Study Guide & Solutions Manual

ORGANIC CHEMISTRY

FIFTH EDITION

Philip S. Bailey Jr

California Polytechnic State University
San Luis Obispo

Christina A. Bailey

California Polytechnic State University
San Luis Obispo

PRENTICE HALL, Englewood Cliffs, NJ 07632

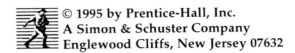 © 1995 by Prentice-Hall, Inc.
A Simon & Schuster Company
Englewood Cliffs, New Jersey 07632

Printed in the United States of America

10 9 8 7 6 5 4 3 2 1

ISBN   0-13-180324-7

Prentice-Hall International (UK) Limited, *London*
Prentice-Hall of Australia Pty. Limited, *Sydney*
Prentice-Hall Canada, Inc. *Toronto*
Prentice-Hall Hispanoamericana, S.A., *Mexico*
Prentice-Hall of India Private Limited, *New Delhi*
Prentice-Hall of Japan, Inc. *Tokyo*
Simon & Schuster Asia Pte. Ltd., *Singapore*
Editora Prentice-Hall do Brasil, Ltda., *Rio de Janeiro*

CONTENTS

HOW TO USE THIS MANUAL

1.   FIRST.......A DO: Listen to and ask questions of your instructor. Read the textbook and question your comprehension as you read.   Work the textbook problems to test your understanding of the material and to apply the concepts you are studying.   Make sure you understand and can communicate the organic chemistry you are studying.

2.   SECOND.......A DON'T: Don't use this manual until you have tried to do the textbook problems yourself!   This is primarily a solutions manual.   We have written it in a way to provide you with understandable answers to the textbook problems.   Your instructor also tries to explain the course material in an understandable way.   Just because you may understand your instructor, and the authors of this manual and your textbook, doesn't mean you can explain the material to others, or do the problems yourself, or perform well on an exam, .

3. Use this manual to check your answers to the textbook problems.   If you get a problem right......feel good; use the success to promote more success.   If you are not entirely correct, try again; you can do it.

HERE ARE SOME TIPS
FOR STUDYING ORGANIC CHEMISTRY

1.   Strive for understanding.   Learn concepts.   Avoid rote memorization.   There are things you will need to remember but it will be easier to remember them if you have a good understanding of the basic concepts.

2.   Organic chemistry builds on itself.   Keep up with the material.   The text begins with atomic structure.   It uses this to explain bond formation which leads to the molecular structure of organic compounds.   We then learn to draw and name organic compounds.   A thorough understanding of organic structure is essential to the understanding of physical and chemical properties.

3.   Ask yourself questions as you read the textbook or your class notes.   Science does not lend itself to light reading.   Read the text with pencil and paper in hand.   Monitor your comprehension.

4.   Work the practice problems in the text.   There are hundreds of problems and worked-out examples in the textbook.   They are there to allow you practice in applying what you know; we are not just pursuing skills.......we are after a lasting and working knowledge of organic chemistry. There are two types of textbook problems. Within a chapter are short exercises often preceded by worked-out examples. Work these as you read the text; subsequent sections will be more meaningful if you understand and actually work with concepts already presented. At the end of each chapter is a more extensive set of problems each labeled according to the topic covered. Work as many of these as you can until you are confident that you fully understand the concept illustrated.

PREPARING FOR AN EXAM

1.   **Test yourself. Never let the instructor be the first to test your knowledge.** If you can talk about something, if you can explain it well to others, there is a good chance you know it. Be honest with yourself. You can always study more but you know when you have done a good job.

2.   **Go for real learning. Insist on true understanding. Avoid superficiality.** Give yourself plenty of time to study and for the opportunity to enjoy learning. Exercise your mind; it needs it as much or more than your body.

3.   **Ask questions of your instructors and teaching assistants.** If it works for you, try to do some of your studying in cooperative groups. Make sure you are a contributor.

4.   **Use the materials available to assist you.** Study the textbook and class notes. Work the practice problems. Use the textbook section headings and glossary terms in the margins as study guides. Use the skill checks to test your comprehension. Use the chapter summaries in the study guide as a review. Check your ability to work problems with the answers in the study guide.

5.   **TEST YOURSELF!** You need to make sure you can remember the material you have studied so hard and have come to understand.

ACKNOWLEDGEMENTS

This manual was produced by the authors using Microsoft Word and CSC ChemDraw with special assistance from Cheryl Schweizer and Doug Brown.

1

Bonding in
Organic Compounds

CHAPTER SUMMARY

Organic chemistry is the study of compounds of carbon. This is a separate branch of chemistry because of the large numbers of organic compounds and their occurrences and applications.

Atoms, the fundamental units of **elements**, are composed of a positively charged **nucleus** which consists of **protons** (charge = +1, mass = 1) and **neutrons** (charge = 0), mass = 1). The nucleus is surrounded by negatively charged **electrons** which have negligible mass. The **atomic number** of an atom is the number of protons in the nucleus; this is equal to the number of electrons surrounding the nucleus in a neutral atom. The **mass number** is the number of protons plus neutrons in the nucleus. **Isotopes** are atoms with the same number of electrons and protons but different numbers of neutrons. The **atomic weight** of an atom is the weighted average of the naturally occurring isotopes.

The space electrons occupy around an atomic nucleus is described by **atomic orbitals**. The most common orbitals in organic chemistry are **s-orbitals**, which are spherical with the atomic nucleus located in the center, and dumbbell shaped **p-orbitals** in which the nucleus is between the lobes. Orbitals exist in **energy levels** or **shells** (numbered 1-7). An atomic orbital can be occupied by 0, 1, or 2 electrons. Atomic orbitals are filled according to the **Aufbau principle** beginning with the lowest energy orbitals and proceeding to higher energy ones.

The **electron configuration** of an atom is described by the orbitals occupied in each shell and the number of electrons in each orbital. The

1

periodic table of elements is organized according to electron configuration. Elements are placed in **periods** which are related to the outermost shell of electrons and in **groups** which are related to the number of electrons in the outer shell. All elements in a group have the same number of outer shell electrons (the same as the group number) and the same electron configuration except for the shell number (for example in Group IV, C is $2s^2 2p^2$ and Si is $3s^2 3p^2$; both outer shells have four electrons). The elements in Group VIII are especially stable; their outer shell configuration is known as a **stable octet**.

     Ionic bonding involves the complete transfer of electrons between two atoms of widely different electronegativities; charged **ions** are formed, both of which usually have a stable octet outer shell. The ionic bond results from the attraction between the **positive cation** and **negative anion**. **Electronegativity** is defined as the attraction of an atom for its outer shell electrons. **Electronegative** elements have strong attraction for electrons and form anions in chemical reactions; **electropositive** elements have relatively weak attractions for electrons and form cations.

     Covalent bonds involve the sharing of electron pairs between atoms of similar electronegativity; in most cases one or both atoms obtains a stable octet outer shell of electrons. Electron dot formulas show **bonding** and **non-bonding** outer-shell electrons in molecules using dots. A **single bond** has one bonding pair of electrons; there are two bonding pairs (four electrons) in a **double bond** and three bonding pairs in a **triple bond**. The number of bonds formed by elements commonly found in organic compounds is: **C - 4; N - 3; O, S - 2; H - 1; F, Cl, Br, I - 1.** A carbon can have four single bonds, two double bonds, a double and two single bonds, or a triple and a single bond; all total four bonds. These bonds can be represented by electron dot or line bond formulas. In drawing **structural formulas**, one must use every atom in the **molecular formula** and satisfy the **valence** (the number of bonds formed) for each.

     Polyatomic ions are charged species in which several atoms are connected by covalent bonds. The magnitude and location of the ion's charge is called the **formal charge**. The formal charge on an atom is equal to the group number of the atom on the periodic table minus the non-bonding electrons and half of the bonding electrons.

A **polar covalent bond** involves atoms with similar but not equal electronegativities. The more electronegative atom is **partially negative** and the other is **partially positive**.

A **molecular orbital** describes a covalent bond; it is formed by the overlap of two atomic orbitals each with one electron. There are two types: sigma and pi. A **sigma molecular orbital** or **sigma bond** involves the overlap of two atomic orbitals head-to-head in one position (such as two s-orbitals, an s and a p-orbital, or two p-orbitals). A **pi-bond** involves the overlap of parallel p-orbitals at both lobes.

The **shapes** of organic molecules are predicted using the following principle: atoms and non-bonding electron pairs bonded to a common central atom are arranged as far apart in space as possible. If there are four surrounding groups, the shape is **tetrahedral**; with three, the groups protrude to the corners of a **triangle**; and with two, the region is **linear**. The four atomic orbitals on carbon (an s and three p's) combine, through a process called **hybridization**, to form new orbitals with the shape and orientation to all the molecular shapes described.

A carbon with four bonded atoms is sp^3-hybridized, tetrahedral, and has approximately 109° bond angles. The four new sp^3 orbitals are raindrop shaped and are oriented to the corners of a tetrahedron. All bonds to the carbon are sigma bonds. **A carbon with three bonded atoms is sp^2-hybridized, trigonal, and has approximately 120° bond angles.** There are three new sp^2 hybrid orbitals directed to the corners of a triangle; these form sigma bonds with other atoms. The remaining p-orbital overlaps with a parallel p-orbital of an adjacent, sigma bonded atom to form a pi-bond and complete the double bond. **A carbon with two bonded atoms is sp hybridized, linear, and has 180° bond angles.** There are two new sp hybrid orbitals that are directed opposite to one another on a straight line; these form sigma bonds. The two remaining p-orbitals overlap with p-orbitals on a similarly hybridized atom to form two pi-bonds and complete the triple bond. Alternatively, the two p-orbitals could overlap with counterparts on two adjacently bonded sp^2-hybridized atoms forming two double bonds. **A single bond is a sigma bond; a double bond is composed of one sigma bond and one pi-bond; a triple bond is one sigma and two pi-bonds.**

3

An **oxygen** with two bonded atoms and two non-bonding electron pairs is sp^3 hybridized and has two sigma bonds (single bonds). With only one bonded atom, the oxygen is sp^2 hybridized and is involved in one sigma and one pi bond, a double bond.

A **nitrogen** with three bonded atoms and one non-bonding electron pair is sp^3 hybridized and has three sigma bonds (single bonds). With two bonded atoms the nitrogen is sp^2 hybridized and involved in two single bonds (sigma) and a double bond (sigma and a pi). A nitrogen with only one bonded atom is sp hybridized, has one single bond (sigma) and one triple bond (one sigma and two pi-bonds).

Connections 1.1 describes three forms of carbon: diamond, graphite, and fullerenes or buckyballs.

SOLUTIONS TO PROBLEMS

1.1   Atomic and Mass Numbers

Subtract the atomic number (the number of electrons; and protons) from the mass number (number of protons and neutrons) to get number of neutrons.

(a-b)	mass number	atomic number	electrons	protons	neutrons
^{12}C	12	6	6	6	6
^{13}C	13	6	6	6	7
^{35}Cl	35	17	17	17	18
^{37}Cl	37	17	17	17	20

(c) F 9, 19.0   S   16, 32.1   Al 13, 27.0

1.2 Electron Configuration

Na $1s^2$ $2s^22p^6$ $3s^1$ Mg $1s^2$ $2s^22p^6$ $3s^2$

Al $1s^2$ $2s^22p^6$ $3s^23p^1$ Si $1s^2$ $2s^22p^6$ $3s^23p^2$

P $1s^2$ $2s^22p^6$ $3s^23p^3$ S $1s^2$ $2s^22p^6$ $3s^23p^4$

Cl $1s^2$ $2s^22p^6$ $3s^23p^5$ Ar $1s^2$ $2s^22p^6$ $3s^23p^6$

1.3 Electron Configuration

He, $1s^2$; Ne, $2s^22p^6$; Ar, $3s^23p^6$; Kr, $4s^24p^6$; Xe, $5s^25p^6$; Rn, $6s^26p^6$

1.4 Ionic Bonding

(a)   Li· + ·F: ⟶ Li$^+$ + :F: (LiF)

           $1s^22s^1$ $1s^22s^22p^5$ $1s^2$ $1s^22s^22p^6$

(b)   Mg: + ·O: ⟶ Mg^{2+} + :O^{2-}:   (MgO)

     $1s^22s^22p^63s^2$ $1s^22s^22p^4$ $1s^22s^22p^6$ $1s^22s^22p^6$

1.5 Electron Dot Formulas

         : Cl:

(a) H: C: Cl: (b)   H·   (c) :Ö::C::Ö: (d) H: C: : : C: Cl:

        : Cl: ·C::O:

                  H·

1.6 Formal Charge

To determine formal charge, subtract the number of non-bonding electrons and have the number of bonding electrons from the group number on the periodic table to which the atom in question belongs.

   H   C   (4) - (0) - 1/2 (8) = 0

  ··   ··   –   H H   +

H:C:O: ·· ··   H's (1) - (0) - 1/2 (2) = 0

  ·· ··   H:C:N:H

   H   ·· ··   O   (6) - (6) - 1/2 (2) = -1

                 H H

                              N   (5) - (0) - 1/2 (8) = +1

1.7 Polar Bonds

a) $\overset{\delta+}{C}\!-\!\overset{\delta-}{Br}$ b) $\overset{\delta+}{C}\!-\!\overset{\delta-}{O}$ c) $\overset{\delta-}{N}\!-\!\overset{\delta+}{H}$ d) $\overset{\delta+}{C}\!-\!\overset{\delta-}{N}$ e) $\overset{\delta+}{C}\!-\!\overset{\delta-}{O}$ f) $\overset{\delta+}{C}\!-\!\overset{\delta-}{S}$

1.8 Propane

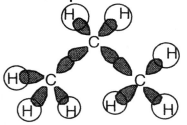

Each carbon is tetrahedral,

has approximately 109° bond angles

and is sp³-hybridized.

1.9 Propene

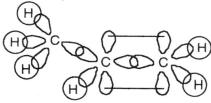

The carbons involved in the double bond are both trigonal, have 120° bond angles, and are sp²-hybridized. The other carbon is tetrahedral, has 109° bond angles, and is sp³-hybridized.

1.10 Propyne

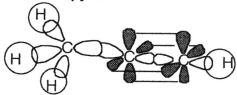

Both carbons in the triple bond have a linear geometry, 180° bond angles, and are sp-hybridized. The other carbon is tetrahed has 109° bond angles, and is sp³-hybridized

1.11 Bonding in Organic Compounds

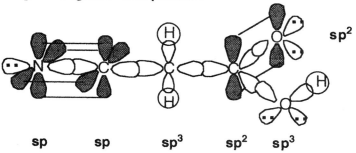

1.12 Atomic and Mass Numbers: Section 1.2A

See problem 1.1 for explanation.

(a) ^{127}I 53 protons, 53 electrons, 74 neutrons; (b) ^{27}Al 13 protons, 13 electrons, 14 neutrons; (c) ^{58}Ni 28 protons, 28 electrons, 30 neutrons; (d) ^{208}Pb 82 protons, 82 electrons, 126 neutrons.

1.13 Electron Configurations: Section 1.2B-D

Groups on Periodic Table

I	II	III	IV	V	VI	VII
H $1s^1$						
Li $2s^1$	Be $2s^2$	B $2s^2sp^1$	C $2s^2sp^2$	N $2s^22p^3$	O $2s^22p^4$	F $2s^22p^5$
Na $3s^1$	Mg $3s^2$	Al $3s^23p^1$	Si $3s^23p^2$	P $3s^23p^3$	S $3s^23p^4$	Cl $3s^23p^5$
K $4s^1$	Ca $4s^2$	Ga $4s^24p^1$	Ge $4s^24p^2$	As $4s^24p^3$	Se $4s^24p^4$	Br $4s^24p^5$
Rb $5s^1$	Sr $5s^2$	In $5s^25p^1$	Sn $5s^25p^2$	Sb $5s^25p^3$	Te $5s^25p^4$	I $5s^25p^5$
Cs $6s^1$	Ba $6s^2$	Tl $6s^26p^1$	Pb $6s^26p^2$	Bi $6s^26p^3$	Po $6s^26p^4$	At $6s^26p^5$
Fr $7s^1$	Ra $7s^2$					

1.14 Electron Configurations: Section 1.2B-D

(a) Na, $3s^1$; (b) Mg, $3s^2$; (c) B, $2s^23p^1$; (d) Ge, $4s^24p^2$; (e) P, $3s^23p^3$; (f) O, $2s^22p^4$; (g) I, $5s^25p^5$; (h) Kr, $4s^24p^6$

1.15 Electron Configurations: Section 1.2B-D

a) Fr b) Sn c) Cl d) Mg e) B f) Se g) He h) Xe i) As

1.16 Outer-Shell Electrons: Section 1.2

The number of outer-shell electrons is the same as the group number on the periodic table. (a) H, 1; (b) Al, 3; (c) C, 4; (d) 5; (e) S, 6; (f) Br, 7.

1.17 Ionic Compounds: Section 1.3

    Put the positive and negative ions in a ratio to form a neutral compound.

    a) NaF   b) $Mg(OH)_2$ c) $CaCO_3$ d) $NaNO_2$ e) $KClO_3$

    f) $PbBr_2$ g) Li_2CO_3 h) CaO   i) $NaHCO_3$ j) $CaSO_4$

    k) $(NH_4)_2SO_4$ l) $Al(OH)_3$

1.18 Ionic Reactions: Section 1.3

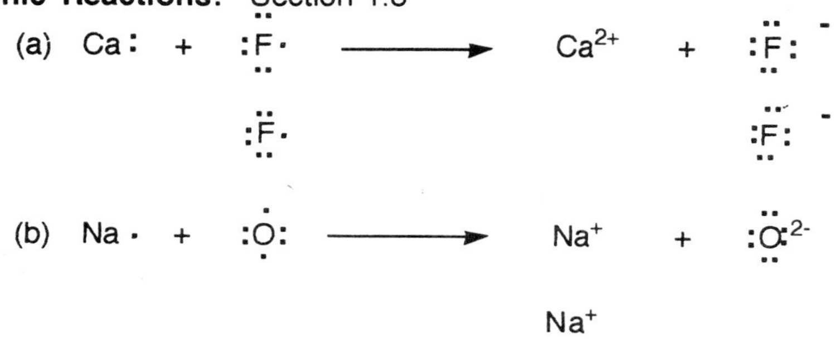

(a) Ca: + :F· ⟶ Ca²⁺ + :F:⁻

:F·   :F:⁻

(b) Na· + :O: ⟶ Na⁺ + :O:²⁻

Na⁺

1.19 Electron Dot Formulas: Section 1.4A-C

Following are electron dot formulas and valences for the elements illustrated in this problem.

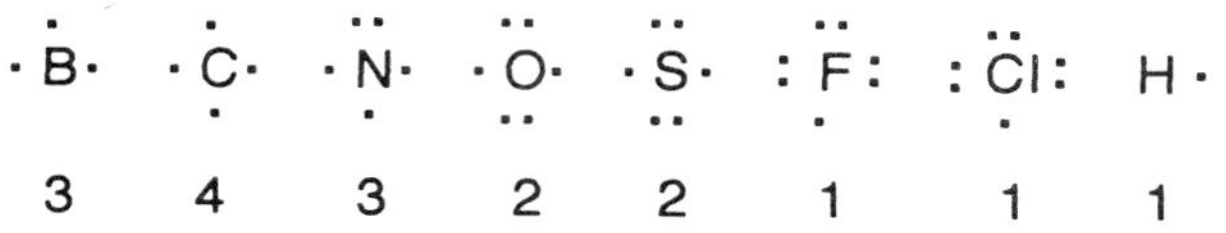

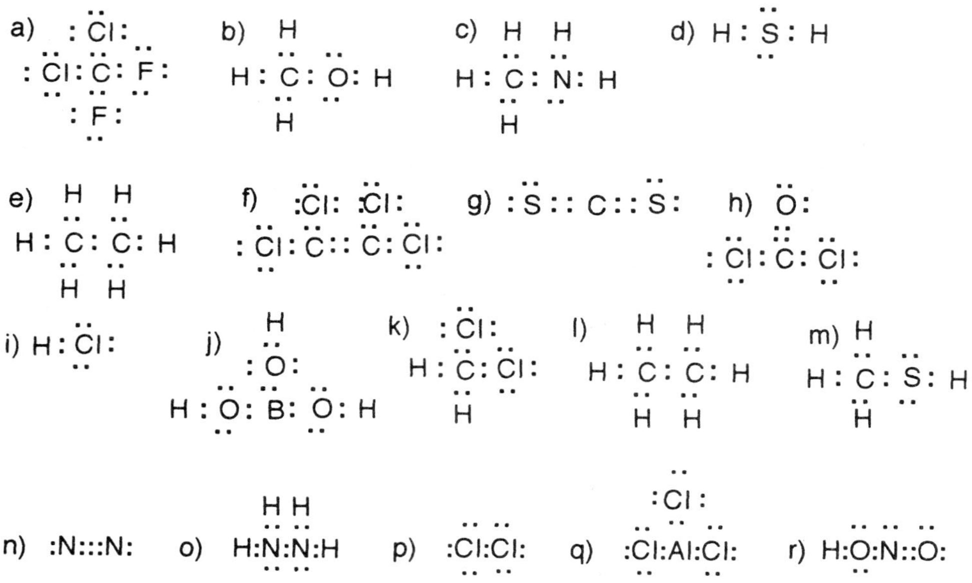

·B·   ·C·   ·N·   ·O·   ·S·   :F: :Cl: H·

3   4   3   2   2   1   1   1

a) :Cl:
    :Cl:C:F:
    :F:

b) H
    H:C:O:H
    H

c) H H
    H:C:N:H
    H

d) H:S:H

e) H H
    H:C:C:H
    H H

f) :Cl: :Cl:
    :Cl:C::C:Cl:

g) :S::C::S:

h) O:
    :Cl:C:Cl:

i) H:Cl:

j)   H
       :O:
    H:O:B:O:H

k) :Cl:
    H:C:Cl:
    H

l) H H
    H:C:C:H
    H H

m) H
    H:C:S:H
    H

n) :N:::N:

o) H H
    H:N:N:H

p) :Cl:Cl:

q) :Cl:
    :Cl:Al:Cl:

r) H:O:N::O:

8

1.20 Electron Dot Formulas: Section 1.4A-C

a)
```
    H   H   H   H                    H              b)   H   H                      H       H
    ..  ..  ..  ..              H  H : C : H  H          ..  ..                     ..      ..
H : C : C : C : C : H           ..            ..    H : C : C : O : H     H : C : O : C : H
    ..  ..  ..  ..          H : C  :   C   :   C : H     ..  ..                     ..      ..
    H   H   H   H               ..      ..      ..       H   H                      H       H
                                H       H       H
```

c)
```
    H   H   H                    H   H   H          d)   H   H                          H  : Cl :
    ..  ..  ..                   ..  ..  ..              ..  ..                         ..   ..
H : C : C : C : Br :        H : C : C : C : H       : Cl : C : C : Cl :     H : C : C : Cl :
    ..  ..  ..                   ..  ..  ..              ..  ..  ..              ..  ..   ..
    H   H   H                    H  : Br : H                 H   H                  H   H
```

e)
```
    H   H                       H       H            f)   H   H     H                    H       H
    ..  ..  ..                  ..  ..  ..                ..  ..                          .. C ..
H : C : C : N : H           H : C : N : C : H        H : C : C :: C : H              H     ../\..     H
    ..  ..                       ..                      ..                             C      C
    H   H   H                    H   H   H                H                       H  ../    \..  H
                                                                                     H          H
```

1.21 Electron Dot Formulas

a)
```
    H   H   ..                b)          H   H       c)      ..            d)   H  H
    ..  ..  ..                    ..      ..  ..              H   O :              ..
H : C : C : O : H            : Cl : C :: C : H              ..  .. ..        H : C :: C : C ::: N :
    ..  ..                       ..                    H : C : C : O : H
    H   H                                                  ..  .. ..
                                                            H
```

e)
```
        .. ..                 f)      H              g)  ..
    H   O : H                         ..                 O :      H   H
    ..  .. ..                 H : C : N :: C :: O :      ..       ..  ..
H : N : C : N : H                 ..                 H : C : O : C : C : H
    ..      ..                    H                      ..  .. ..  ..
                                                            H   H
```

h)
```
    H   H     H   H
    ..  ..    ..  ..  ..
H : C : C :: C : C : S : H
    ..            ..  ..
    H         H
```

i)
```
                ..
    H   H   .. O :
    ..  ..  ..  ..
H : C : C : C : O : H
    ..  ..  ..
    H : O : H
        ..
```

j)
```
    H   H   H
    ..  ..  ..  ..
H : C : C : C : S : H
    ..  ..  ..  ..
    H   H   H
```

1.22 Electron Dot Formulas: Section 1.4A-C

Place five carbons in a row connected by single bonds. Twelve hydrogens are needed if all bonds are single. If three double bonds are inserted, the six hydrogens will satisfy the remaining valences of the five carbons. Likewise, one triple bond and one double bond will allow the valences to be satisfied.

C:C:C:C:C

$$H \quad H\ H\ H$$
$$H:\ddot{C}::C:\ddot{C}:\ddot{C}::\ddot{C}:H$$

$$H\ H\ H$$
$$H:C:::C:\ddot{C}:\ddot{C}::\ddot{C}:H$$
$$H$$

1.23 Formal Charge: Section 1.4E

a) $+CH_3$ b) $\cdot CH_3$ c) $-: CH_3$ d) $CH_3 \overset{H}{\underset{}{—}} \overset{+}{\underset{..}{O}} —CH_3$

e) $(CH_3)_4N^+$ f) $CH_3 —\overset{+}{N}\equiv\overset{-}{N}:$ g) $CH_3\overset{..}{\underset{..}{Cl}}$

1.24 Polar Covalent Bonds: Section 1.4G

Most bonds in organic compounds are considered polar except carbon-hydrogen and carbon-carbon bonds.

$$\partial+ \quad \partial- \quad \overset{H}{\underset{H}{}} \quad \overset{H}{\underset{N}{}} \quad O\ \partial-$$

$$H-S-C-C-C$$

1.25-1.27 Bonding in Organic Compounds: Section 1.5

See section 1.5; there is a summary in 1.5G. Also see problems 1.8-1.10 and example 1.4.

Problem 1.25: The carbons involved only in single bonds have four bonded groups and are sp^3 hybridized, are tetrahedral, and have 109° bond angles. The carbons that have one double bond have three bonded groups and are sp^2 hybridized, trigonal, and have 120° bond angles. The carbons involved in triple bonds or two double bonds are sp hybridized, linear, and have 180° bond angles.

Problem 1.26: All single bonds are sigma bonds. A double bond is a sigma and a pi bond. A triple bond is constructed of a sigma and two pi bonds.

Problem 1.27: Bonding picture.

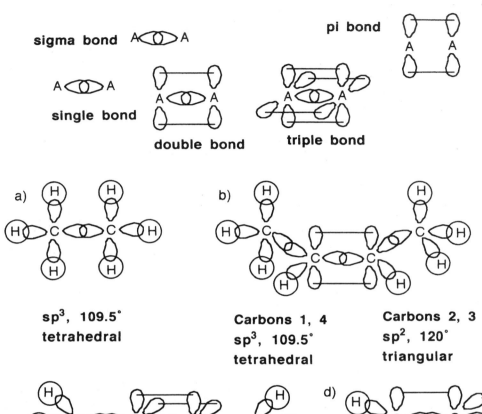

sigma bond   A◁◯▷A

A◁◯▷A
single bond

pi bond

double bond   **triple bond**

a)   sp³, 109.5°
tetrahedral

b)   Carbons 1, 4
sp³, 109.5°
tetrahedral

Carbons 2, 3
sp², 120°
triangular

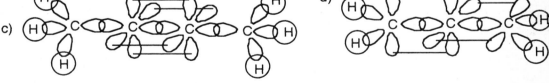

c)   carbons 1, 4 sp³, 109.5°
tetrahedral
carbons 2, 3 sp, 180°
linear

d)   Inside C - sp, 180°, linear;
Outside C's - sp², 120°, triangular

e)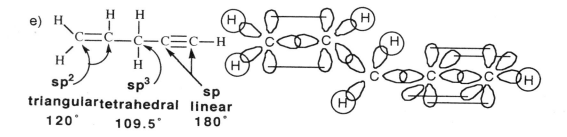

sp² — triangular — 120°

sp³ — tetrahedral — 109.5°

sp — linear — 180°

1.28-1.30 Bonding with Oxygen and Nitrogen: Section 1.6

Problem 1.28: Carbons, nitrogens, and oxygens involved only in single bonds have four groups that occupy space; four bonded groups with carbon, three bonded groups and a non-bonding electron pair with nitrogen, and two bonded groups and two non-bonding pairs with oxygen. All are sp³ hybridized, tetrahedral, and have 109° bond angles. Carbons, nitrogens, and oxygens with one double bond have three groups that occupy space; three bonded groups for carbon, two bonded groups and a non-bonding pair for nitrogen and one bonded group and two non-bonding pairs for oxygen. All are sp² hybridized, trigonal, and have 120° bond angles. Carbons and nitrogens with a triple bond have only two groups that occupy space; two bonded groups for carbon and one bonded group and a non-bonding electron pair for nitrogen. Both are sp hybridized, linear, and have 180° bond angles.

Problem 1.29: Single bonds are sigma bonds; double bonds are one sigma and one pi bond. Triple bonds are one sigma and two pi bonds.

Problem 1.30: Bonding pictures.

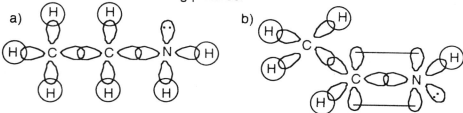

All atoms are sp³, tetrahedral, approximate bond angles 109°.

C H₃ is tetrahedral, sp³, 109°. Other C and N are sp², triangular, 120°.

c) 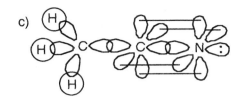   d)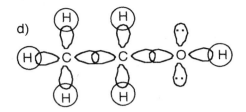

Carbon-2 is sp³, tetrahedral, 109° bond angles. The carbon and nitrogen in the triple bond are sp, linear, 180°.

All carbons and the oxygen are sp³, tetrahedral, with approximately 109° bond angles.

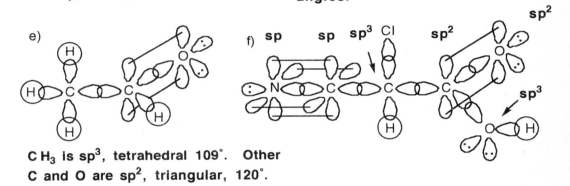

C H₃ is sp³, tetrahedral 109°. Other C and O are sp², triangular, 120°.

1.31 Bonding and Molecular Orbitals: Section 1.5

Element	Compound	
· Ċ ·	$s^1p^1p^1p^1$	$(sp)^1(sp)^1p^1p^1$	
· Ö ·	$s^2p^2p^1p^1$	$(sp^2)^2(sp^2)^2(sp^2)^1p^1$	$:\ddot{S}::C::\ddot{O}:$
· Ṡ ·	$s^2p^2p^1p^1$	$(sp^2)^2(sp^2)^2(sp^2)^1p$	

1.32 Silicon: Section 1.4A-C

Silicon was a logical choice of an element for the Star Trek episode about a very different life form. Silicon is just below carbon in group IV of the periodic table, has the same number of outer shell electrons, and has some properties that are similar. It can bond to itself (though not as extensively as carbon) and like carbon it is tetravalent.

13

1.33 Molecular Shape: Section 1.5

In NH_3, nitrogen has four groups that occupy space, three bonding pairs (hydrogens) and one non-bonding pair of electrons. As such the preferred geometry is tetrahedral and the nitrogen is sp^3 hybridized.

N s^2 p^1 p^1 p^1 hybridizes to $(sp^3)^2$ $(sp^3)^1$ $(sp^3)^1$ $(sp^3)^1$ in NH_3

Surrounding boron are three space occupying groups; the three fluorines. Boron does not have an octet of electrons. Therefore it assumes a trigonal shape and is sp^2 hybridized.

B s^1 p^1 p^1 p^0 hybridizes to $(sp^2)^1$ $(sp^2)^1$ $(sp^2)^1$ p^0

1.34 Bond Angles: Section 1.5

All four compounds have four pairs of electrons around the central atom. In CH_4 they are all bonding pairs and relatively confined to the carbon-hydrogen bonds. Methane is a classic example of a tetrahedral molecule with $109°$ bond angles.   In ammonia, NH_3, there are three bonding pairs of electrons and one non-bonding pair. The non-bonding pair tends to occupy more space and repel the bonding pairs thus slightly compressing the bond angles; the result is $107°$ bond angles. In water there are two non-bonding electron pairs. These spacious pairs repel each other and the two bonding pairs thus further compressing the bond angles to $105°$.

1.35 Molecular Shape: Section 1.5

14

1.36 Hybridization: Section 1.5

Be has two outer shell electrons and thus forms only two bonds; there are no non-electron pairs. Thus Be in BeH_2 is linear, sp hybridized, and has 180° bond angles.

Be s^1 p^1 p^0 p^0 **hybridizes to** $(sp)^1$ $(sp)^1$ p^0 p^0 **in** BeH_2

The bonding picture below shows the two sp-s sigma bonds. In addition, the Be has two empty p-orbitals that are not shown.

1.37 Reactivity: Section 1.4A-C

The carbon in CH_4 has a stable octet and all eight electrons are expressed as four bonding electron pairs. In NH_3, the nitrogen has a stable octet but, since nitrogen is in group V and has five outers shell electrons, there is a non-bonding electron pair remaining following formation of three bonds. This pair of electrons is quite available for sharing with electron-deficient species relative to the bonding pairs of CH_4. Boron is in group III of the periodic table. Since it has only three outer shell electrons it forms three bonds. However it does not achieve a stable octet. Consequently, it is attracted to species that have electron pairs available for bonding (such as the N in NH_3) since in reacting, it can achieve a stable octet.

```
    H
    ..
H : B
    ..
    H
```

```
    H
    ..
H : C : H
    ..
    H
```

```
    H
    ..
H : N :
    ..
    H
```

2

THE ALKANES:

STRUCTURE AND NOMENCLATURE

OF SIMPLE HYDROCARBONS

CHAPTER SUMMARY

Organic compounds are classified according to common structural features that impart similar chemical and physical properties to the compounds within each group or family. **Hydrocarbons** are composed only of carbon and hydrogen and fall into two major classes - saturated and unsaturated. **Saturated hydrocarbons** or the **alkanes** are are entirely constructed of single bonds and have the general formula C_nH_{2n+2}. **Unsaturated hydrocarbons** include the **alkenes** (C_nH_{2n}) in which there is at least one carbon-carbon double bond; the **alkynes** (C_nH_{2n-2}) where there is at least one carbon-carbon triple bond; and **aromatic hydrocarbons** which appear to have double bonds but actually have a special structure that is discussed in Chapter 6.

Compounds are described by **molecular formulas** which provide the kinds of atom and numbers of each in a molecule and **structural formulas** which give the bonding arrangements of the atoms; that is, what atoms are bonded to one another and by what kind of bond. **Isomers** are compounds with the same molecular formula but different structural formulas. **Structural isomers (skeletal, positional, and functional)** differ in the bonding arrangement of atoms; different atoms are attached to one another. In **stereoisomerism (geometric, conformational, and optical)** the same atoms are bonded to one another but their orientation in space differs. In drawing structural formulas, every atom in the molecular formula must be used and in a way to satisfy its **valence (C-4; N-3; O,S-2; H,F,Cl,Br,I - 1).**

Cycloalkanes are hydrocarbons possessing at least one ring and differ from **open-chain alkanes** in this way.

Structural formulas can be represented in a variety of ways from full formulas that show every individual atom and bond to very condensed line or stick formulas.

Positional isomers differ in the position of a noncarbon group or of a double bond or triple bond.

Organic compounds can be named systematically by **IUPAC nomenclature.** The base name of alkanes is derived from the Greek for the number of carbons in the **longest continuous carbon chain** (Table 2.1 of the text) followed by the suffix **ane. Alkyl groups** (Table 2.2 of the text) are shorter chains and named by changing the **ane** to **yl.** The positions of alkyl groups are described with numbers; the longest carbon chain of an alkane is numbered starting at the end that gives the lowest number to the first substituent. Multiple numbers of identical alkyl groups are indicated with **di, tri, tetra**, etc. The base name of cycloalkanes is usually based on the Greek for the number of carbons in the ring with the prefix **cyclo** and the suffix **ane.** The prefixes **fluoro, chloro, bromo, and iodo** are used to indicate the presence of halogen in a molecule.

Conformational isomers are isomers in which the spatial relationship of atoms differs because of rotation around a carbon-carbon double bond. Because the rotation is unrestricted in most cases, conformational isomers are constantly interconverting and are not isolatable. There are two extreme conformations. In the **eclipsed conformation,** atoms on adjacent carbons are lined up with one another and are as close together as possible; this is the least stable conformational arrangement. In the **staggered conformation,** atoms on adjacent atoms are staggered with one another and are as far apart as possible; this is the most stable conformational arrangement. Staggered and eclipsed forms are represented with **sawhorse diagrams** or **Newman projections.**

Cycloalkanes are generally depicted with regular polygons though they actually exist in three-dimensional conformations. **Cyclopropane** and **cyclobutane** are less stable than other cycloalkanes since they are planar or nearly so and have internal bond angles significantly smaller (60° and 90° respectively) than the preferred tetrahedral angle (109°).

Cyclohexane exists in two puckered conformations, the **boat and chair forms**, that have tetrahedral bond angles. The boat form is less stable and not preferred because of interactions between the two end or **flagpole** carbons and because the hydrogens on the other adjacent carbons are eclipsed. In the preferred chair form, atoms on adjacent carbons are staggered and there are no flagpole type interactions. There are two orientations of hydrogens in the chair conformation. **Axial hydrogens** are oriented directly above or below the "plane" of the ring in an alternating arrangement. **Equatorial hydrogens** protrude out along the perimeter of the ring. Equatorial positions are more spacious and in substituted cyclohexanes they are preferred. Cyclohexane rings flip between chair forms and establish an equilibrium. In the process of flipping, all equatorial positions become axial and all axial positions become equatorial. The equilibrium favors the chair in which substituents are equatorial. In disubstituted cyclohexanes where one group is axial and one equatorial, the equilibrium favors the chair form where the larger group occupies the more spacious equatorial position.

Disubstituted cycloalkanes exhibit **geometric isomerism**, a type of stereoisomerism. If both groups are on the same "side" of the ring (both up or both down) the isomer is termed **cis.** If they are on opposite "sides" (one up and one down) the isomer is **trans.**

Physical properties of hydrocarbons are related to structure. Melting points and boiling points generally increase with molecular weight within a **homologous series** (a series of compounds in which each succeeding member differs from the previous one by a CH_2 group). **Branched chain hydrocarbons** have less surface area and thus less opportunity for intermolecular attractions; as a result, their boiling points are lower than the straight chain isomers. However, their compact nature causes them to fit more easily in a crystal lattice and thus they generally have higher melting points. Hydrocarbons are **non-polar** and insoluble in water and, because they are less dense, they float on the surface of water.

Connections 2.1 describes the origin of coal and petroleum, their importance and applications, and the fractionation of petroleum into the following fractions: gas, gasoline, kerosene, gas-oil, wax-oil, wax, and residue.

SOLUTIONS TO PROBLEMS

2.1 Skeletal Isomerism

Start with a chain of seven carbons.

$$CH_3CH_2CH_2CH_2CH_2CH_2CH_3$$

Now draw isomers with six carbons in the longest chain and vary the position of a one-carbon chain.

$$\underset{\displaystyle CH_3CHCH_2CH_2CH_2CH_3}{\overset{\displaystyle CH_3}{|}} \qquad\qquad \underset{\displaystyle CH_3CH_2CHCH_2CH_2CH_3}{\overset{\displaystyle CH_3}{|}}$$

Now draw five carbon chains and place one two-carbon chain or two one-carbon chains on it. If the two-carbon side chain is placed on either the first or second carbon, it merely extends the longest chain. However, placing it on the third carbon gives us an isomer.

$$\underset{\displaystyle CH_3CH_2CHCH_2CH_3}{\overset{\displaystyle CH_2CH_3}{|}}$$

Now attach two one-carbon chains to the carbon skeleton.

$$\overset{\displaystyle CH_3}{\underset{\displaystyle CH_3}{|}}\;CH_3CCH_2CH_2CH_3 \quad \overset{\displaystyle CH_3}{\underset{\displaystyle CH_3}{|}}\;CH_3CH_2CCH_2CH_3 \quad \overset{\displaystyle CH_3}{\underset{\displaystyle CH_3}{|}}\;CH_3CHCHCH_2CH_3 \quad \overset{\displaystyle CH_3\;\;CH_3}{|\;\;\;\;|}\;CH_3CHCH_2CHCH_3$$

Finally, draw a four carbon chain with three one-carbon side chains.

$$\underset{\displaystyle CH_3}{\overset{\displaystyle CH_3\;\;CH_3}{CH_3C\!\!-\!\!-\!\!CHCH_3}}$$

2.2 Skeletal Isomers of Cycloalkanes: Five cyclic compounds of C_5H_{10}.

Start with a five-membered ring. Then use a four-membered ring with a one-carbon side chain. Finally, draw a three-membered ring with either one two-carbon side chain or two one-carbon side chains.

2.3 Positional Isomers: Five isomers of C_3H_6BrCl.

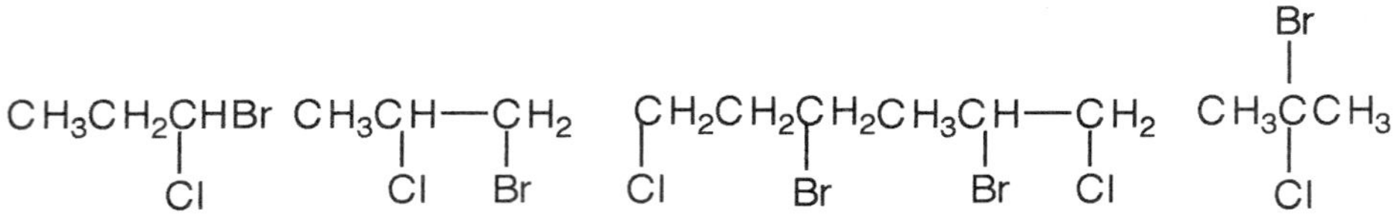

2.4 Alkane Nomenclature

a. Names of compounds in Example 2.1 as they appear.
hexane, 2-methylpentane, 3-methylpentane, 2,2-dimethylbutane, and 2,3-dimethylbutane.

b. Structures from names

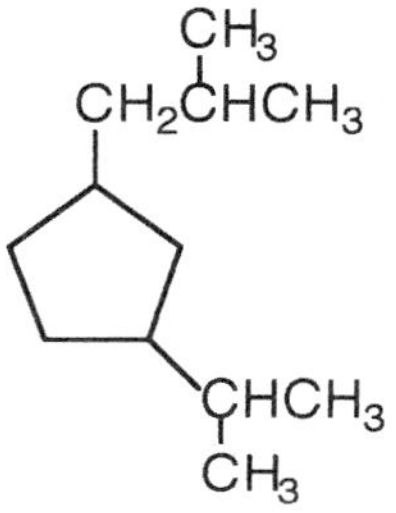

1-isobutyl-3-isopropylcyclopentane

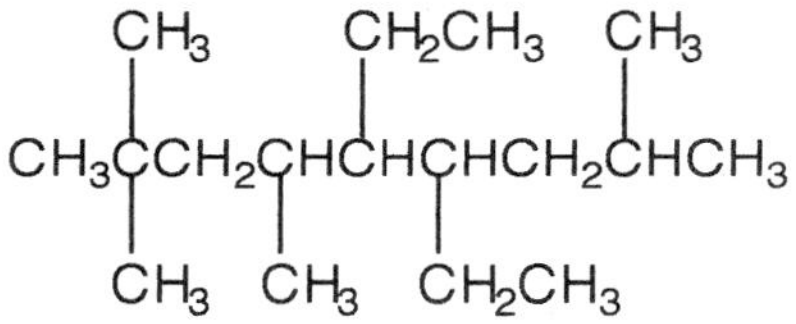

5,6-diethyl-2,2,4,8-tetramethylnonane

2.5 Alkyl Halide Nomenclature

Names of compounds in section 2.5 as they appear.
1-bromobutane, 2-bromobutane,
1-bromo-2-methylpropane, and 2-bromo-2-methylpropane.

2.6 Alkyl Halide Nomenclature

Names of compounds from problem 2.3 as they appear.
1-bromo-1-fluoropropane, 1-bromo-2-chloropropane,
1-bromo-3-chloropropane, 2-bromo-1-chloropropane, and
2-bromo-2-chloropropane.

2.7 Skeletal and Positional Isomerism

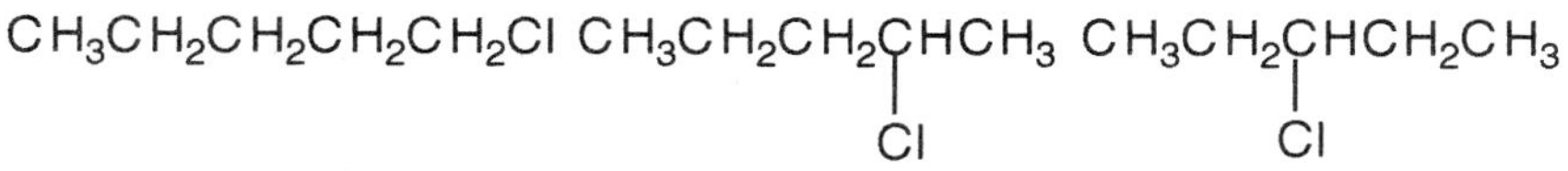

1-chloropentane 2-chloropentane 3-chloropentane

$$CH_3\underset{\underset{Cl}{|}}{\overset{\overset{CH_3}{|}}{C}}HCH_2CH_2 \qquad CH_3\underset{\underset{Cl}{|}}{\overset{\overset{CH_3}{|}}{C}}HCHCH_3 \qquad CH_3\underset{\underset{Cl}{|}}{\overset{\overset{CH_3}{|}}{C}}CH_2CH_3$$

1-chloro-3-methylbutane   2-chloro-3-methylbutane   2-chloro-2-methylbutane

$$CH_2\underset{\underset{Cl}{|}}{\overset{\overset{CH_3}{|}}{C}}HCH_2CH_3 \qquad\qquad CH_3\underset{\underset{Cl}{|}}{\overset{\overset{CH_3}{|}}{C}}CH_2CH_3$$

1-chloro-2-methylbutane   2-chloro-2-methylbutane

2.8 Conformational Isomers of Butane

Follow the procedure outlined in Example 2.7 of the text. In this problem both the front and back carbons have two hydrogens and one methyl group.

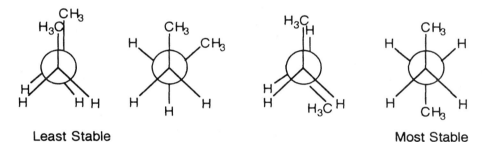

Least Stable   Most Stable

2.9 Cyclohexane Chair: Axial and Equatorial Positions

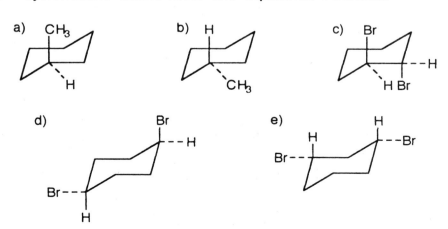

2.10 Equilibrium between Chair Forms

(a) The equatorial position is more spacious and the isomer more stable.

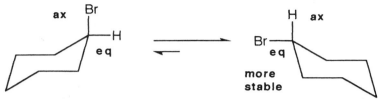

(b) The equatorial positions are more spacious and preferred in the equilibrium.

(c) The equilibrium favors the isomer in which the larger isopropyl group
     is in the more spacious equatorial position.

2.11 Geometric Isomerism in Cyclic Compounds

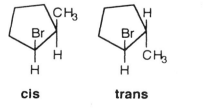

cis   trans

1-bromo-2-methylcyclopentane

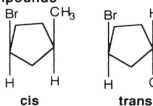

cis   trans

1-bromo-3-methylcyclopentane

2.13 and 2.14   Skeletal Isomerism and Nomenclature Sections 2.3 and 2.6

(a) See problem 2.1. Names in order of appearance:

heptane

2-methylhexane; 3-methylhexane

3-ethylpentane

2,2-dimethylpentane; 3,3-dimethylpentane; 2,3-dimethylpentane;

2,4-dimethylpentane

2,2,3-trimethylbutane

b) C_8H_{18}: Start with an eight-carbon chain. Then systematically make the longest chain one carbon shorter. Arrange the remaining carbons in each case on the chain in as many different ways as possible without extending the length of the base chain. The following isomers are drawn in a logical, systematic order.

$CH_3CH_2CH_2CH_2CH_2CH_2CH_2CH_3$

octane

$$\underset{\textstyle CH_3CHCH_2CH_2CH_2CH_2CH_3}{\overset{\textstyle CH_3}{|}}$$

2-methylheptane

$$\underset{\textstyle CH_3CH_2CHCH_2CH_2CH_2CH_3}{\overset{\textstyle CH_3}{|}}$$

3-methylheptane

$$\underset{\textstyle CH_3CH_2CH_2CHCH_2CH_2CH_3}{\overset{\textstyle CH_3}{|}}$$

4-methylheptane

$$\underset{\textstyle CH_3CH_2CHCH_2CH_2CH_3}{\overset{\textstyle CH_2CH_3}{|}}$$

3-ethylhexane

$$\underset{\textstyle CH_3}{\overset{\textstyle CH_3}{|}}\ CH_3CCH_2CH_2CH_2CH_3$$

2,2-dimethylhexane

$$\underset{\textstyle CH_3}{\overset{\textstyle CH_3}{|}}\ CH_3CH_2CCH_2CH_2CH_3$$

3,3-dimethylhexane

$$\overset{\textstyle CH_3\ \ CH_3}{CH_3CH-CHCH_2CH_2CH_3}$$

2,3-dimethylhexane

$$\overset{\textstyle CH_3\ \ CH_3}{CH_3CHCH_2CHCH_2CH_3}$$

2,4-dimethylhexane

$$\overset{\textstyle CH_3\ \ CH_3}{CH_3CHCH_2CH_2CHCH_3}$$

2,5-dimethylhexane

CH₃ CH₃
CH₃CH₂CH—CHCH₂CH₃

CH₃ CH₂—CH₃
CH₃CH—CHCH₂CH₃

CH₃
CH₃CH₂—C—CH₂CH₃
CH₂
CH₃

3,4-dimethylhexane   3-ethyl-2-methylhexane   3-ethyl-3-methylhexane

CH₃ CH₃
CH₃C—CHCH₂CH₃
CH₃

CH₃ CH₃
CH₃CCH₂CHCH₃
CH₃

CH₃ CH₃
CH₃CH—C—CH₂CH₃
CH₃

2,2,3-trimethylpentane   2,2,4-trimethylpentane   2,3,3-trimethylpentane

CH₃ CH₃ CH₃
CH₃CH—CH—CHCH₃

CH₃ CH₃
CH₃C—CCH₃
CH₃ CH₃

2,3,4-trimethylpentane   2,2,3,3-tetramethylbutane

(c) three isomers of C_9H_{20} with eight carbons in the longest chain: write the longest chain straight across. Don't put anything on the chain that would make it longer.

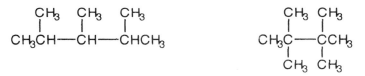

CH₃
CH₃CHCH₂CH₂CH₂CH₂CH₂CH₃
2-methyloctane

CH₃
CH₃CH₂CHCH₂CH₂CH₂CH₂CH₃
3-methyloctane

CH₃
CH₃CH₂CH₂CHCH₂CH₂CH₂CH₃
4-methyloctane

(d) 11 isomers of C_9H_{20} with seven carbons in the longest chain:

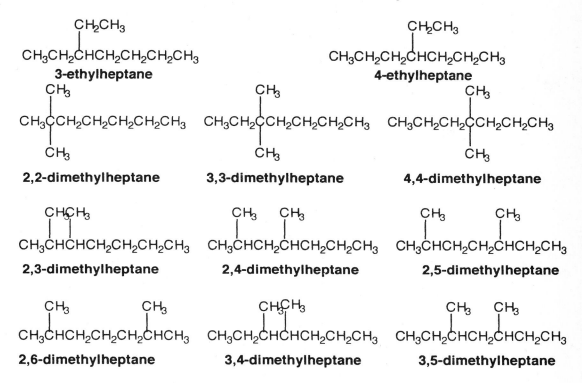

$$CH_2CH_3$$
$$CH_3CH_2CHCH_2CH_2CH_2CH_3$$
3-ethylheptane

$$CH_2CH_3$$
$$CH_3CH_2CH_2CHCH_2CH_2CH_3$$
4-ethylheptane

$$CH_3$$
$$CH_3CCH_2CH_2CH_2CH_2CH_3$$
$$CH_3$$
2,2-dimethylheptane

$$CH_3$$
$$CH_3CH_2CCH_2CH_2CH_2CH_3$$
$$CH_3$$
3,3-dimethylheptane

$$CH_3$$
$$CH_3CH_2CH_2CCH_2CH_2CH_3$$
$$CH_3$$
4,4-dimethylheptane

$$CH_3CH_3$$
$$CH_3CHCHCH_2CH_2CH_2CH_3$$
2,3-dimethylheptane

$$CH_3 \quad CH_3$$
$$CH_3CHCH_2CHCH_2CH_2CH_3$$
2,4-dimethylheptane

$$CH_3 \quad CH_3$$
$$CH_3CHCH_2CH_2CHCH_2CH_3$$
2,5-dimethylheptane

$$CH_3 \qquad CH_3$$
$$CH_3CHCH_2CH_2CH_2CHCH_3$$
2,6-dimethylheptane

$$CH_3CH_3$$
$$CH_3CH_2CHCHCH_2CH_2CH_3$$
3,4-dimethylheptane

$$CH_3 \quad CH_3$$
$$CH_3CH_2CHCH_2CHCH_2CH_3$$
3,5-dimethylheptane

e)   Eight isomers of C_9H_{20} with five carbons in longest chain: Draw a five-carbon chain across the paper in a straight line. Arrange the remaining four carbons in as many ways as possible and see how many ways you can put these variations on the chain without making it longer. The ways to arrange the remaining four carbons are: a) one four-carbon chain; b) a three- and a one-carbon chain, c) 2 two-carbon chains; d) 1 two- and 2 one-carbon chains; and e) 4 one-carbon chains. Variations a) and b) are not usable as there is no way to place a four- or three-carbon unit on a five-carbon chain without extending the longest chain.

$$\begin{array}{c} CH_3 \\ | \\ CH_2 \\ | \\ CH_3CH_2-C-CH_2CH_3 \\ | \\ CH_2 \\ | \\ CH_3 \end{array}$$

$$\begin{array}{c} CH_3 \\ | \\ CH_2 \quad CH_3 \\ | \qquad | \\ CH_3CH_2CH-CCH_3 \\ | \\ CH_3 \end{array}$$

$$\begin{array}{c} CH_3 \\ | \\ CH_2 \\ | \\ CH_3CHCHCHCH_3 \\ | \quad | \\ CH_3 \quad CH_3 \end{array}$$

3,3-diethylpentane   **2,2-dimethyl-3-ethylpentane**   **2,4-dimethyl-3-ethylpentane**

$$\begin{array}{c} CH_3 \\ | \\ CH_2 \\ | \\ CH_3CH-CCH_2CH_3 \\ | \quad | \\ CH_3 \quad CH_3 \end{array}$$

$$\begin{array}{c} CH_3 \quad CH_3 \\ | \qquad | \\ CH_3C-CCH_2CH_3 \\ | \qquad | \\ CH_3 \quad CH_3 \end{array}$$

$$\begin{array}{c} CH_3 \qquad CH_3 \\ | \qquad\quad | \\ CH_3C-CH_2-CCH_3 \\ | \qquad\quad | \\ CH_3 \qquad CH_3 \end{array}$$

2,3-dimethyl-3-ethylpentane   **2,2,3,3-tetramethylpentane**   **2,2,4,4-tetramethylpentane**

$$\begin{array}{c} CH_3 \quad CH_3 \quad CH_3 \\ | \qquad | \qquad | \\ CH_3C-CH-CHCH_3 \\ | \\ CH_3 \end{array}$$

$$\begin{array}{c} CH_3 \quad CH_3 \quad CH_3 \\ | \qquad | \qquad | \\ CH_3CH-C-CHCH_3 \\ | \\ CH_3 \end{array}$$

2,2,3,4-tetramethylpentane   **2,3,3,4-tetramethylpentane**

(f) four isomers of $C_{10}H_{22}$ with nine carbons in the longest chain:

$$\begin{array}{c} CH_3 \\ | \\ CH_3CHCH_2CH_2CH_2CH_2CH_2CH_2CH_3 \end{array}$$

2-methylnonane

$$\begin{array}{c} CH_3 \\ | \\ CH_3CH_2CHCH_2CH_2CH_2CH_2CH_2CH_3 \end{array}$$

3-methylnonane

$$\begin{array}{c} CH_3 \\ | \\ CH_3CH_2CH_2CHCH_2CH_2CH_2CH_2CH_3 \end{array}$$

4-methylnonane

$$\begin{array}{c} CH_3 \\ | \\ CH_3CH_2CH_2CH_2CHCH_2CH_2CH_2CH_3 \end{array}$$

5-methylnonane

(g) two isomers of $C_{10}H_{22}$ with only two alkyl groups on a six carbon chain:

$$CH_3CH_2\overset{\displaystyle CH_2CH_3}{\underset{\displaystyle CH_2CH_3}{C}}CH_2CH_2CH_3$$

3,3-diethylhexane

$$CH_3CH_2\overset{\displaystyle CH_2CH_3}{\underset{\displaystyle CH_2CH_3}{CH}}CHCH_2CH_3$$

3,4-diethylhexane

(h) six isomers of $C_{10}H_{22}$ with five carbons in the longest chain:

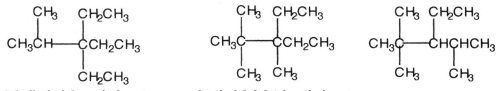

3,3-diethyl-2-methylpentane   **3-ethyl-2,2,3-trimethylpentane**

3-ethyl-2,2,4-trimethylpentane

3-ethyl-2,3,4-trimethylpentane **2,2,3,3,4-pentamethylpentane**

2,2,3,4,4-pentamethylpentane

(i) the isomer of $C_{13}H_{28}$ with the shortest longest chain possible

$$CH_3\overset{\displaystyle CH_3}{\underset{\displaystyle CH_3}{C}}\overset{\displaystyle \overset{\displaystyle CH_3}{CH_2}}{\underset{\displaystyle \overset{\displaystyle CH_2}{CH_3}}{C}}\overset{\displaystyle CH_3}{\underset{\displaystyle CH_3}{C}}CH_3$$

3,3-diethyl-2,2,4,4-tetramethylpentane

(j) five cyclic compounds of C_5H_{10}

Names in order: cyclopentane; methylcyclobutane; ethylcyclopropane
1,1-dimethylcyclopropane; 1,2-dimethylcyclopropane

27

(k) twelve cyclic compounds of C_6H_{12}

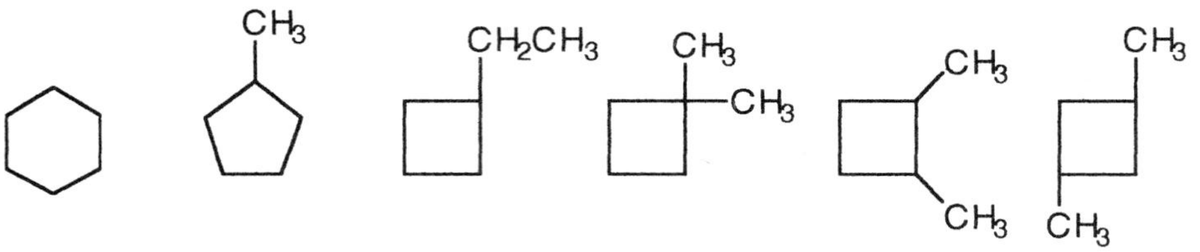

Names in order: **cyclohexane; methylcyclopentane; ethylcyclobutane; 1,1-dimethylcyclobutane; 1,2-dimethylcyclobutane; 1,3-dimethylcyclobutane.**

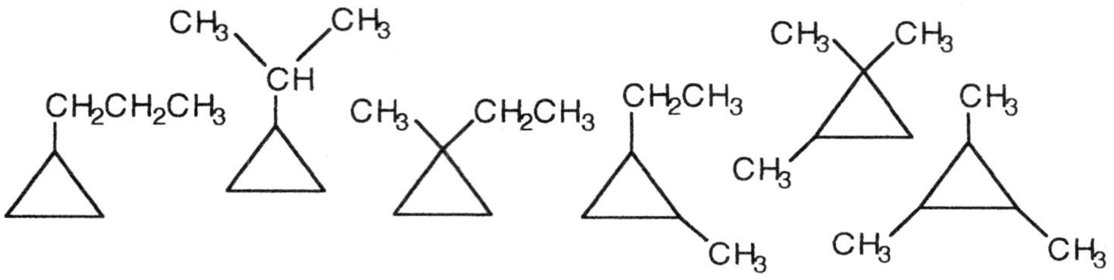

Names in order: **propylcyclopropane; isopropylcyclopropane; 1-ethyl-1-methylcyclopropane; 1-ethyl-2-methylcyclopropane; 1,1,2-trimethylcyclopropane; 1,2,3-trimethylcyclopropane.**

2.14 Nomenclature of Alkanes: Section 2.6

See problem 2.13 for compound names.

2.15 Positional Isomers: Section 2.5

Each carbon that can have a hydrogen replaced with a chlorine is numbered. Identically numbered carbons produce the same isomer upon chlorination.

a) $CH_3CH_2CH_2CH_2CH_2CH_3$
        1   2   3   3   2   1

        3 isomers

b) 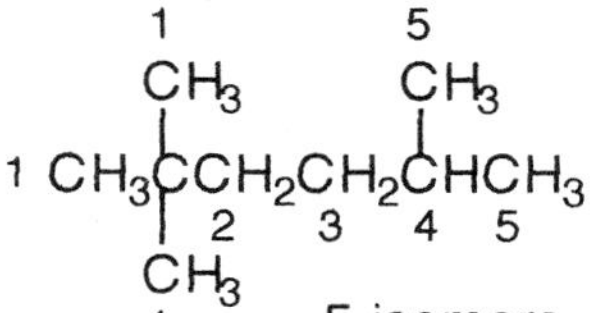   1 $CH_3CCH_2CH_2CHCH_3$
            2   3   4   5

with CH_3 (1) at top position 1 and CH_3 (5) at top position 5, and CH_3 (1) at bottom.

        5 isomers

c)
```
    1        5        1
   CH₃      CH₃      CH₃
    |        |        |
 CH₃CHCH₂CHCH₂CHCH₃
   1   2   3   4   3   2   1
```

5 isomers

d)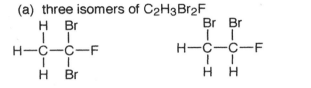

1 isomer

2.16 Skeletal and Positional Isomerism: Sections 2.3 and 2.5

(a) three isomers of $C_2H_3Br_2F$

```
    H   Br                Br  Br               Br   H
    |    |                 |   |                |    |
 H—C—C—F              H—C—C—F             H—C—C—F
    |    |                 |   |                |    |
    H   Br                 H   H               Br   H
```

1,1-dibromo-1-fluoroethane   1,2-dibromo-1-fluoroethane   1,1-dibromo-2-fluoroethane

(b) four isomers of $C_3H_6Br_2$

```
                         Br
                         |
 CH₃CH₂CHBr      CH₃CCH₃      CH₃CH—CH₂      CH₂CH₂CH₂
         |               |            |    |        |        |
         Br              Br           Br   Br       Br       Br
```

1,1-dibromopropane   2,2-dibromopropane   1,2-dibromopropane   1,3-dibromopropane

(c) twelve isomers of C_4H_8BrF

```
         Br            Br
         |             |
 CH₃CH₂CH₂CH   CH₃CH₂CHCH₃   CH₃CH₂CHCH₂   CH₃CHCH₂CH₂
         |             |             |   |           |    |
         F             F             F   Br          F    Br
```

Names in order: 1-bromo-1-fluorobutane; 2-bromo-2-fluorobutane;
1-bromo-2-fluorobutane; 1-bromo-3-fluorobutane

```
 CH₂CH₂CH₂CH₂   CH₃CH₂CH—CH₂   CH₃CHCH₂CH₂   CH₃CH—CHCH₃
  |         |            |    |          |   |          |    |
  F         Br           Br   F          Br  F          Br   F
```

Names in order: 1-bromo-4-fluorobutane; 2-bromo-1-fluorobutane;
3-bromo-1-fluorobutane; 2-bromo-3-fluorobutane

$$CH_3CHCHBr \quad CH_3-C-CH_2 \quad CH_3C-CH_2 \quad CH_2-C-CH_2$$

Names in order: **1-bromo-1-fluoro-2-methylpropane; 1-bromo-2-fluoro-2-methylpropane; 2-bromo-1-fluoro-2-methylpropane; 1-bromo-3-fluoro-2-methylpropane**

(d) nine isomers of $C_5H_{10}Br_2$ with four carbons in the longest chain

For simplicity let us just show the required carbon skeleton and move the two bromines around systematically. First, place two bromines on the same carbon.

1.1-dibromo-3-methylbutane   2,2-dibromo-3-methylbutane   1,1-dibromo-2-methylbutane

Now put a bromine on carbon-1 and vary the position of the other bromine. Finally, draw isomers in which the bromines are on the middle two carbons and the two carbons on the other end.

1,2-dibromo-2-methylbutane   1,3-dibromo-2-methylbutane   1,4-dibromo-2-methylbutane

1,3-dibromo-2-ethylpropane   2,3-dibromo-2-methylbutane   1,2-dibromo-3-methylbutane

(e) five isomers of $C_6H_{13}Cl$ with four carbons in the longest chain.

Again let us draw the carbon skeletons, there are two, and vary the position of the chlorine.

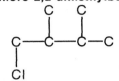

1-chloro-2,2-dimethylbutane   **3-chloro-2,2-dimethylbutane**   **1-chloro-3,3-dimethylbutane**

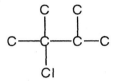

1-chloro-2,3-dimethylbutane   **2-chloro-2,3-dimethylbutane**

2.17 Nomenclature of Halogenated Alkanes: Section 2.6D
See problem 2.16 for the names.

2.18 Nomenclature of Alkanes: Section 2.6
(a) propane   (b) decane   (c) octane

2.19 Nomenclature of Alkanes: Section 2.6
(a) 4-methylnonane;   (b) 5-propyldecane;   (c) 2,5-dimethylhexane;
(d) 4-ethyl-2-methylhexane;   (e) 2,2,4,6-tetramethylheptane;
(f) 6-butyl-2,4,4-trimethylnonane;   (g) 2-cyclobutylhexane;
(h) 3,5,7-trimethyldecane (note that the longest chain is not written straight
across the page; the two fragments below are part of the longest chain.)
(i) 2,2,3,3-tetramethylbutane;   (j) ethylcyclopropane;
(k) isopropylcyclopentane;   (l) 1-butyl-4-t-butylcyclohexane;
(m) 2,2-dimethylbutane;   (n) 2,4-dimethylhexane;
(o) 4-ethyl-2,2-dimethylhexane

2.20 Nomenclature of Halogenated Alkanes: Section 2.6D
(a) triiodomethane;   (b) 2-bromo-4-methylbutane;
(c) 1,3-dibromo-4-fluoro-2,4-dimethylpentane

2.21 IUPAC Nomenclature: Section 2.6

(a) CCl_2F_2 (b) 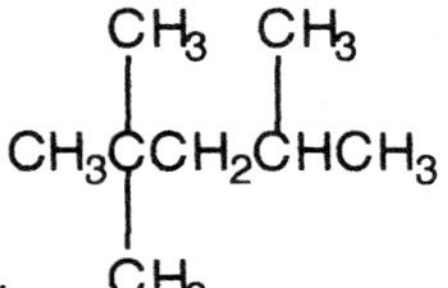 $CH_3CCH_2CHCH_3$ (with CH_3, CH_3 substituents)   (c)

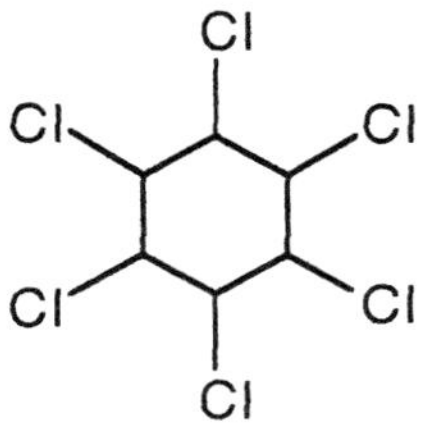

(d) 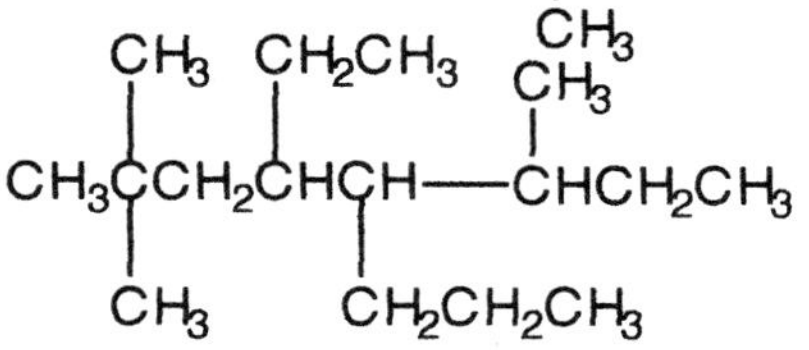 $CH_3CCH_2CHCH \text{---} CHCH_2CH_3$ (with CH_3, CH_2CH_3, CH_3, CH_3, CH_3, $CH_2CH_2CH_3$ substituents)

2.22 Conformational Isomers: Section 2.7

In each case, draw the compound, determine what three groups are on each of the carbons to be placed in the Newman projection, draw the Newman projection with the bonds for the front carbon emanating from the center of the circle and those of the back carbon coming from the perimeter, and put the three groups on each carbon. Rotate between staggered and eclipsed conformations to get the extreme forms.

a) $CH_3CH_2CH_3$

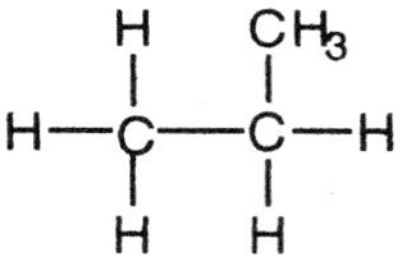

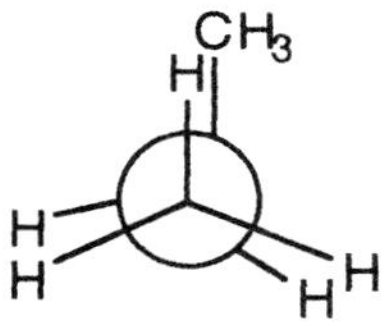

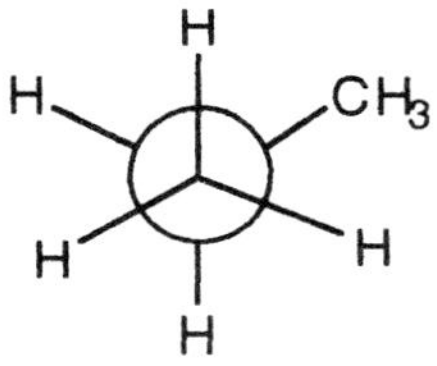

b) $HOCH_2 \text{---} CH_2OH$

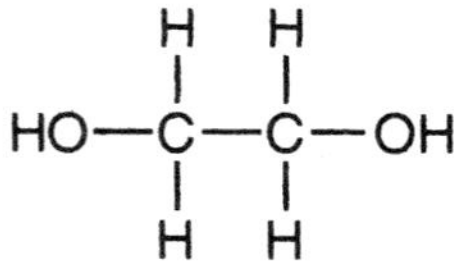

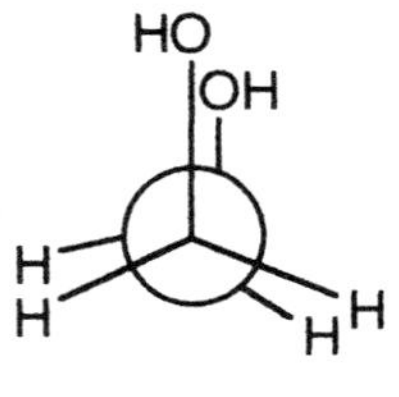

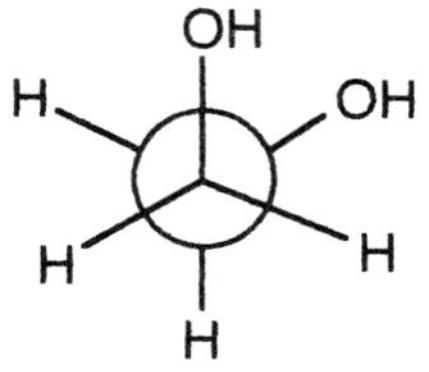

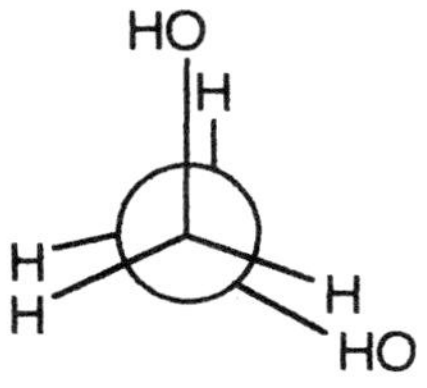

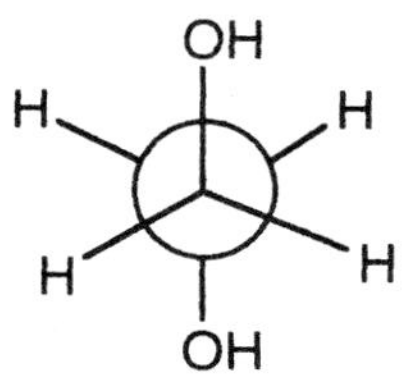

c)

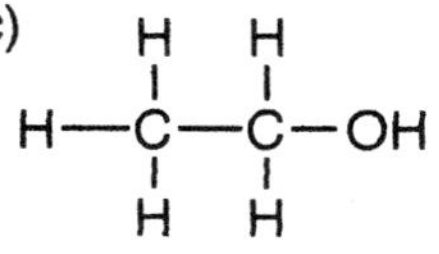

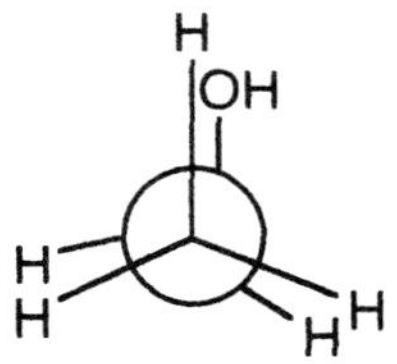

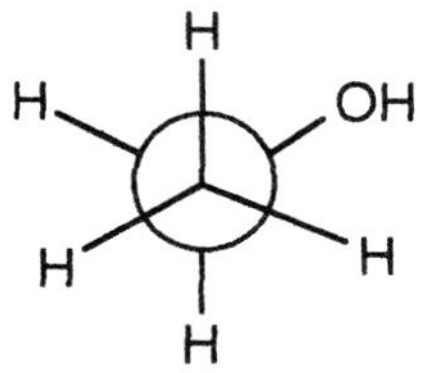

d)

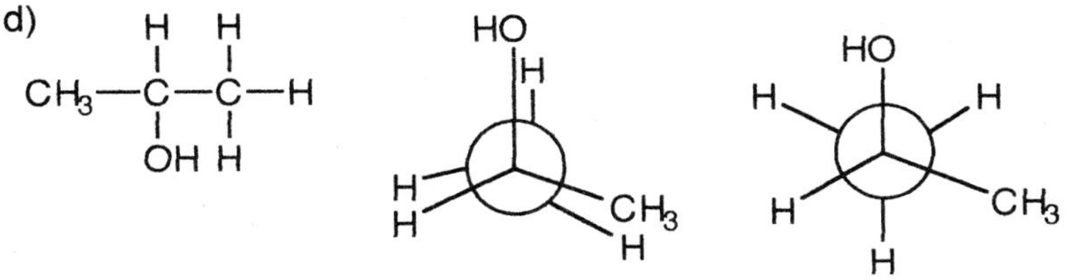
H H
CH3—C—C—H
OH H
HO
H
H
H
CH3
H
HO
H
H
H
CH3
H

2.23 Conformational Isomerism in Substituted Cyclohexanes:
Section 2.8 B-D

In doing these problems, remember that equatorial positions are roomier than axial positions and substituents occupy equatorial positions preferentially when possible. If there are two groups on the cyclohexane chair, the conformation in which the larger group is equatorial, or, if possible, both groups are equatorial, is preferred.

(a)

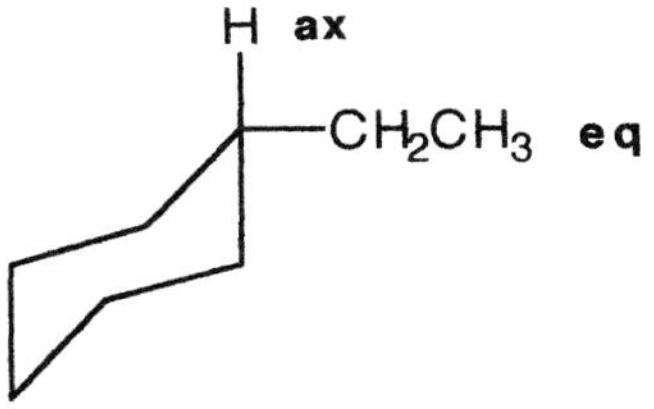
H ax
CH2CH3 eq

ethylcyclohexane

(b) most stable chair forms of 1,2-; 1,3-; and 1,4-dimethylcyclohexane

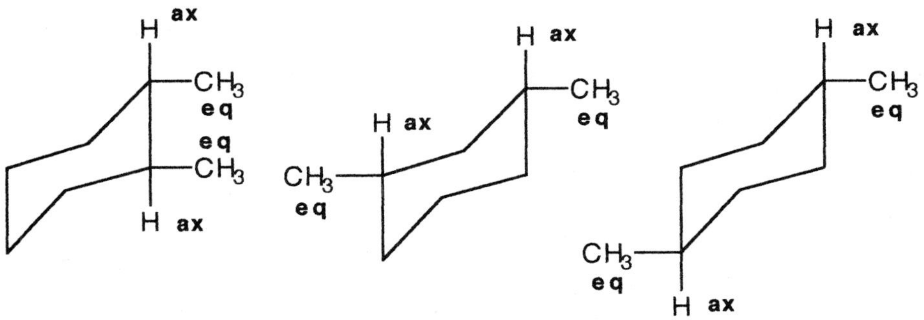
H ax
CH3 eq
eq CH3
H ax
H ax
CH3 eq
H ax
CH3 eq
H ax
CH3 eq
CH3 eq
H ax

(c)   least stable chair forms of compounds in part b

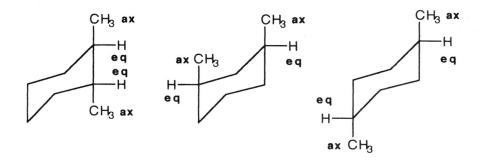

(d)   1,2-dimethylcyclohexane with one group axial and one equatorial

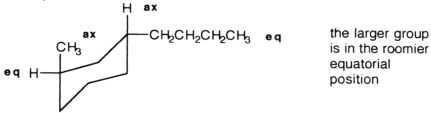

(e) most stable chair form of 1-butyl-3-methylcyclohexane with one group axial and one equatorial

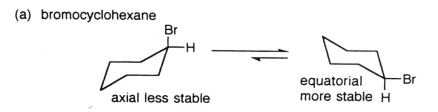

the larger group
is in the roomier
equatorial
position

2.24 Conformational Isomerism in Substituted Cyclohexanes:
Section 2.8D

(a) bromocyclohexane

(b)

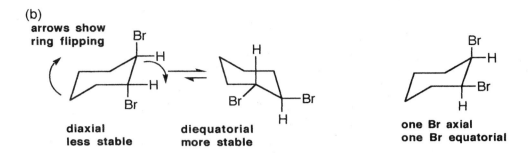

arrows show
ring flipping

diaxial
less stable

diequatorial
more stable

one Br axial
one Br equatorial

(c)   1-ethyl-3-methylcyclohexane

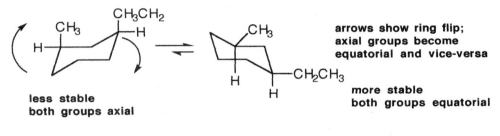

arrows show ring flip;
axial groups become
equatorial and vice-versa

less stable
both groups axial

more stable
both groups equatorial

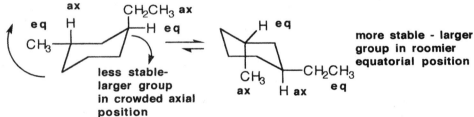

more stable - larger
group in roomier
equatorial position

less stable-
larger group
in crowded axial
position

(d)   1-ethyl-4-methylcyclohexane

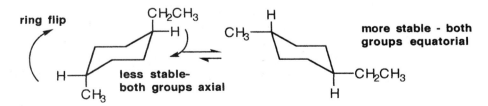

ring flip

more stable - both
groups equatorial

less stable-
both groups axial

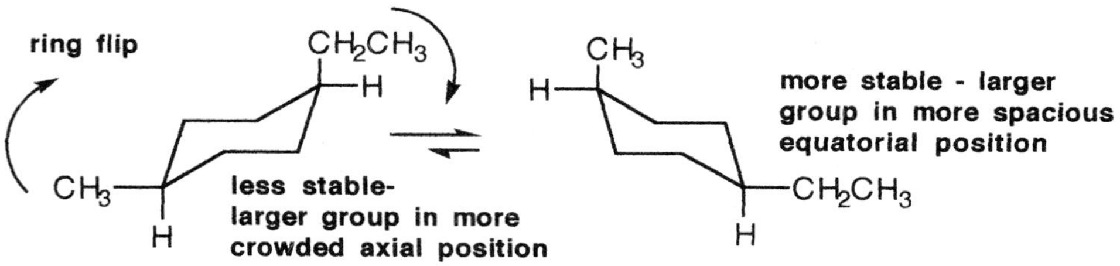

ring flip

CH₂CH₃
—H

**less stable-
larger group in more
crowded axial position**

CH₃—
H

CH₃
H—

**more stable - larger
group in more spacious
equatorial position**

—CH₂CH₃
H

(e) 1,3,5-tribromocyclohexane

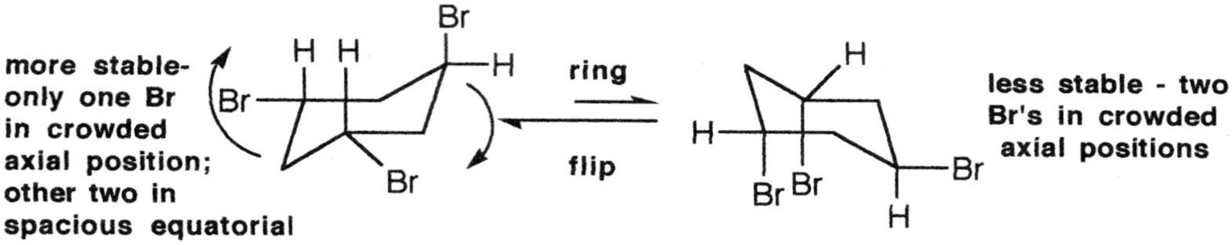

**more stable-
only one Br
in crowded
axial position;
other two in
spacious equatorial**

H H
Br— —H

Br

Br

ring

flip

H

H—

Br Br

Br

H

**less stable - two
Br's in crowded
axial positions**

2.25 Conformational Isomerism in Cyclohexane: Section 2.8B

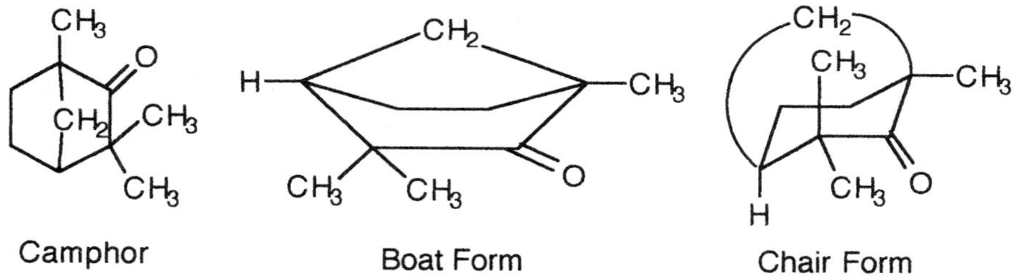

CH₃
O
CH₂ CH₃
CH₃

Camphor

CH₂
H—   —CH₃
CH₃ CH₃
O

Boat Form

CH₂
CH₃
—CH₃
CH₃ O
H

Chair Form

The structures show a one-carbon bridge between the first and fourth carbons of the ring. In the boat form, the first and fourth carbons are directed toward one another and are easily tied together by the single bonds to the —CH2— bridge. However, in the chair form, these two carbons are so far removed from one another that they cannot be bridged by a single carbon. The two single bonds are not long enough nor can they be conveniently directed in the necessary geometry.

2.26 Geometric Isomerism: Section 2.8E

(a)   1,2-dimethylcyclopropane

cis 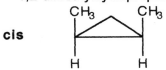   **trans**

(b)   1-bromo-3-chlorocyclobutane

cis 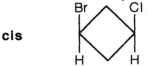   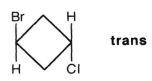 **trans**

2.27 Geometric Isomerism: Section 2.8E

(a)   1,2-dimethylcyclohexane

cis 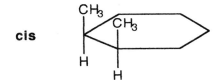   **trans**

(b)   1,3-dimethylcyclohexane

cis 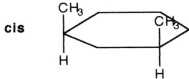   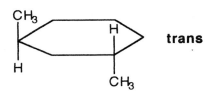 **trans**

(c)   1,4-dimethylcyclohexane

cis 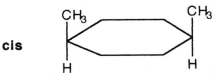   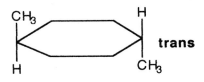 **trans**

2.28 Geometric Isomerism: Section 2.8E

(a)   1,2-dibromo-3-chlorocyclopropane

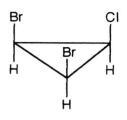 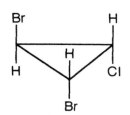 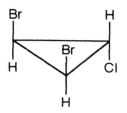

(b)   1,2,3-tribromocyclopropane

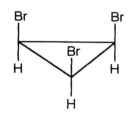 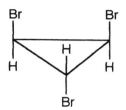

(c)   1,2,4-tribromocyclopentane

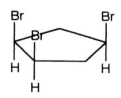 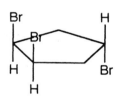 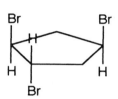

(d)   1,4-dibromo-4-chlorocyclopentane

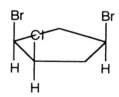 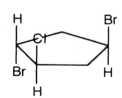 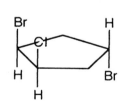 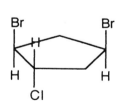

2.29 Geometric Isomerism in the Cyclohexane Chair: Section 2.8

To understand cis and trans on the
cyclohexane chair, first draw the chair
and insert bonds for the axial and
equatorial positions (these are labeled
in the diagram). Then note on each
carbon, one bond can be considered
up (**u**) and one down (**d**) relative to one
another. In disubstituted cyclohexanes,
if both groups are up or both down the
isomer is cis. If they are up/down or
down/up, the isomer is trans. Rationalize
this with the chart in the text. For example,
in 1,2-disubstituted up/up or down/down is
ax/eq or eq/ax and up/down or down/up is
ax/ax or eq/eq.

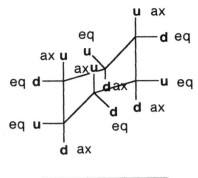

| cis | u/u | or | d/d |
| trans | u/d | or | d/u |

(a)   cis 1,2-dimethylcyclohexane

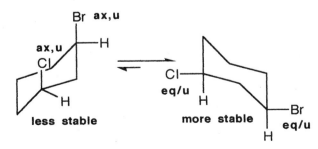

cis is up/up or down/down
which is ax/eq or eq/ax in
this case. The two
conformations are essentially
identical and of equal
stability.

(b)   cis 1-bromo-3-chlorocyclohexane

Cis is ax/ax or eq/eq in
1,3-disubstituted chairs
since this is u/u or d/d.
The eq/eq is more stable
since the two large groups
are in the more spacious
equatorial positions.

(c)   trans 1,4-diethylcyclohexane

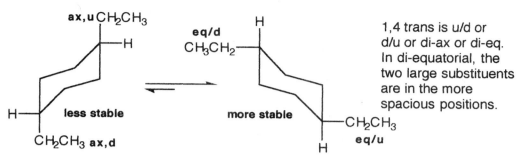

1,4 trans is u/d or
d/u or di-ax or di-eq.
In di-equatorial, the
two large substituents
are in the more
spacious positions.

(d)   cis 1-ethyl-4-methylcyclohexane

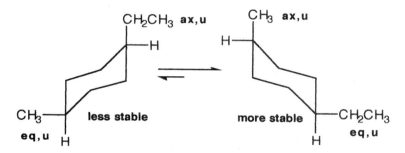

Cis is u/u or d/d
which as shown is
ax/eq or eq/ax for 1,4
disubstitution. The
conformer in which the
larger ethyl group is
equatorial is the more
stable since equatorial
positions are more sp

(e)   trans 1-ethyl-3-methylcyclohexane

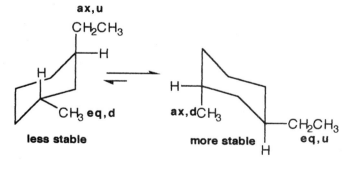

Trans is u/d as shown
here. The more stable
conformer has the
larger ethyl group
in the more spacious
equatorial position and
the smaller methyl
group in the more
crowded axial position.

2.30 Stereoisomerism: Section 2.8

1,2-dibromocyclopropane

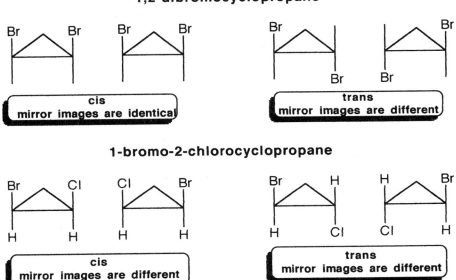

In the first example the two substituents are the same and thus the molecule has symmetry. As a result the cis isomer has only one form; the trans isomer has non-superimposable mirror images. In 1-bromo-2-chlorocyclopropane, the two substituents are different and both the cis and trans isomer have non-superimposable mirror images.

2.31 Physical Properties: Section 2.9

    a)   Boiling points increase with molecular weight within a homologous series. Both examples are alkanes; ethane has the greater molecular weight.

b)   These two compounds are isomers and have the same molecular weight. The unbranched isomer has greater surface area and thus there are more opportunities for intermolecular attraction. The greater the attractions between molecules, the more energy necessary to break these attractions and the higher is the boiling point. The branched isomer has less surface area and less intermolecular interactions as a result.

    c) CBr_4 has the higher molecular weight and thus the higher boiling point.

d)   Cyclohexane is more symmetrical and compact and consequently forms a more stable crystal lattice. Such a lattice requires more energy to disrupt and cyclohexane has a higher melting point.

e)   Melting points generally increase with molecular weight (all other factors being equal) since it requires more energy (higher temperatures) to give the heavier molecules enough motion to break out of a stationary crystal lattice.

2.32 Combustion: Connection 2

(a) CH_4 + $2O_2$ $\longrightarrow$ CO_2 + $2H_2O$

(b) C_3H_8 + $5O_2$ $\longrightarrow$ $3CO_2$ + $4H_2O$

(c) $2C_8H_{18}$ + $25O_2$ $\longrightarrow$ $16CO_2$ + $18H_2O$

2.33 Petroleum Fractions: Connection 2

a)   Gas - any hydrocarbon with 1-4 carbons; for example CH_4, the main component of natural gas.

b)   Gasoline - any hydrocarbon with 5-10 carbons; for example

$$\begin{array}{cc} CH_3 & CH_3 \\ | & | \\ CH_3CCH_2CHCH_3 \\ | \\ CH_3 \end{array}$$

c)   Kerosene - any hydrocarbon with 11-18 carbons; for example $CH_3(CH_2)_{10}CH_3$

d)   Gas-Oil - any hydrocarbon with 15-18 carbons such as $CH_3(CH_2)_{16}CH_3$

e)   Wax-Oil - hydrocarbons with 18-20 carbons; $CH_3(CH_2)_{17}CH_3$

f)   Wax - high molecular weight hydrocarbons usually with 20 or more carbons; for example $CH_3(CH_2)_{22}CH_3$

3

Alkenes and Alkynes:
Structure and Nomenclature

CHAPTER SUMMARY

Alkenes are hydrocarbons in which there is at least one **carbon-carbon double bond**; **alkynes** have at least one **carbon-carbon triple bond**. Both are termed **unsaturated** because the carbons involved in the multiple bonds do not have the maximum number of bonded atoms. Alkenes have the general formula C_nH_{2n} and alkynes are C_nH_{2n-2}.

In **IUPAC nomenclature** double bonds are described with an **-ene** suffix attached to the name of the longest chain of carbons; the suffix is **-yne** for alkynes. The carbon chain is numbered to give the lowest possible number to the multiple bond nearest the end; when there is a choice, double bonds take precedence.

Alkenes and alkynes exhibit **skeletal isomerism** in which the carbon chain is varied and **positional isomerism** where the position of the multiple bond is different. **Functional isomers** differ in the class of compounds to which they belong. Common functional groups in organic chemistry include: **alkanes, alkenes, alkynes, aromatic hydrocarbons, carboxylic acids, aldehydes, ketones, alcohols, ethers, amines.**

Alkenes in which there are two different groups on each of the double-bonded carbons are capable of exhibiting **geometric isomerism.** In the **cis isomer** two identical or comparable groups are on the same side of the double bond and in the **trans isomer** they are on opposite sides. The **pi-bond**

43

restricts rotation around the carbon-carbon double bond and prevents interconversion of the two isomeric forms.

One unit of unsaturation is expressed as a **double bond or a ring**. A **triple bond** is **two units of unsaturation.**

Connections 3.1 is about oral contraceptives.

Connections 3.2 is about chemical communication and pheromones.

Connections 3.3 is about geometric isomerism and vision.

SOLUTIONS TO PROBLEMS

3.1   General Molecular Formulas

Alkanes: C_nH_{2n+2}; Cycloalkanes: C_nH_{2n}; Alkenes: C_nH_{2n}; Alkynes: C_nH_{2n-2} Cycloalkenes: C_nH_{2n-2}; Dienes: C_nH_{2n-2}; Cycloalkynes: C_nH_{2n-4}

3.2   Nomenclature of Alkenes and Alkynes

(a) 1-heptene;   (b) 2-heptyne;   (c) 3-methylcyclohexene;
(d) 2,4-heptadiyne

3.3.   Common Nomenclature

$CH_3CH_2CH{=}CH_2$ $CH_2{=}CHBr$ $BrCH_2CH{=}CH_2$ $CH_3C{\equiv}CCH_2CH_3$
Butylene   Vinyl bromide   Allyl bromide   Ethylmethylacetylene

3.4   Functional Isomers

$CH_3C{\equiv}CH$ $CH_2{=}C{=}CH_2$
Propyne   Propadiene   Cyclopropene

3.5   Skeletal and Positional Isomers

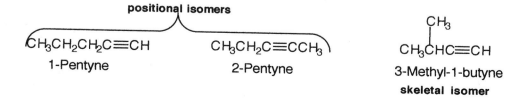

positional isomers

$CH_3CH_2CH_2C{\equiv}CH$ $CH_3CH_2C{\equiv}CCH_3$
1-Pentyne   2-Pentyne

$CH_3\overset{\displaystyle CH_3}{\underset{|}{C}}HC{\equiv}CH$
3-Methyl-1-butyne
skeletal isomer

3.6   Skeletal and Positional Isomers

Each horizontal row of compounds represents a group of positional isomers. The top two compounds are skeletal isomers of the bottom three. Any cyclic compound such as cyclopentane would be a functional isomer as there would no longer be a double bond.

$$CH_3CH_2CH_2CH{=}CH_2 \qquad\qquad CH_3CH_2CH{=}CHCH_3$$

1-Pentene   2-Pentene

$$\overset{\displaystyle CH_3}{\underset{\displaystyle}{|}}$$
CH₃CHCH=CH₂   CH₃C=CHCH₃   CH₃CH₂C=CH₂

3-Methyl-1-butene   2-Methyl-2-butene   2-Methyl-1-butene

3.7   Functional Groups

$$CH_3CH{=}CH_2 \qquad CH_3C{\equiv}CH \qquad CH_3CH_2\overset{O}{\overset{\|}{C}}OH \qquad CH_3CH_2\overset{O}{\overset{\|}{C}}H$$

alkene   **alkyne**   **carboxylic acid**   **aldehyde**

$$CH_3\overset{O}{\overset{\|}{C}}CH_3 \qquad CH_3CH_2CH_2OH \qquad CH_3CH_2CH_2NH_2 \qquad CH_3OCH_2CH_3$$

ketone   **alcohol**   **amine**   **ether**

3.8   Functional Groups

See Connections 3.2 in the text.

3.9   Geometric Isomerism

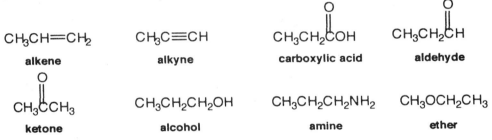

a) cis: / trans:

(b)   For geometric isomerism, each carbon in the double bond must have two different attached groups. In part (a) each carbon has a hydrogen and methyl. In 1-butene, the first carbon has two identical groups, hydrogens, and thus cis-trans isomers do not exist.

3.10 Geometric Isomerism

(a)   1,2-dibromocyclobutane   (b)   1-bromo-3-chlorocyclobutane

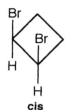
cis

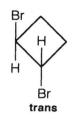
trans

cis

trans

3.11 Units of Unsaturation
   (a) 1   (b) 2   (c) 4   (d) 5

3.12 Skeletal and Positional Isomerism: Section 3.3
   (a)   thirteen alkenes with the formula C_6H_{12} that are skeletal or positional isomers. Note the systematic method for drawing the isomers.

$CH_3CH_2CH_2CH_2CH=CH_2$ $CH_3CH_2CH_2CH=CHCH_3$ $CH_3CH_2CH=CHCH_2CH_3$
1-hexene   **2-hexene**   **3-hexene**

CH_3
$CH_3CH_2CH_2C=CH_2$
2-methyl-1-pentene

CH_3
$CH_3CH_2CHCH=CH_2$
3-methyl-1-pentene

CH_3
$CH_3CHCH_2CH=CH_2$
4-methyl-1-pentene

CH_3
$CH_3CH_2CH=CCH_3$
2-methyl-2-pentene

CH_3
$CH_3CH_2C=CHCH_3$
3-methyl-2-pentene

CH_3
$CH_3CHCH=CHCH_3$
4-methyl-2-pentene

CH_2CH_3
$CH_3CH_2C=CH_2$
2-ethyl-1-butene

$CH_3 CH_3$
$CH_3CH-C=CH_2$
2,3-dimethyl-1-butene

CH_3
$CH_3CCH=CH_2$
CH_3
3,3-dimethyl-1-butene

$CH_3 CH_3$
$CH_3C=CCH_3$
2,3-dimethyl-2-butene

(b) twelve cycloalkanes with the formula C_6H_{12}

Names in order: cyclohexane; methylcyclopentane; ethylcyclobutane; 1,1-dimethylcyclobutane; 1,2-dimethylcyclobutane;1,3-dimethylcyclobutane

Names in order: propylcyclopropane; isopropylcyclopropane; 1-ethyl-1-methylcyclopropane; 1-ethyl-2-methylcyclopropane; 1,1,2-trimethylcyclopropane; 1,2,3-trimethylcyclopropane

(c) the six alkynes with the formula C_6H_{10}

$CH_3CH_2CH_2CH_2C{\equiv}CH$ $CH_3CH_2CH_2C{\equiv}CCH_3$ $CH_3CH_2C{\equiv}CCH_2CH_3$
1-hexyne   **2-hexyne**   **3-hexyne**

3-methyl-1-pentyne   **4-methyl-1-pentyne**   **4-methyl-2-pentyne**

3.13 Nomenclature of Alkenes, Alkynes, and Cycloalkanes
Section 3.2; Please see names in problem 3.12 solutions.

3.14 Nomenclature of Alkenes: Section 3.2
(a) 1-heptene;   (b) 3,4-dimethyl-3-heptene;   (c) 4,4-dimethyl-2-pentene;   (d) 4-ethyl-1-cyclopentene;   (e) 2-cyclopropyl-5-propyl-3-octene;

(f) 3,5-diethyl-8-methyl-3-nonene;   (g) 2,4-octadiene;   (h) 1,2,5,7-
cyclooctatetraene;   (i) 4,5-dibromo-2-methyl-2,4,6-nonatriene.
 Let us illustrate the procedure with the last example:

(i)   1. Nine carbons in the longest chain: **non**
   2. Three triple bonds: **nonatriene**
   3. Number the chain left to right; complete suffix: **2,4,6-nonatriene**
   4. Name all other groups with prefixes. The complete name is:
   4,5-dibromo-2-methyl-2,4,6-nonatriene

3.15 Nomenclature of Alkynes: Section 3.2

 (a) 1-butyne;   (b) 2,2-dibromo-7-methyl-3-octyne;   (c) 4-methyl-2-
pentyne;   (d) 3-ethyl-3-methyl-1-pentyne;   (e) cyclooctyne;
(f) 9-methyl-2,4,6-decatriyne

3.16 Nomenclature of Alkenes and Alkynes: Section 3.2

 (a) 2,9,9-trimethyl-2,5-decadiene;   (b) 2,4,6,8-decatetrayne;
(c) 1-hexyn-4-ene;   (d) 2-hexen-4-yne; (e) 2-methyl-1,3-decadien-5,7,9-triyne
A few comments: In (c) the triple bond is the first multiple bond reached in
numbering and thus numbering is right to left. In (d) whichever way you number
you encounter a multiple bond at carbon-2. In these cases, the double bond
takes precedence and the numbering is left to right. The last problem is
illustrated in detail below.

(e)   1. The longest chain is ten carbons: **dec**
   2. There are two double bonds: **decadien**
    and three triple bonds: **decadien triyne**
   3. Either way you number the chain, a multiple bond is at carbon-1;
    give precedence to the double bond: **1,3-decadien-5,7,9-triyne**
   4. The methyl is named with a prefix; the complete name is:
   2-methyl-1,3-decadien-5,7,9-triyne

3.17 IUPAC Nomenclature: Section 3.2

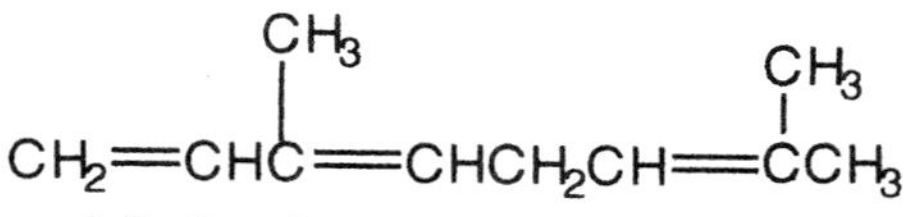

CH_2=CHC=CHCH$_2$CH=CCH$_3$ **3,7-dimethyl-1,3,6-octatriene**

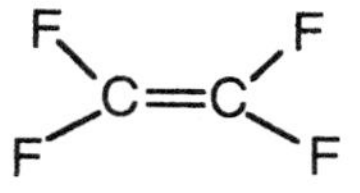

tetrafluoroethene 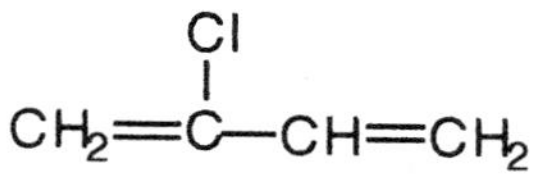   **2-chloro-1,3-butadiene**

$$CH_2\!\!=\!\!CH\!-\!CH\!\!=\!\!CH\!-\!C\!\!\equiv\!\!C\!-\!C\!\!\equiv\!\!C\!-\!C\!\!\equiv\!\!C\!-\!CH\!\!=\!\!CHCH_3$$

1,3,11-tridecatrien-5,7,9-triyne

$$CH_2\!\!=\!\!CCl_2$$

1,1-dichloroethene

$$\underset{\displaystyle CH_3}{\underset{|}{CH_3}}\overset{\displaystyle \overset{CH_3}{|}\;\;\overset{CH_3}{|}}{CCH_2CHCH_3}$$

2,2,4-trimethylpentane

3.18 Positional Isomerism: Section 3.3

(a)

$$CH_3CH_2CH_2CH_2CH_2CH_2CH\!\!=\!\!CH_2 \qquad CH_3CH_2CH_2CH_2CH_2CH\!\!=\!\!CHCH_3$$

$$CH_3CH_2CH_2CH_2CH\!\!=\!\!CHCH_2CH_3 \qquad CH_3CH_2CH_2CH\!\!=\!\!CHCH_2CH_2CH_3$$

(b)

$$\underset{|}{\overset{CH_3}{CH_3CHCH_2CH_2OH}} \quad \underset{\underset{OH}{|}}{\overset{CH_3}{CH_3CHCHCH_3}} \quad \underset{\underset{OH}{|}}{\overset{CH_3}{CH_3CCH_2CH_3}} \quad \underset{\underset{OH}{|}}{\overset{CH_3}{CH_2CHCH_2CH_3}}$$

(c)

$$\overset{\displaystyle O}{\overset{\|}{CH_3CCH_2CH_2CH_2CH_2CH_3}} \quad \overset{\displaystyle O}{\overset{\|}{CH_3CH_2CCH_2CH_2CH_2CH_3}} \quad \overset{\displaystyle O}{\overset{\|}{CH_3CH_2CH_2CCH_2CH_2CH_3}}$$

(d)

$$\underset{\underset{CH_3}{|}}{\overset{\overset{CH_3}{|}}{CH_3CCH_2CH_2CH_2C\!\equiv\!CH}} \qquad\qquad \underset{\underset{CH_3}{|}}{\overset{\overset{CH_3}{|}}{CH_3CCH_2CH_2C\!\equiv\!CCH_3}}$$

$$\underset{\underset{CH_3}{|}}{\overset{\overset{CH_3}{|}}{CH_3CCH_2C\!\equiv\!CCH_2CH_3}} \qquad\qquad \underset{\underset{CH_3}{|}}{\overset{\overset{CH_3}{|}}{CH_3CC\!\equiv\!CCH_2CH_2CH_3}}$$

(e)

(f)

$$CH_3\overset{\overset{\displaystyle CH_3}{|}}{C}HCH_2\overset{\overset{\displaystyle CH_3}{|}}{C}HCH_2Br \qquad CH_3\overset{\overset{\displaystyle CH_3}{|}}{C}HCH_2\underset{\underset{\displaystyle Br}{|}}{\overset{\overset{\displaystyle CH_3}{|}}{C}}CH_3 \qquad CH_3\overset{\overset{\displaystyle CH_3}{|}}{C}H\underset{\underset{\displaystyle Br}{|}}{\overset{\overset{\displaystyle CH_3}{|}}{C}}HCHCH_3$$

3.19 Skeletal and Positional Isomerism: Section 3.3

(a)   four isomers of C_3H_9N

$$CH_3CH_2CH_2NH_2 \quad CH_3\underset{\underset{\displaystyle NH_2}{|}}{C}HCH_3 \quad CH_3NHCH_2CH_3 \quad CH_3\overset{\overset{\displaystyle CH_3}{|}}{N}CH_3$$

(b)   eight isomers of $C_4H_{11}N$

$$CH_3CH_2CH_2CH_2NH_2 \quad CH_3CH_2\underset{\underset{\displaystyle NH_2}{|}}{C}HCH_3 \quad CH_3\overset{\overset{\displaystyle CH_3}{|}}{C}HCH_2NH_2 \quad CH_3\underset{\underset{\displaystyle NH_2}{|}}{\overset{\overset{\displaystyle CH_3}{|}}{C}}CH_3$$

$$CH_3NHCH_2CH_2CH_3 \quad CH_3NH\underset{\underset{\displaystyle CH_3}{|}}{C}HCH_3 \quad CH_3CH_2NHCH_2CH_3 \quad CH_3\overset{\overset{\displaystyle CH_3}{|}}{N}CH_2CH_3$$

3.20 Functional Isomers: Section 3.4

Functional isomers of $C_4H_8O_2$

a) $CH_3CH_2CH_2\overset{\overset{\displaystyle O}{||}}{C}OH$ b) $HOCH_2CH_2CH_2\overset{\overset{\displaystyle O}{||}}{C}H$ c) $HOCH_2CH_2\overset{\overset{\displaystyle O}{||}}{C}CH_3$

    carboxylic acid   **alcohol-aldehyde**   **alcohol-ketone**

d) $CH_3OCH_2CH_2\overset{\overset{\displaystyle O}{||}}{C}H$ e) $CH_3OCH_2\overset{\overset{\displaystyle O}{||}}{C}CH_3$ f) $HOCH_2CH=CHCH_2OH$

    ether-aldehyde   **ether-ketone**   **alkene-dialcohol**

g) $CH_3OCH=CHOCH_3$ h)   i) HO–⬦–OH j)

    alkene-diether   **alcohol-ether**   **dialcohol**   **diether**

50

3.21 Skeletal, Positional, and Functional Isomers: Section 3.3

(a) aldehydes with the formula C_4H_8O:

$$CH_3CH_2CH_2\overset{\overset{\displaystyle O}{\|}}{C}H \qquad CH_3\overset{\overset{\displaystyle CH_3}{|}}{C}H\text{—}\overset{\overset{\displaystyle O}{\|}}{C}H$$

(b) ketones with the formula $C_6H_{12}O$

$$CH_3\overset{\underset{\displaystyle O}{\|}}{C}CH_2CH_2CH_2CH_3 \qquad CH_3CH_2\overset{\underset{\displaystyle O}{\|}}{C}CH_2CH_2CH_3 \qquad CH_3\overset{\overset{\displaystyle CH_3}{|}}{C}H\overset{\underset{\displaystyle O}{\|}}{C}CH_2CH_3$$

$$CH_3\overset{\overset{\displaystyle CH_3}{|}}{C}HCH_2\overset{\underset{\displaystyle O}{\|}}{C}CH_3 \qquad CH_3\overset{\overset{\displaystyle CH_3}{|}}{\underset{\underset{\displaystyle O}{\|}}{C}}HCH_2CH_3 \qquad CH_3\overset{\overset{\displaystyle CH_3}{|}}{C}\text{—}\overset{\underset{\displaystyle CH_3}{|}}{\underset{\underset{\displaystyle O}{\|}}{C}}CH_3$$

(c) aldehydes or ketones with the formula $C_5H_{10}O$

$$CH_3CH_2CH_2CH_2\overset{\overset{\displaystyle O}{\|}}{C}H \qquad CH_3CH_2CH_2\overset{\overset{\displaystyle O}{\|}}{C}CH_3 \qquad CH_3CH_2\overset{\overset{\displaystyle O}{\|}}{C}CH_2CH_3$$

$$CH_3\overset{\overset{\displaystyle CH_3}{|}}{C}HCH_2\overset{\overset{\displaystyle O}{\|}}{C}H \qquad CH_3\overset{\overset{\displaystyle CH_3}{|}}{C}H\text{—}\overset{\overset{\displaystyle O}{\|}}{C}CH_3 \qquad H\overset{\overset{\displaystyle O}{\|}}{C}\text{—}\overset{\overset{\displaystyle CH_3}{|}}{C}HCH_2CH_3 \qquad CH_3\overset{\overset{\displaystyle CH_3}{|}}{C}\text{—}\overset{\overset{\displaystyle O}{\|}}{C}H$$

(d) carboxylic acids with the formula $C_6H_{12}O_2$

$$CH_3CH_2CH_2CH_2CH_2\overset{\overset{\displaystyle O}{\|}}{C}OH \qquad CH_3CH_2CH_2\overset{\underset{\underset{\displaystyle CH_3}{|}}{C}}{H}\overset{\overset{\displaystyle O}{\|}}{C}OH \qquad CH_3CH_2\overset{\underset{\underset{\displaystyle CH_3}{|}}{C}}{H}CH_2\overset{\overset{\displaystyle O}{\|}}{C}OH$$

$$CH_3\overset{\underset{\underset{\displaystyle CH_3}{|}}{C}}{H}CH_2CH_2\overset{\overset{\displaystyle O}{\|}}{C}OH \qquad CH_3CH_2\overset{\overset{\displaystyle CH_3}{|}}{\underset{\underset{\displaystyle CH_3}{|}}{C}}\text{—}\overset{\overset{\displaystyle O}{\|}}{C}OH \qquad CH_3\overset{\underset{\underset{\displaystyle CH_3}{|}}{C}}{H}\text{—}\overset{\underset{\underset{\displaystyle CH_3}{|}}{C}}{H}\overset{\overset{\displaystyle O}{\|}}{C}OH$$

$$CH_3\overset{\overset{\displaystyle CH_3}{|}}{\underset{\underset{\displaystyle CH_3}{|}}{C}}CH_2\overset{\overset{\displaystyle O}{\|}}{C}OH \qquad CH_3CH_2\overset{\underset{\underset{\displaystyle CH_3CH_2}{|}}{C}}{H}\overset{\overset{\displaystyle O}{\|}}{C}OH$$

(e) alcohols or ethers with the formula $C_4H_{10}O$

$$CH_3CH_2CH_2CH_2OH \qquad CH_3CH_2\underset{\underset{\displaystyle OH}{|}}{C}HCH_3 \qquad CH_3\underset{\underset{\displaystyle CH_3}{|}}{C}HCH_2OH \qquad CH_3\underset{\underset{\displaystyle OH}{|}}{\overset{\overset{\displaystyle CH_3}{|}}{C}}CH_3$$

$$CH_3OCH_2CH_2CH_3 \qquad CH_3O\underset{\underset{\displaystyle CH_3}{|}}{C}HCH_3 \qquad CH_3CH_2OCH_2CH_3$$

(f) alcohols or ethers with the formula $C_5H_{12}O$

$$CH_3CH_2CH_2CH_2CH_2OH \qquad CH_3CH_2CH_2\underset{\underset{\displaystyle OH}{|}}{C}HCH_3 \qquad CH_3CH_2\underset{\underset{\displaystyle OH}{|}}{C}HCH_2CH_3$$

$$CH_3\overset{\overset{\displaystyle CH_3}{|}}{C}HCH_2\underset{\underset{\displaystyle OH}{|}}{C}H_2 \qquad CH_3\overset{\overset{\displaystyle CH_3}{|}}{C}H\underset{\underset{\displaystyle OH}{|}}{C}HCH_3 \qquad CH_3\overset{\overset{\displaystyle CH_3}{|}}{\underset{\underset{\displaystyle OH}{|}}{C}}CH_2CH_3 \qquad \underset{\underset{\displaystyle OH}{|}}{C}H_2\overset{\overset{\displaystyle CH_3}{|}}{C}HCH_2CH_2$$

$$CH_3\overset{\overset{\displaystyle CH_3}{|}}{\underset{\underset{\displaystyle CH_3}{|}}{C}}CH_2OH \qquad CH_3OCH_2CH_2CH_2CH_3 \qquad CH_3O\overset{\overset{\displaystyle CH_3}{|}}{C}HCH_2CH_3$$

$$CH_3OCH_2\overset{\overset{\displaystyle CH_3}{|}}{C}HCH_3 \qquad CH_3O\overset{\overset{\displaystyle CH_3}{|}}{\underset{\underset{\displaystyle CH_3}{|}}{C}}CH_3 \qquad CH_3CH_2OCH_2CH_2CH_3 \qquad CH_3CH_2O\overset{\overset{\displaystyle CH_3}{|}}{C}HCH_3$$

3.22 Functional Isomerism: Section 3.4

Six functional isomers with the formula $C_5H_{10}O$

$$CH_3CH_2CH_2CH_2\overset{\overset{\displaystyle O}{\|}}{C}H \qquad CH_3CH_2CH_2\overset{\overset{\displaystyle O}{\|}}{C}CH_3 \qquad CH_2{=}CHCH_2CH_2CH_2OH$$

      aldehyde   **ketone**   **alkene-alcohol**

$$CH_2{=}CHCH_2CH_2OCH_3$$

        alkene-ether   **alcohol**   **ether**

3.23 Geometric Isomerism in Alkenes: Section 3.5

See Example 3.5 in the text for a procedure for drawing geometric isomers.

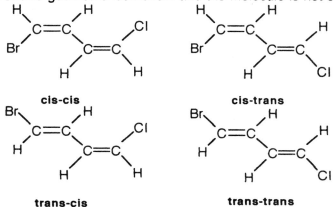

3.24 Geometric Isomerism in Alkenes: Section 3.5

Consider each double bond individually and be sure to draw the trigonal geometry carefully.

(a) This molecule is capable of exhibiting four geometric isomers since each double bond shows geometric isomerism and the molecule is not symmetrical.

cis-cis   **cis-trans**

trans-cis   **trans-trans**

(b)   This molecule has two double bonds both capable of geometric isomerism. However, each double bond has the same attached groups and the molecule is symmetrical. As a result the cis/trans and trans/cis isomers are the same and the total number of geometric isomers is only three.

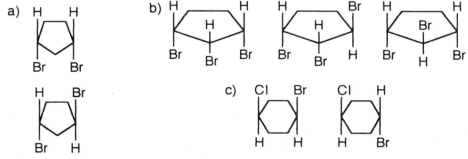

cis-cis   **trans-cis or cis-trans**   **trans-trans**

(c)   The compound has three double bonds capable of geometric isomerism and it is not symmetrical; there are eight possible isomers:

cis cis cis	cis cis trans	cis trans trans	trans trans trans
	cis trans cis	trans cis trans	
	trans cis cis	trans trans cis	

cis-cis-cis   **trans-trans-trans**

3.25 Geometric Isomerism in Cyclic Compounds: Section 2.8E

a)

b)

c)

3.26 Geometric Isomerism: Section 3.5

(a)

(b)

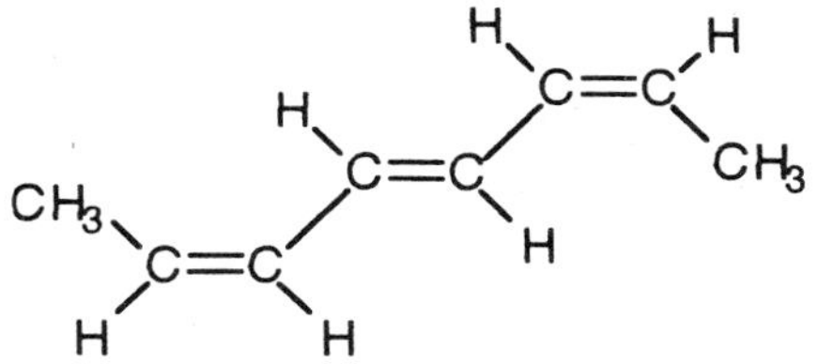

(c)

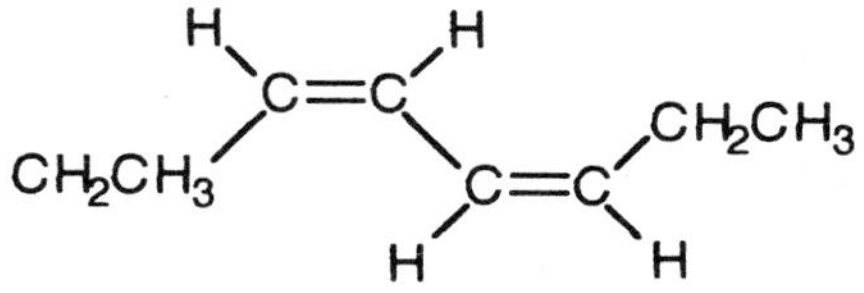

3.27 Expressing Units of Unsaturation: Section 3.6

    a) C_8H_{10}: One isomer with as many triple bonds as possible, one with as many double bonds as possible, and one with as many rings as possible.

$$HC \equiv CC \equiv CCH_2CH_2CH_2CH_3 \quad CH_2 = CHCH = CHCH = CHCH = CH_2$$

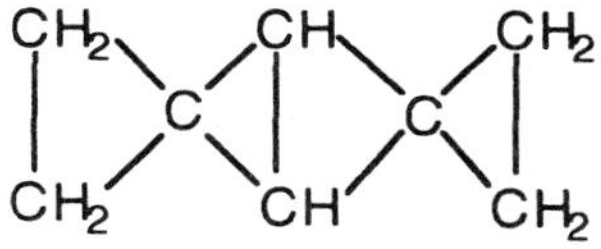

    b) C_6H_8: Six isomers that differ in the expression of the three units of unsaturation.

$$CH_2 = CH - C \equiv C - CH_2CH_3 \qquad CH_2 = CH - CH = CH - CH = CH_2$$

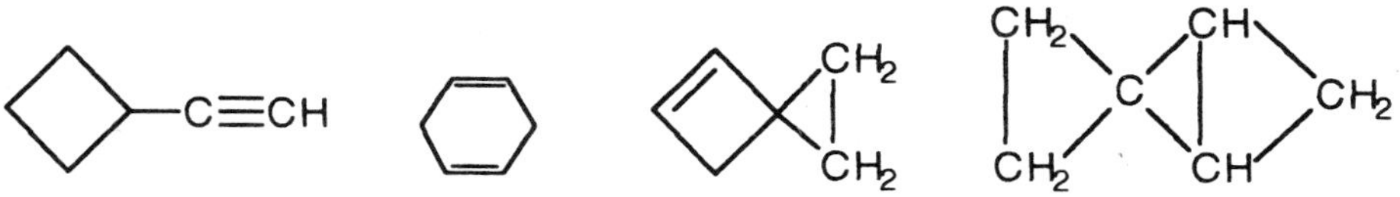

    c) Expression of Four Units of Unsaturation
        Two triple bonds
        One triple and two double bonds
        One triple bond, one double bond, and one ring
        One triple bond and two rings
        Four double bonds
        Three double bonds and one ring
        Two double bonds and two rings
        One double bond and three rings
        Four rings

3.28 Units of Unsaturation: Section 3.6

(a) By several methods you can determine that for a hydrocarbon with 11 carbons there need to be 2n+2 hydrogens (24 H's) for the compound to be saturated. This formula is deficient 10 hydrogens and has 5 units of unsaturation. A triple bond is two units of unsaturation and a double bond is one. This compound can have a maximum of two triple bonds.

(b) A compound with five units of unsaturation can have a maximum of five double bonds since a double bond represents one unit.

(c) This compound with one triple bond, two units of unsaturation, theoretically can have three rings since each represents one unit of unsaturation.

3.29 Units of Unsaturation: Section 3.6

(a) A compound with 13 carbons must have 2n+2 or 28 monovalent atoms to be saturated. This one has a triple and double bond, three units of unsaturation and thus needs only 22 monovalent atoms to satisfy valences. It has three bromines and thus needs 19 hydrogens.

(b) A compound with seven carbons and one oxygen needs 16 monovalent atoms to be saturated. With two double bonds, one triple bond, and one ring, this compound has five units of unsaturation and needs only six monovalent atoms to satisfy valences. Since it has five hydrogens already, it must have only one chlorine.

3.30 Units of Unsaturation: Section 3.6

(a) four double bonds, one triple bond, and one ring: **seven units**

(b) three double bonds and three rings: **six units**

(c) seven double bonds and three rings: **ten units**

3.31 Isomers: Sections 3.3-3.5

(a) six isomers of C_4H_8

$CH_3CH_2CH=CH_2$

$$\underset{\underset{\displaystyle CH_2}{\big\|}}{CH_3\overset{\displaystyle \overset{CH_3}{|}}{C}}$$

(square) (triangle with CH₃)

(b) five isomers of C_3H_5Br

$BrCH_2CH{=\!=}CH_2$ $CH_3\underset{\underset{\displaystyle Br}{|}}{C}{=\!=}CH_2$

$$\underset{H}{\overset{CH_3}{\diagdown}}C{=\!=}\underset{H}{\overset{Br}{\diagup}}C \qquad \underset{H}{\overset{CH_3}{\diagdown}}C{=\!=}\underset{Br}{\overset{H}{\diagup}}C$$

(triangle with Br)

(c) twelve isomers of C_5H_{10}

$CH_3CH_2CH_2CH{=\!=}CH_2$

$$\underset{H}{\overset{CH_3CH_2}{\diagdown}}C{=\!=}\underset{H}{\overset{CH_3}{\diagup}}C \qquad \underset{H}{\overset{CH_3CH_2}{\diagdown}}C{=\!=}\underset{CH_3}{\overset{H}{\diagup}}C$$

$$\underset{\underset{\displaystyle}{}}{CH_3\overset{\overset{CH_3}{|}}{C}HCH{=\!=}CH_2} \qquad CH_3\overset{\overset{CH_3}{|}}{C}{=\!=}CHCH_3 \qquad CH_2{=\!=}\overset{\overset{CH_3}{|}}{C}CH_2CH_3$$

(pentagon) (square with CH₃)

(triangle with CH₂CH₃) (triangle with CH₃, CH₃) (bicyclic with H, H, CH₃, CH₃) (bicyclic with H, CH₃, CH₃, H)

(d) Sixteen isomers of C_3H_4BrF

$$\underset{H}{\overset{CH_3}{\diagdown}}C{=\!=}\underset{F}{\overset{Br}{\diagup}}C \quad \underset{H}{\overset{CH_3}{\diagdown}}C{=\!=}\underset{Br}{\overset{F}{\diagup}}C \quad \underset{Br}{\overset{CH_3}{\diagdown}}C{=\!=}\underset{H}{\overset{F}{\diagup}}C \quad \underset{Br}{\overset{CH_3}{\diagdown}}C{=\!=}\underset{F}{\overset{H}{\diagup}}C$$

$$\underset{F}{\overset{CH_3}{\diagdown}}C{=\!=}\underset{H}{\overset{Br}{\diagup}}C \quad \underset{F}{\overset{CH_3}{\diagdown}}C{=\!=}\underset{Br}{\overset{H}{\diagup}}C \quad \underset{H}{\overset{FCH_2}{\diagdown}}C{=\!=}\underset{Br}{\overset{H}{\diagup}}C \quad \underset{H}{\overset{FCH_2}{\diagdown}}C{=\!=}\underset{H}{\overset{Br}{\diagup}}C$$

$$\underset{H}{\overset{BrCH_2}{\diagdown}}C{=\!=}\underset{H}{\overset{F}{\diagup}}C \quad \underset{H}{\overset{BrCH_2}{\diagdown}}C{=\!=}\underset{F}{\overset{H}{\diagup}}C \quad BrCH_2\overset{\overset{}{|}}{\underset{\underset{F}{|}}{C}}{=\!=}CH_2 \quad FCH_2\overset{}{\underset{\underset{Br}{|}}{C}}{=\!=}CH_2$$

57

BrCHCH=CH₂
  |
  F

$BrCHCH{=}CH_2$ with F below

3.32 Types of Isomerism

(a) skeletal;   (b) functional;   (c) geometric;   (d) positional;
(e) skeletal; (f) functional; (g) conformational; (h) positional;
(i) functional; (j) geometric; (k) conformational

3.33 Isomers

a) $CH_3C{-}CCH_3$ with CH_3, CH_3 above and CH_3, CH_3 below

b) $CH_3CH_2CH_2CH_2CH_2CH_2CH_2CH_2CH_2CH_2CH_2CH_3$

c) $CH_3CH_2CH_2CH_2CH_2CH_2CH_2CH_2CH_2\overset{O}{\overset{\|}{C}}OH$

d) $HC{\equiv}C\overset{O}{\overset{\|}{C}}CH{=}CH_2$

e) $HC{\equiv}C$ ... $C{=}C$ ... $\overset{O}{\overset{\|}{C}}H$

f)

g) cyclopentene with OH

h) $HC{\equiv}C{-}C{\equiv}C{-}C{\equiv}C{-}C{=}CHCH_3$

i) $CH_3OCH_2OCH_3$

j) $CH_3{-}N{-}CH_3$ with CH_3 above

k) cyclohexane with CH_3 and CH_2CH_3 substituents

3.34-3.37   Functional, Positional, and Skeletal Isomerism

(a)

$$CH_3\overset{\displaystyle O}{\overset{\|}{C}}CH_2CH_2CH_3 \qquad CH_3CH_2CH_2CH_2\overset{\displaystyle O}{\overset{\|}{C}}H \qquad CH_3CH_2\overset{\displaystyle O}{\overset{\|}{C}}CH_2CH_3 \qquad CH_3\overset{\displaystyle CH_3O}{\overset{|}{C}H}\text{-}\overset{}{C}CH_3$$

     ketone   **functional-aldehyde**   **positional**   **skeletal**

(b)

$$CH_3CH_2CH_2OCH_3 \qquad CH_3CH_2CH_2CH_2OH \qquad CH_3CH_2OCH_2CH_3 \qquad CH_3\overset{\displaystyle CH_3}{\overset{|}{C}H}OCH_3$$

     ether   **functional- alcohol**   **positional**   **skeletal**

(c)

$$CH_3CH_2CH\!=\!CH_2 \qquad \square \qquad CH_3CH\!=\!CHCH_3 \qquad CH_3\overset{\displaystyle CH_3}{\overset{|}{C}}\!=\!CH_2$$

       alkene   **functional-cycloalkane**   **positional**   **skeletal**

(d)

$$CH_3\overset{\displaystyle CH_3}{\overset{|}{C}H}CH_2\overset{\displaystyle O}{\overset{\|}{C}}H \qquad CH_3\overset{\displaystyle CH_3O}{\overset{|}{C}H}\text{-}\overset{}{C}CH_3 \qquad H\overset{\displaystyle O}{\overset{\|}{C}}\overset{\displaystyle CH_3}{\overset{|}{C}H}CH_2CH_3 \qquad CH_3CH_2CH_2CH_2\overset{\displaystyle O}{\overset{\|}{C}}H$$

  aldehyde   **functional-ketone**   **positional**   **skeletal**

(e)

$$HO\!-\!\!\bigcirc\!\!-\!CH_3 \qquad CH_3(CH_2)_5\overset{\displaystyle O}{\overset{\|}{C}}H \qquad \overset{HO}{\bigcirc}\!\!-\!CH_3 \qquad CH_3CH_2\!\!-\!\!\bigcirc\!\!-\!OH$$

    alcohol   **functional-aldehyde**   **positional**   **skeletal**

(f)

$$HO\overset{\displaystyle O}{\overset{\|}{C}}CH_2CH_2\overset{\displaystyle CH_3}{\overset{|}{C}H}CH_3 \qquad H\overset{\displaystyle O}{\overset{\|}{C}}CH_2OCH_2\overset{\displaystyle CH_3}{\overset{|}{C}H}CH_3 \qquad CH_3CH_2CH_2\overset{\displaystyle CH_3O}{\overset{|}{C}H}\overset{}{C}OH \qquad CH_3(CH_2)_4\overset{\displaystyle O}{\overset{\|}{C}}OH$$

  carboxylic acid   **functional-aldehyde/ether**   **positional**   **skeletal**

(g)

$$CH_3CH_2CH_2C\!\!\equiv\!\!CH \qquad CH_2\!=\!CHCH\!=\!CHCH_3 \qquad CH_3C\!\!\equiv\!\!CCH_2CH_3 \qquad CH_3\overset{\displaystyle CH_3}{\overset{|}{C}H}C\!\!\equiv\!\!CCH_3$$

    alkyne   **functional-alkene**   **positional**   **skeletal**

3.38 Geometric Isomerism: Section 3.5

This compound can exhibit geometric isomerism because the nitrogen is sp^2 hybridized and trigonal.   The carbon has a methyl and a hydrogen and the nitrogen has an OH and an electron-pair.

$$CH_3 \diagdown_{C=N} \diagup^{OH} \qquad CH_3 \diagdown_{C=N} \diagup^{:}$$
$$H \diagup \qquad \diagdown : \qquad\qquad H \diagup \qquad \diagdown OH$$

3.39 Cycloalkanes

The four carbons attached to or directly involved in the triple bond are in a linear arrangement. This is difficult to accomplish in a small ring compound due to the preferred internal angles of the polygon. It isn't until an eight membered ring that the flexibility is sufficient to accommodate a triple bond.

3.40 Geometric Isomerism in Cycloalkenes

The flexibility to have a trans configuration in a ring does not occur until cyclooctane. As an extreme, consider the impossibilty of having a trans configuration in cyclopropane.

3.41 Geometric Isomerism in Cycloalkanes

Because of the nature of a ring, the cis configuration is preferred. It isn't until eight membered rings that the trans can even exist. In larger rings it is easier for the trans configuration to exist without creating undue angle strain. Thus trans cyclodecene is more stable than trans cyclooctene.

3.42 Geometric Isomerism

$$CH_3 \diagdown C = C \diagup CH_3$$
$$H \diagup \qquad \diagdown H$$

$$CH_3 \diagdown \overset{CH_3 \; CH_3}{\underset{CH_3 \; H}{C}} \diagup C = C \diagup \overset{CH_3}{\underset{H \; CH_3}{C}}$$

Cis 2-butene is more stable because the two methyl groups are relatively small and do not interfere with each other as much in the cis configuration as the two tertiary butyl groups in 2,2,5,5-tetramethyl-3-hexene.

61

4

An Introduction to
Organic Reactions

CHAPTER SUMMARY

A **reaction equation** describes what happens in a chemical reaction by displaying the reactants and products. There are three main types of organic reactions. In **substitution reactions,** an atom or group of atoms is replaced by another species. **Elimination reactions** involve the removal of a pair of atoms or groups from two adjacent atoms to form a multiple bond. In **addition reactions** atoms or groups add to adjacent atoms involved in a multiple bond; the multiple bond is reduced.

A **reaction mechanism** is a step-by-step description of how a reaction occurs. The reaction equation describes what happens; the mechanism describes how the reaction happens. Multistep reaction mechanisms proceed through **reaction intermediates.** There are three major reaction intermediates involving carbon. A **carbocation** has a carbon with only three bonds, six outer-shell electrons and a positive charge. A **free radical** is a neutral carbon with only three bonds and seven outer shell electrons, one of which is unpaired. A **carbanion** has only three bonds but has eight outer shell electrons, one of which is a non-bonding pair, and a negative charge.

In chemical reactions, bonds break in the reactants and new bonds form in the products. In **homolytic** bond cleavage, the bonding electrons are evenly divided among the two parting atoms; neutral free radicals are often the result.

In **heterolytic** bond cleavage, the bonding electrons are unevenly divided between the two parting atoms; charged species, such as carbocations and carbanions usually result.

Organic reactions usually occur at sites within molecules where there is a special availability or deficiency of electrons. **Electrophiles** are regions of a molecule or ion that are positive or deficient in electrons and which tend to attract electron-rich species and accept electrons in a chemical reaction. **Nucleophiles** are electron-rich, provide electrons in a chemical reaction, and tend to attract electron-deficient or positive species. **Double and triple bonds** are active reaction sites because they are rich in electrons and the electrons are accessible due to the nature of pi-bonds. Because of the charge separation in **polar covalent bonds**, they are common reaction sites since they attract charged species.

Nucleophiles and electrophiles are also described as Lewis bases and acids. A **Lewis base** is a species that has a non-bonding pair of outer-shell electrons that can be shared in a chemical reaction. A **Lewis acid** is a substance that can accept a pair of elctrons for sharing in a chemical reaction. Nitrogen compounds, such as **amines**, and oxygen derivatives, such as **alcohols and ethers**, are often Lewis bases because the nitrogen and oxygen have non-bonding electron pairs. **Hydrogen ion and simple boron and aluminum compounds** are examples of Lewis acids. **Carbocations** are Lewis acids and **carbanions** are Lewis bases.

Chlorination or bromination of alkanes is an example of a substitution reaction. A hydrogen on an alkane is replaced by a halogen; hydrogen halide is the by-product. The reaction is initiated by light or heat. To promote monohalogenation, a high ratio of alkane to halogen is used. Polyhalogenation is caused using a high ratio of halogen to alkane. Halogenation proceeds by a **free-radical chain reaction** mechanism. In the **initiation step**, light or heat causes a halogen molecule to dissociate into free radicals. There are two **propagation steps.** In one the halogen free radical abstracts a hydrogen from the alkane leaving a carbon free radical. In the other, the carbon free radical reacts with a halogen molecule to form a carbon-halogen bond and a new halogen free radical. The two propagation steps alternate. The chain reaction can be slowed or halted by **chain termination steps** in which free radicals combine to form compounds without producing a new free radical to continue the chain reaction process.

Alkenes and alkynes are prepared by **elimination reactions** in which a carbon-carbon single bond is converted to a double or triple bond. In elimination reactions, atoms or groups are eliminated from adjacent carbons. Elimination once produces double bonds; twice produces triple bonds. In **dehydrohalogenation reactions,** hydrogen and halogen are the atoms eliminated from adjacent carbons. Bases such as potassium hydroxide and sodium amide are the reagents. Both alkenes and alkynes can be synthesized by dehydrohalogenation. In **dehydration reactions**, the elements of water, H and OH are eliminated from adjacent carbon atoms; sulfuric acid is used as a catalyst. Generally the reaction is only effective in producing carbon-carbon double bonds.

The **Saytzeff rule** is used to predict the product of elimination when more than one product is possible. According to the Saytzeff rule, the most stable alkene is formed; this is the one in which the double bond is most highly substituted with alkyl groups.

The **dehydration reaction** proceeds via a **carbocation mechanism.** The three step mechanism starts with the protonation of the alcohol oxygen with a hydrogen ion from sulfuric acid by a Lewis acid-Lewis base reaction. Water departs in the second step leaving a carbocation intermediate. In the final step, a hydrogen ion leaves the adjacent carbon and the double bond forms.

Connections 4.1 describes the effects of chlorofluorocarbons on the earth's ozone layer. The ozone layer is damaged by a free-radical chain reaction involving ozone and halogen atoms.

Connections 4.2 introduces organo halogen general anesthetics.

SOLUTIONS TO PROBLEMS

4.1   Types of Reactions

(a) substitution;   (b) elimination;   (c) addition;   (d) addition
(e) elimination;   (f) substitution

4.2   Polar Covalent Bonds

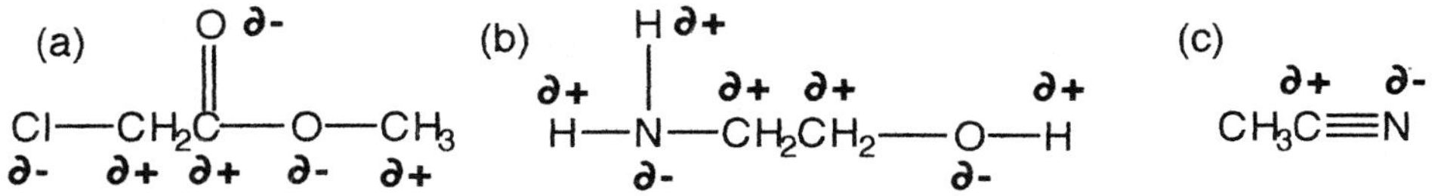

(a)

$$Cl\!-\!CH_2\overset{\overset{\displaystyle O \;\; \partial -}{\|}}{C}\!-\!O\!-\!CH_3$$
$$\partial - \;\; \partial + \;\; \partial + \;\; \partial - \;\; \partial +$$

(b)

$$H\!-\!\overset{\overset{\displaystyle H \;\; \partial +}{|}}{\underset{\underset{\displaystyle \partial -}{}}{N}}\!-\!CH_2CH_2\!-\!O\!-\!H$$
$$\partial + \qquad \partial + \;\; \partial + \qquad \partial + \qquad \partial -$$

(c)

$$\partial + \;\;\; \partial -$$
$$CH_3C\!\equiv\!N$$

4.3   Lewis Bases

In a-b, the oxygen has two non-bonding pairs of electrons in each molecule and is the Lewis base site.   In c-d, the nitrogens have one non-bonding electron pair and are Lewis base sites.

4.4   Lewis Acid-Lewis Base Reaction

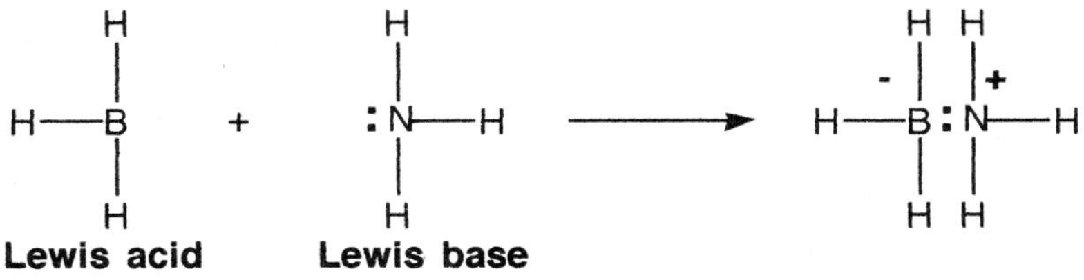

Lewis acid   **Lewis base**

4.5   Reaction Sites

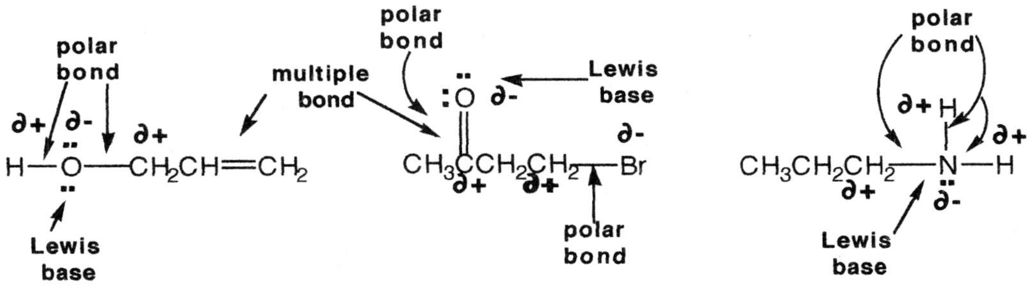

4.6   Chlorination of Ethane

After one hydrogen is replaced by a chlorine, that molecule can compete for chlorine with the ethane. This can continue to give products ranging from one hydrogen being replaced to all replaced by chlorine. HCl is the inorganic by-product.

CH_3CH_2Cl, CH_3CHCl_2, $ClCH_2CH_2Cl$, CH_3CCl_3, $ClCH_2CHCl_2$,

$ClCH_2CCl_3$, $Cl_2CHCHCl_2$, Cl_2CHCCl_3, Cl_3CCCl_3

4.7   Monochlorination of 2-methylpentane

$$
\begin{array}{cccc}
\underset{\displaystyle |}{CH_3} & \underset{\displaystyle |}{CH_3} & \underset{\displaystyle |}{CH_3} & \underset{\displaystyle |}{CH_3} \\[2pt]
\underset{\displaystyle |}{CH_2CHCH_2CH_3} & \underset{\displaystyle |}{CH_3CCH_2CH_3} & \underset{\displaystyle |}{CH_3CHCHCH_3} & \underset{\displaystyle |}{CH_3CHCH_2CH_2} \\[2pt]
Cl & Cl & Cl & Cl
\end{array}
$$

4.8   Free-radical Chain Reaction Mechanism

Initiation $Br_2 \xrightarrow{\text{light}} 2\,Br^{\bullet}$

Propagation $CH_3CH_3 + Br^{\bullet} \longrightarrow CH_3CH_2^{\bullet} + HBr$

Propagation $CH_3CH_2^{\bullet} + Br_2 \longrightarrow CH_3CH_2Br + Br^{\bullet}$

4.9   Elimination Reactions

(a) $CH_3CH_2CH_2OH \xrightarrow{H_2SO_4} CH_3CH=CH_2 + H_2O$

(b) $CH_3CH_2CH_2CH_2Br + KOH \xrightarrow[\text{alc.}]{\text{aq}} CH_3CH_2CH=CH_2 + KBr + H_2O$

(c) $ClCH_2CH_2Cl + 2\,NaNH_2 \longrightarrow HC\equiv CH + 2NaCl + 2\,NH_3$

(d) $CH_3CH_2CH_2CHBr_2 + 2\,NaNH_2 \longrightarrow CH_3CH_2C\equiv CH + 2\,NaBr + 2\,NH_3$

4.10 Orientation of Elimination: See Example 4.2

In (a), two alkenes are possible. Concentrate on the carbon with the OH. If the H on the carbon to the right eliminates, the product is monosubstituted; if the H on the carbon to the left is eliminated the major product, trisubstituted, is formed. In (b) a disubstituted alkene is formed if a hydrogen from the carbon to the left of the carbon-chlorine bond is eliminated.   The product shown, which is tetrasubstituted, is formed if the H on the carbon to the right is eliminated.

$$
\begin{array}{ll}
\text{a)}\ \ \underset{\displaystyle |}{CH_3} & \text{b)}\ \ \underset{\displaystyle |}{CH_3}\ \ \underset{\displaystyle |}{CH_3} \\[2pt]
CH_3C=CHCH_3 & CH_3C=CCH_3
\end{array}
$$

4.11 Mechanism of Dehydration

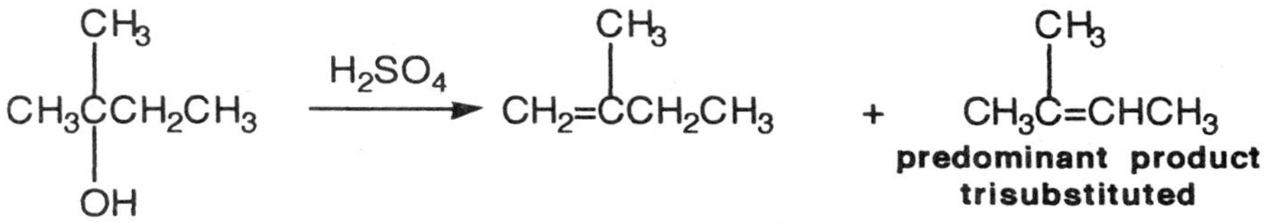

$$CH_3CCH_2CH_3 \xrightarrow{H_2SO_4} CH_2=CCH_2CH_3 + CH_3C=CHCH_3$$

with CH_3 and OH substituents on the starting material, CH_3 on the products

predominant product
trisubstituted

Reaction Mechanism

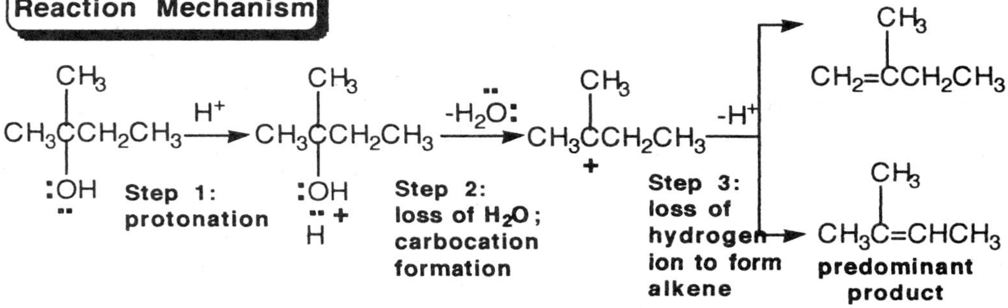

$CH_3CCH_2CH_3 \xrightarrow{H^+} CH_3CCH_2CH_3 \xrightarrow{-H_2O:} CH_3CCH_2CH_3 \xrightarrow{-H^+}$

:OH
Step 1:
protonation

:OH +
H
Step 2:
loss of H_2O ;
carbocation
formation

+
Step 3:
loss of
hydrogen
ion to form
alkene

$CH_2=CCH_2CH_3$ (with CH_3)

$CH_3C=CHCH_3$ (with CH_3)
**predominant
product**

4.12 Reaction Mechanisms

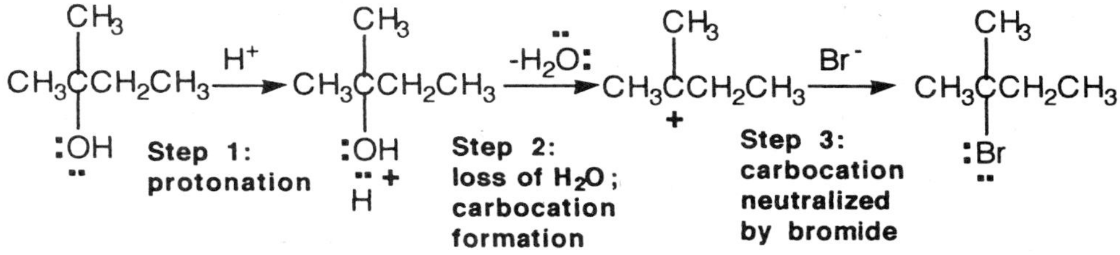

$CH_3CCH_2CH_3 \xrightarrow{H^+} CH_3CCH_2CH_3 \xrightarrow{-H_2O:} CH_3CCH_2CH_3 \xrightarrow{Br^-} CH_3CCH_2CH_3$

:OH
Step 1:
protonation

:OH +
H
Step 2:
loss of H_2O ;
carbocation
formation

+
Step 3:
carbocation
neutralized
by bromide

:Br

4.13 Polar Bonds: Section 4.2B

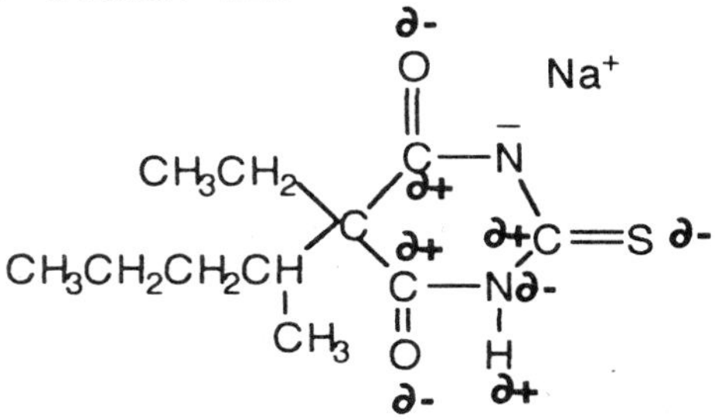

CH_3CH_2, $CH_3CH_2CH_2CH$, CH_3 bonded to central C with C, and groups: $\partial-$ O, $\partial+$ C—N$^-$ Na$^+$, $\partial+$ C=S $\partial-$, $\partial+$ C—N $\partial-$ with H $\partial+$, $\partial-$ O

4.14 Lewis Acids and Bases: Section 4.2C

Lewis bases have a lone pair of electrons, not used in bonding, which can be shared with a Lewis acid. Lewis acids usually have an incomplete outer shell and thus can accept the non-bonding electron pair of a Lewis base.

a) $\underset{\text{acid}}{Br-Al-Br}$
with Br above Al

b) $\underset{\text{base}}{CH_3-N-CH_3}$
with H above N

c) $\underset{\text{base}}{CH_3-\overset{..}{\underset{..}{O}}-H}$

d) $\underset{\text{acid}}{H-B-H}$
with H above B

4.15 Lewis Acids and Bases: Section 4.2C

a) $CH_3CH_2-\overset{..}{\underset{..}{O}}-H + H^+ \longrightarrow CH_3CH_2-\overset{H\ +}{\underset{..}{O}}-H$

b) $\underset{Cl}{\overset{Cl}{Cl-Al}} + :\overset{..}{O}-H \longrightarrow Cl-\overset{Cl}{\underset{Cl}{Al}}:\overset{-\ ..\ +}{O}-H$
(with H below O)

c) $CH_3-\overset{H}{\underset{H}{N}}: + HCl \longrightarrow CH_3-\overset{H\ +}{\underset{H}{N}}:H \quad Cl^-$

d) $CH_3-\overset{CH_3}{\underset{CH_3}{N}}: + \overset{F}{\underset{F}{B}}-F \longrightarrow CH_3-\overset{+CH_3}{\underset{CH_3}{N}}:\overset{F\ -}{\underset{F}{B}}-F$

4.16 Reaction Sites: Section 4.2

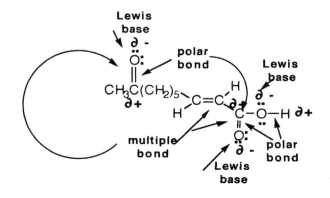

4.17 Electrophiles and Nucleophiles: Section 4.2

(a) nucleophile - negative, non-bonding electron pairs; (b) electrophile - positive, incomplete octet; (c) nucleophile - non-bonding electron pairs, Lewis base; (d) nucleophile - non-bonding electron pair; (e) nucleophile - negative, non-bonding electron pairs; (f) electrophile - positive, carbocation with incomplete octet; (g) nucleophile - negative, non-bonding electron pair; (h) nucleophile - negative, non-bonding electron pairs.

4.18 Reactive Intermediates: Section 4.1C

$$CH_3\overset{\underset{\displaystyle CH_3}{|}}{\underset{\underset{\displaystyle CH_3}{|}}{C}}\!:\!A \longrightarrow CH_3\overset{\underset{\displaystyle CH_3}{|}}{\underset{\underset{\displaystyle CH_3}{|}}{C}}\!+ \;+\; A\!:^-$$

carbocation

$$CH_3\overset{\underset{\displaystyle CH_3}{|}}{\underset{\underset{\displaystyle CH_3}{|}}{C}}\!\cdot\!\!|\;A \longrightarrow CH_3\overset{\underset{\displaystyle CH_3}{|}}{\underset{\underset{\displaystyle CH_3}{|}}{C}}\!\cdot \;+\; \cdot A$$

free radical

$$CH_3\overset{\underset{\displaystyle CH_3}{|}}{\underset{\underset{\displaystyle CH_3}{|}}{C}}\!:\!|\;A \longrightarrow CH_3\overset{\underset{\displaystyle CH_3}{|}}{\underset{\underset{\displaystyle CH_3}{|}}{C}}\!:^- \;+\; A^+$$

carbanion

4.19 Reactive Intermediates: Section 4.1C

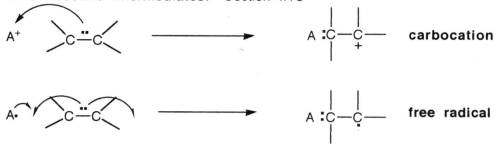

4.20 Halogenation of Alkanes: Section 4.4C

a)

$$
\underset{\underset{Cl}{|}}{CH_2}\overset{\overset{CH_3}{|}}{CHCH_2CH_2}\overset{\overset{CH_3}{|}}{CHCH_3}
\qquad
CH_3\overset{\overset{CH_3}{|}}{\underset{\underset{Cl}{|}}{C}}CH_2CH_2\overset{\overset{CH_3}{|}}{CHCH_3}
\qquad
CH_3\overset{\overset{CH_3}{|}}{CH}\overset{}{\underset{\underset{Cl}{|}}{CH}}CH_2\overset{\overset{CH_3}{|}}{CHCH_3}
$$

b)

c)

$$
\underset{\underset{Cl}{|}}{CH_2}\overset{\overset{CH_3}{|}}{CHCH_2CH_2}\overset{\overset{CH_3}{|}}{\underset{\underset{CH_3}{|}}{C}}CH_3
\qquad
CH_3\overset{\overset{CH_3}{|}}{\underset{\underset{Cl}{|}}{C}}CH_2CH_2\overset{\overset{CH_3}{|}}{\underset{\underset{CH_3}{|}}{C}}CH_3
\qquad
CH_3\overset{\overset{CH_3}{|}}{CH}CHCH_2\overset{\overset{CH_3}{|}}{\underset{\underset{CH_3}{|}}{C}}CH_3
$$

$$
CH_3\overset{\overset{CH_3}{|}}{CH}CH_2\overset{\overset{CH_3}{|}}{\underset{\underset{Cl}{|}}{CH}}\overset{}{\underset{\underset{CH_3}{|}}{C}}CH_3
\qquad
CH_3\overset{\overset{CH_3}{|}}{CH}CH_2CH_2\overset{\overset{CH_3}{|}}{\underset{\underset{CH_3}{|}}{C}}CH_2Cl
$$

70

4.21 Halogenation of Alkanes: Section 4.4C

a) C_5H_{12}

$$CH_3-\underset{\underset{CH_3}{|}}{\overset{\overset{CH_3}{|}}{C}}-CH_3$$

b) C_8H_{18}

$$CH_3\underset{\underset{CH_3}{|}\;\;\underset{CH_3}{|}}{\overset{\overset{CH_3}{|}\;\;\overset{CH_3}{|}}{C—C}}CH_3$$

4.22 Halogenation of Alkanes: Section 4.4C

To obtain predominantly bromoethane use a large excess of ethane relative to the bromine. Statistically, the bromine is more likely to encounter an ethane molecule than a bromoethane.

$$CH_3CH_3 \text{ (excess)} + Br_2 \xrightarrow{\text{light}} CH_3CH_2Br + HBr$$

To obtain hexabromoethane provide enough bromine (6 moles) to ensure that every hydrogen (six) can be replaced.

$$CH_3CH_3 + 6\,Br_2 \longrightarrow Br_3CCBr_3 + 6\,HBr$$

4.23 Halogenation of Alkanes - Reaction Mechanism

Section 4.4D

Initiation $Br \!:\! Br \xrightarrow{\text{light}} 2Br\cdot$

Propagation $Br\cdot + CH_4 \longrightarrow CH_3\cdot + HBr$

                $CH_3\cdot + Br_2 \longrightarrow CH_3Br + Br\cdot$

4.24 Halogenation of Alkanes - Reaction Mechanism   Section 4.4D
The ethyl radicals formed from the decomposition of tetraethyllead can react with methane to form methyl radicals or with chlorine to form chlorine radicals. Both of these are part of the propagation steps.

Initiation

$$Pb(CH_2CH_3)_4 \xrightarrow{140°C} Pb + 4\,CH_3CH_2\cdot$$

Initiation	**Initiation**
$CH_3CH_2\cdot + CH_4 \longrightarrow CH_3CH_3 + CH_3\cdot$	$CH_3CH_2\cdot + Cl_2 \longrightarrow CH_3CH_2Cl + Cl\cdot$
Propagation	**Propagation**
$CH_3\cdot + Cl_2 \longrightarrow CH_3Cl + Cl\cdot$	$Cl\cdot + CH_4 \longrightarrow CH_3\cdot + HCl$
$Cl\cdot + CH_4 \longrightarrow CH_3\cdot + HCl$	$CH_3\cdot + Cl_2 \longrightarrow CH_3Cl + Cl\cdot$

4.25 Dehydration of Alcohols: Section 4.5C

$$CH_3CH_2\underset{\underset{\overset{\displaystyle |}{:\!\underset{\cdot\cdot}{O}H}}{}}{C}HCH_2CH_3 \xrightarrow{\;H^+\;} CH_3CH_2\underset{\underset{\overset{\displaystyle |}{\underset{+}{:O}-H}}{\overset{\displaystyle |}{\underset{}{}}}}{C}HCH_2CH_3 \xrightarrow{\;-H_2\ddot{O}:\;} CH_3CH_2\overset{+}{C}HCH_2CH_3$$

1. Protonation
2. Loss of water to
   form carbocation
3. Loss of proton to
   form alkene

$$CH_3CH_2CH=CHCH_3$$

$- H^+$
from
adjacent
carbon

4.26 Elimination Reactions to Produce Alkenes: Section 4.5A-B

Examples a and c are dehydrohalogenation reactions and the others are dehydrations. The predominant product, when more than one product is possible, is the most highly substituted alkene.

a) $CH_3CH=CH_2$ b) $CH_3CH_2CH=CHCH_3$ c) $CH_3CH=\underset{\overset{\displaystyle |}{CH_3}}{C}CH_3$

d) $CH_3\underset{\overset{\displaystyle |}{CH_3}}{\overset{\overset{\displaystyle CH_3}{|}}{C}}=CCH_3$ e) (cyclohexene with two CH₃ groups)   (f) $CH_2=CH-CH=CH_2$

4.27 Elimination Reactions to Produce Alkynes: Section 4.5A-B

a) $CH_3C\equiv CH$ b) $CH_3\underset{\overset{\displaystyle |}{CH_3}}{\overset{\overset{\displaystyle CH_3}{|}}{C}}C\equiv CH$

4.28 Preparation of Alkenes and Alkynes: Section 4.5A-B

a) $CH_3\underset{\overset{\displaystyle |}{X}}{\overset{\overset{\displaystyle CH_3}{|}}{C}}HCH_2CH_2 \xrightarrow{\;\text{Reagent}\;} CH_3\overset{\overset{\displaystyle CH_3}{|}}{C}HCH=CH_2$

X = OH   Reagent = H2SO4
X = Cl, Br, I   Reagent = KOH

The above is the preferred method for preparing the desired product since having X on the next carbon would give the most substituted product predominantly as shown in part b.

b)
$$
\underset{X}{\overset{CH_3}{CH_3CHCHCH_2CH_3}} \xrightarrow{\text{Reagent}} \underset{\text{major product}}{\overset{CH_3}{CH_3C=CHCH_2CH_3}} + \underset{\text{minor product}}{\overset{CH_3}{CH_3CHCH=CHCH_3}}
$$

X = OH   Reagent = H_2SO_4

X = Cl, Br, I   Reagent = KOH

c)
$$
\underset{X}{\overset{X}{CH_3CH_2CH_2CH}} + 2\,KOH \longrightarrow CH_3CH_2C\equiv CH + 2\,KX + 2\,H_2O
$$

X = Cl, Br, I

The other two possible dihalide starting materials are less desirable as they can give a diene product as well as the alkyne or 2-butyne.

$$
\underset{X\ \ \ X}{CH_3CH_2CH-CH_2} \xrightarrow{\text{KOH}} CH_3CH_2C\equiv CH + CH_3CH=C=CH_2
$$

$$
\underset{X}{\overset{X}{CH_3CH_2CCH_3}} \xrightarrow{\text{KOH}} CH_3CH_2C\equiv CH + CH_3C\equiv CCH_3 + CH_3CH=C=CH_2
$$

4.29 Preparation of Alkenes and Alkynes: Section 4.5A-B

(a) Only one product is possible upon dehydrohalogenation of 1-bromopentane and it is not the desired product. Two products can form from 2-chloropentane; the most substituted is the desired 2-pentene and is formed predominantly.

$$
CH_3CH_2CH_2CH_2CH_2Br \longrightarrow \underset{\textbf{only product}}{CH_3CH_2CH_2CH=CH_2}
$$

$$
\underset{Cl}{CH_3CH_2CH_2CHCH_3} \longrightarrow \underset{\textbf{major product}}{CH_3CH_2CH=CHCH_3} + \underset{\textbf{minor product}}{CH_3CH_2CH_2CH=CH_2}
$$

(b)   1,1-dichloropropane can only form propyne upon dehydrohalogenation. 2,2-dichloropropane could possibly eliminate in two different directions to give a diene.

$$CH_3\overset{\boxed{H\ Cl}}{\underset{\boxed{H\ Cl}}{C}}-\overset{}{C}H \longrightarrow CH_3C\equiv CH \longleftarrow \overset{\boxed{H\ C\ H}}{\underset{\boxed{H\ C\ H}}{HC-C}}\overset{}{C}H \longrightarrow H_2C=C=CH_2$$

(c)   Either compound can produce the desired product. The first one can produce two alkenes whereas the second only one, the target compound.

$$(CH_3)_2\underset{OH}{\overset{}{C}}CH_2CH_2CH_3 \xrightarrow{\ H_2SO_4\ } (CH_3)_2C=CHCH_2CH_3 \ + \ H_2O$$

4.30 Carbocations: Section 4.1C

A carbocation has three bonded groups and is trigonal, sp^2 hybridized, and has $120°$ bond angles. The empty orbital is the unhybridized p-orbital.

4.31 Carbanions: Section 4.1C

In a carbanion, there are four space occupying groups - three bonded groups and the non-bonding electron pair. As a result, it is tetrahedral, sp^3 hybridized, and has $109°$ bond angles. The non-bonding pair is in an sp^3 hybrid orbital.

4.32 Reactions

Propene is a possible product from a "dehydration" type process.

$$(CH_3)_2CH\text{-}\overset{..}{\underset{..}{O}}\text{-}CH(CH_3)_2 \xrightarrow{\ H^+\ } (CH_3)_2CH\text{-}\overset{\overset{H^+}{..}}{\underset{..}{O}}\text{-}CH(CH_3)_2$$

$$CH_3CH=CH_2 \xleftarrow{\ \text{-}H^+\ } (CH_3)_2CH+ \xleftarrow{} $$

$$\text{-}(CH_3)_2CHOH$$

$$\text{-}H_2O \qquad H^+$$

5

$$H_2C{=}CH_2 \xrightarrow{\;Br_2\;} \underset{\underset{Br \quad Br}{|\quad\;|}}{H_2C{-}CH_2} \qquad \underset{\underset{Br \quad Br}{|\quad\;|}}{\overset{\overset{Br \quad Br}{|\quad\;|}}{HC{-}CH}} \xleftarrow{\;2\,Br_2\;} HC{\equiv}CH$$

Reactions of
Alkenes and Alkynes

CHAPTER SUMMARY

Addition is the characteristic reaction of **alkenes and alkynes.**
Since the carbons of a double or triple bond do not have the maximum number
of attached atoms, they can add additional groups or atoms; **double bonds**
undergo addition once and **triple bonds** can undergo addition twice. The
reactivity of alkenes and alkynes is due to the presence of **pi bonds.** Unlike
sigma bonds, pi-bonds are directed away from the carbons; the electrons are
loosely held, very accessible, and quite attractive to an electron-deficient
species **(electrophile)** seeking an electron source.

Alkenes add **hydrogen halides, halogens** (chlorine and bromine),
water (sulfuric acid catalyst), and **hydrogen** (metal catalyst). One part of the
adding reagent adds to each carbon of the double bond; the double bond
becomes a single bond during the process. With the exception of
hydrogenation, these reactions occur by an **electrophilic addition
mechanism.** The **electrophile** (H^+ or X^+) attacks the electron-rich pi-bond
of the double bond. The electrons are used to form a single bond between the
carbon and attacking species; the other carbon becomes a **carbocation.** The
carbocation is then neutralized by halide ion or water; the addition is complete.
In **bromination** reactions, the bromine adds in a **trans** fashion.

When an **unsymmetrical reagent** adds to an **unsymmetrical
alkene,** two addition products are possible. The one resulting from the more

76

stable carbocation intermediate is the favored product.   Following is the order of **carbocation stability: 3º > 2º > 1º > methyl.**   A **tertiary carbocation** has three bonded alkyl groups. **Secondary carbocations** have two alkyl groups bonded directly to the carbocation carbon and in **primary carbocations** there is only one. Since **alkyl groups** are **electron-releasing groups** they stabilize the positive carbocation.   Tertiary carbocations have the greatest number of alkyl groups and are the most stable.

Alkynes add hydrogen, hydrogen halides, and halogens (chlorine and bromine). They can add one mole of reagent to produce a double bond or two moles to form a single bond. The mechanism is **electrophilic addition** as with alkenes and **orientation of addition** of unsymmetrical reagents to unsymmetrical alkynes is determined by the stability of the intermediate carbocation.   Alkynes add water to form aldehydes and ketones.

Hydrogenation of alkenes and alkynes is accomplished in the presence of a metal catalyst which attracts both the hydrogen and hydrocarbon to its surface. As a result of the reactants being adsorbed onto the same surface, the reaction occurs with **cis addition.**

A **polymer** is a giant molecule composed of a repeating structural unit called a **monomer. Addition polymers** result from the addition of alkene molecules to one another. The **polymerization** occurs by cationic, free-radical, and anionic reaction mechanisms. Examples of addition polymers include polyethylene, polystyrene, PVC, and Teflon. **Rubber** can be produced by the polymerization of **isoprene** by a **1,4-addition reaction.**

Conjugated dienes are compounds in which two carbon-carbon double bonds are separated by a single bond.   Upon treatment with adding reagents, conjugated dienes undergo **1,2-addition**, in which the reagent adds to one of the double bonds and **1,4-addition** in which the reagent adds to the first and fourth carbons with the remaining double bond shifting between carbons 2 and 3. This is caused by the formation of an allylic intermediate such as an **allylic carbocation.**   An allylic carbocation is one in which the carbocation carbon is attached directly to a carbon-carbon double bond. Such a carbocation engages in resonance allowing neutralization at the second and fourth carbons of the original conjugated diene.

Resonance forms are classical structures used to describe a more complex system; they do not actually exist. The species is more accurately described by a **resonance hybrid** which can be imagined as an average of the resonance forms. Each atom in a resonance stabilized system has a p-orbital. **Allylic carbocations, free radicals, and carbanions** are resonance stabilized.

Treatment of alkenes with **potassium permanganate** produces 1,2-diols. **Ozonolysis** cleaves the carbon-carbon double bond to form aldehydes and ketones. **Terminal alkynes** have weakly acidic hydrogens that can be abstracted by strong bases such as sodium amide.

Connections 5.1 relates stories of serendipity in the discovery of polymers.

Connections 5.2 is about recycling plastics.

Connections 5.3 is about terpenes, naturally occurring compounds characterized by carbon skeletons constructed of isoprene units.

Connections 5.4 is about cholesterol and the treatment of atherosclerosis.

SOLUTIONS TO PROBLEMS

5.1   Addition Reactions of Alkenes

a) $CH_3CH{=}CHCH_3 + H_2 \xrightarrow{\text{Pt}} CH_3CH_2CH_2CH_3$

b) $CH_3CH{=}CHCH_3 + Cl_2 \longrightarrow CH_3\underset{Cl}{CH}{-}\underset{Cl}{C}HCH_3$

c) $CH_3CH{=}CHCH_3 + HBr \longrightarrow CH_3\underset{Br}{CH}{-}\underset{H}{C}HCH_3$

d) $CH_3CH{=}CHCH_3 + H_2O \xrightarrow{H_2SO_4} CH_3\underset{HO}{CH}{-}\underset{H}{C}HCH_3$

5.2   Electrophilic Addition Mechanisms

In each mechanism there is initial attack by an electrophile on the electron-rich pi-bond followed by neutralization of the resulting carbocation by a Lewis base, either halide ion or water.

a) 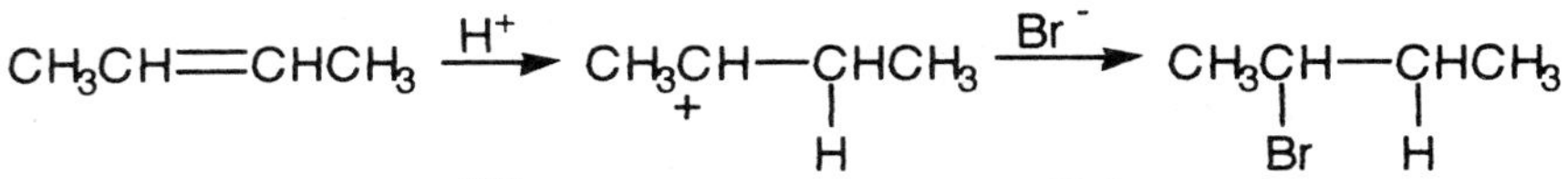 $CH_3CH{=}CHCH_3 \xrightarrow{H^+} \underset{\underset{H}{|}}{CH_3CH{-}\overset{+}{C}HCH_3} \xrightarrow{Br^-} \underset{\overset{|}{Br}\ \overset{|}{H}}{CH_3CH{-}CHCH_3}$

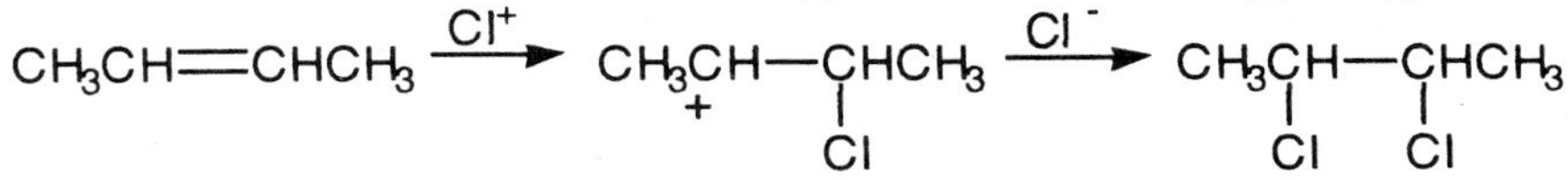

b) $CH_3CH{=}CHCH_3 \xrightarrow{Cl^+} \underset{\underset{Cl}{|}}{CH_3CH{-}\overset{+}{C}HCH_3} \xrightarrow{Cl^-} \underset{\overset{|}{Cl}\ \overset{|}{Cl}}{CH_3CH{-}CHCH_3}$

c) $CH_3CH{=}CHCH_3 \xrightarrow{H^+} \underset{\underset{H}{|}}{CH_3CH{-}\overset{+}{C}HCH_3} \searrow^{H_2O}$

$\underset{\overset{|}{OH}\ \overset{|}{H}}{CH_3CH{-}CHCH_3} \xleftarrow{-H^+} \underset{\underset{H}{\overset{+}{O}}\ \ \overset{|}{H}}{CH_3CH{-}CHCH_3}$

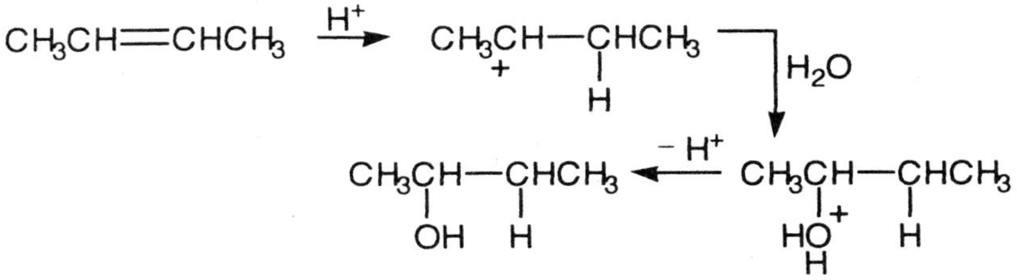

5.3   Orientation of Addition

(a)

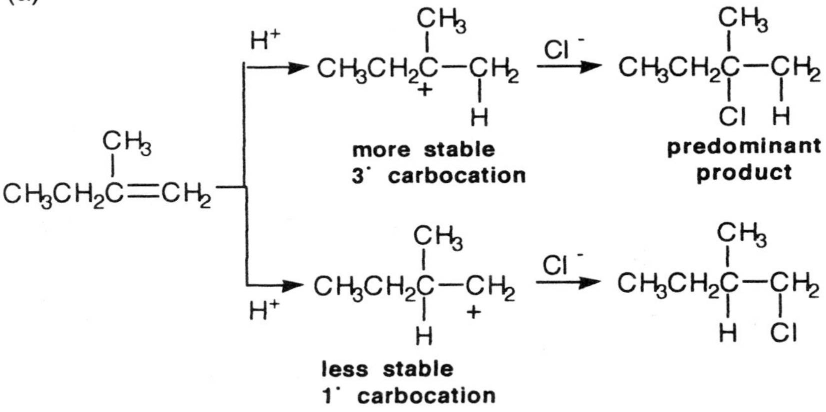

CH_3CH_2C=CH_2 (with CH_3 branch)

$\xrightarrow{H^+}$ more stable 3˙ carbocation $\xrightarrow{Cl^-}$ predominant product

$\xrightarrow{H^+}$ less stable 1˙ carbocation $\xrightarrow{Cl^-}$

(b)

$\xrightarrow{H^+}$ more stable 3˙ carbocation $\xrightarrow{H_2O}\xrightarrow{-H^+}$ predominant product

$\xrightarrow{H^+}$ less stable 2˙ carbocation $\xrightarrow{H_2O}\xrightarrow{-H^+}$

79

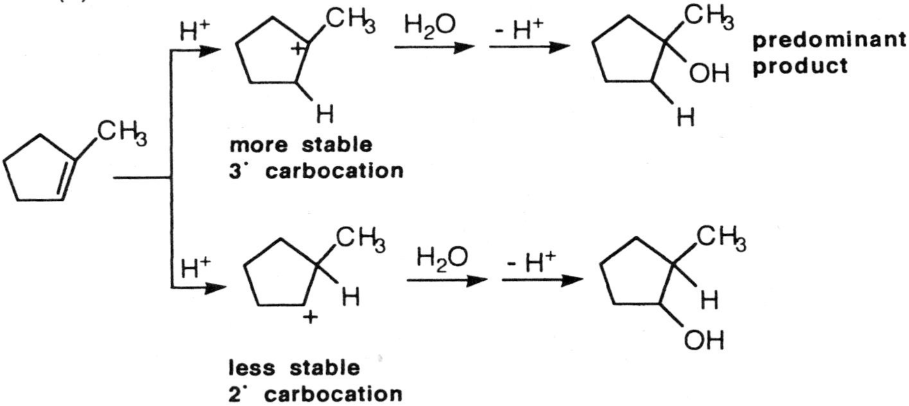

5.4   Addition Reactions of Alkynes

(a) $CH_3C{\equiv}CCH_3$ + 1Br$_2$ $\longrightarrow$ $CH_3\underset{Br}{C}{=}\underset{Br}{C}CH_3$

(b) $CH_3C{\equiv}CCH_3$ + 2Br$_2$ $\longrightarrow$ $CH_3\overset{Br}{\underset{Br}{C}}{-}\overset{Br}{\underset{Br}{C}}CH_3$

(c) $CH_3C{\equiv}CCH_3$ + 1Cl$_2$ $\longrightarrow$ $CH_3\underset{Cl}{C}{=}\underset{Cl}{C}CH_3$

(d) $CH_3C{\equiv}CCH_3$ + 2H$_2$ $\xrightarrow{\text{Ni}}$ $CH_3\overset{H}{\underset{H}{C}}{-}\overset{H}{\underset{H}{C}}CH_3$

5.5   Hydrogenation of Alkynes

$CH_3CH_2C{\equiv}CCH_3$

$\xrightarrow[\text{Pt}]{\text{1H}_2}$ $\underset{H}{\overset{CH_3CH_2}{\diagdown}}C{=}C\underset{H}{\overset{CH_3}{\diagup}}$ **cis addition**

$\xrightarrow[\text{Pt}]{\text{2H}_2}$ $CH_3CH_2CH_2CH_2CH_3$

5.6   Electrophilic Addition to Alkynes

$CH_3CH_2CH_2C{\equiv}CH$ + 2HBr $\longrightarrow$ $CH_3CH_2CH_2\overset{Br}{\underset{Br}{C}}{-}\overset{H}{\underset{H}{C}}H$

Reaction Mechanism

$$CH_3CH_2CH_2C\equiv CH \xrightarrow{H^+} CH_3CH_2CH_2C\overset{+}{=}CH \xrightarrow{:\overset{..}{\underset{..}{Br}}:^-} CH_3CH_2CH_2C=CH$$

HBr adds to triple bond and then to the resulting double bond. In each case the more stable carbocation is formed.

$$CH_3CH_2CH_2\overset{Br}{\underset{Br}{\overset{|}{C}}}-\overset{H}{\underset{H}{\overset{|}{CH}}} \longleftarrow :\overset{..}{\underset{..}{Br}}:^- CH_3CH_2CH_2\overset{+}{\underset{Br}{\overset{|}{C}}}-\overset{H}{\underset{H}{\overset{|}{CH}}} \longleftarrow H^+$$

5.7   1,2 and 1,4 Addition

$$CH_2=CH-CH=CH_2 + 1HBr \longrightarrow$$

$$CH_2=CH-\underset{Br}{\overset{|}{CH}}-\underset{H}{\overset{|}{CH_2}}$$

1,2 Addition

$$\underset{Br}{\overset{|}{CH_2}}-CH=CH-\underset{H}{\overset{|}{CH_2}}$$

1,4 Addition

Reaction Mechanism

STEP 1: Electrophile, H^+ is attracted to pi-cloud and uses two pi-electrons to bond. More stable allylic carbocation results.

$$CH_2=CH-CH=CH_2$$
$$\downarrow H^+$$

$$\left\{ CH_2=CH-\underset{+}{CH}-\underset{H}{\overset{|}{CH_2}} \longleftrightarrow \underset{+}{CH_2}-CH=CH-\underset{H}{\overset{|}{CH_2}} \right\}$$

Resonance Forms

STEP 2: The allylic carbocation is resonance stabilized. Resonance forms show the two places it can be neutralized by bromide ion.

$$\downarrow Br^-$$

$$CH_2=CH-\underset{Br}{\overset{|}{CH}}-\underset{H}{\overset{|}{CH_2}} \quad + \quad \underset{Br}{\overset{|}{CH_2}}-CH=CH-\underset{H}{\overset{|}{CH_2}}$$

1,2 Addition   **1,4 Addition**

5.8   Free-radical Chain Polymerization

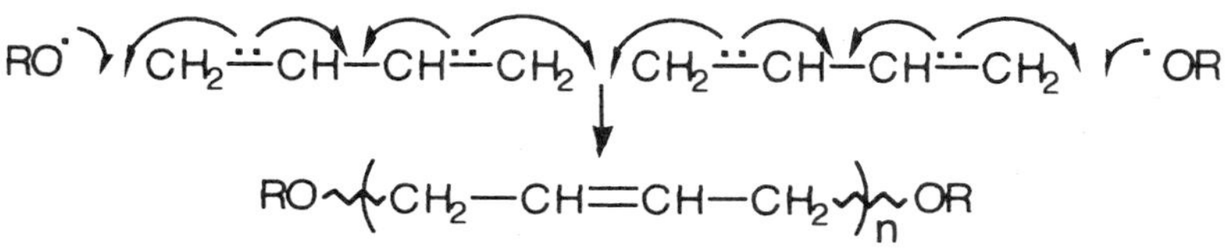

5.9   Hydroxylation with Potassium Permanganate

(a) 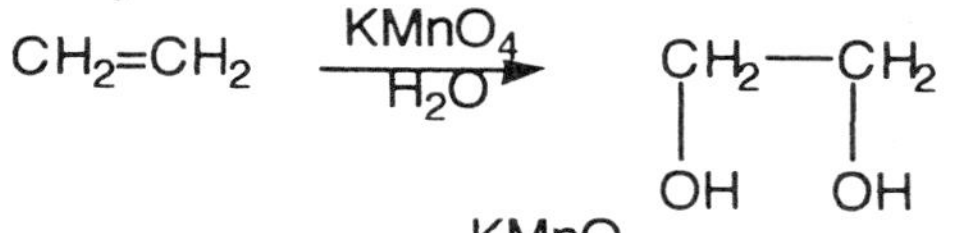 $CH_2=CH_2 \quad \xrightarrow[H_2O]{KMnO_4}$

(b) 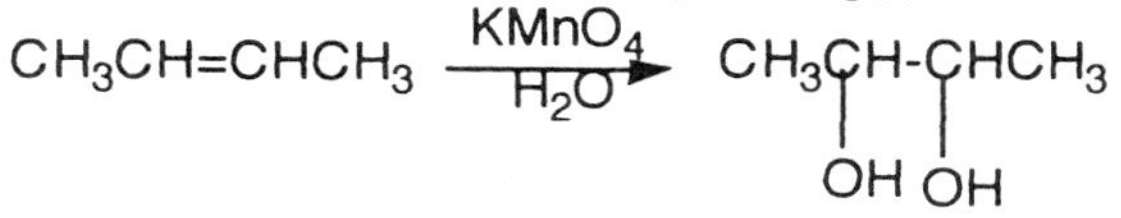 $CH_3CH=CHCH_3 \quad \xrightarrow[H_2O]{KMnO_4} \quad CH_3CH\text{-}CHCH_3$

(c)

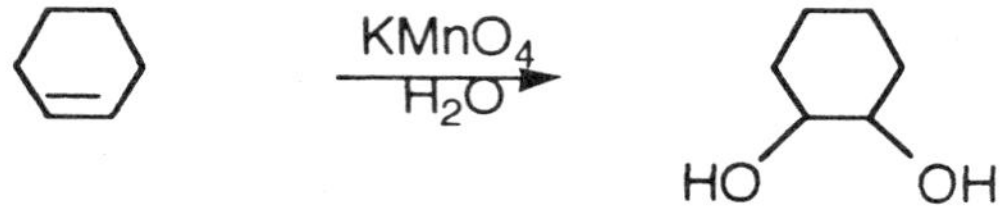

5.10 Ozonolysis

Each double bond is cleaved; the carbons become carbon-oxygen double bonds.

(a)

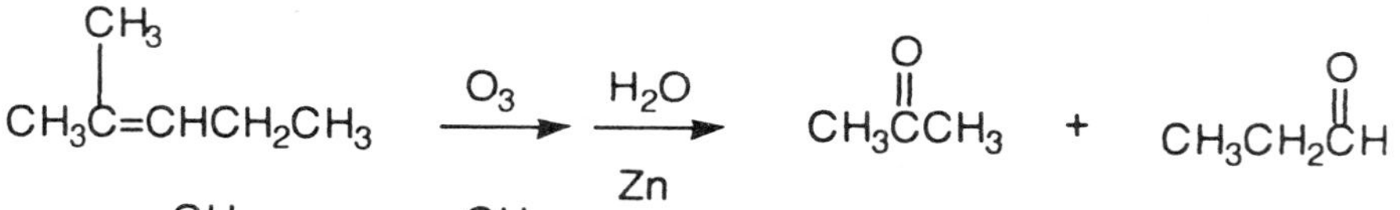

(b)

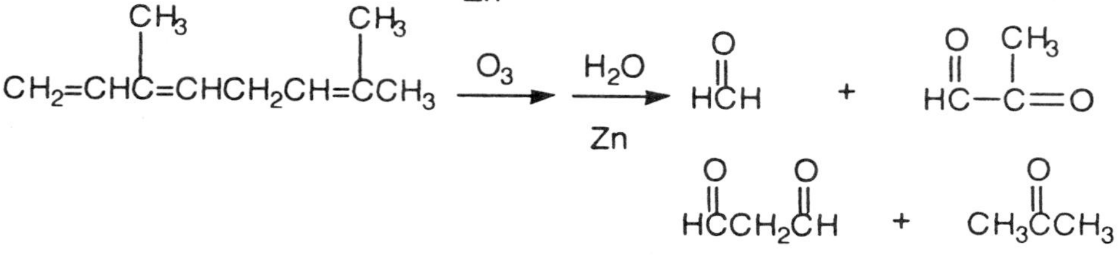

c)

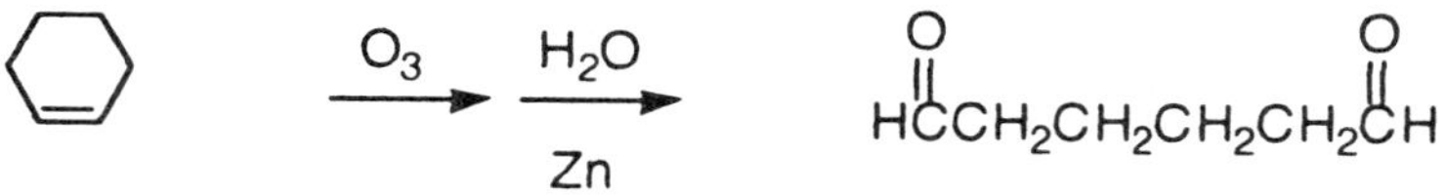

82

5.11 Ozonolysis

Where ever you see a carbon-oxygen double bond, there was originally a carbon-carbon double bond. Since there are only two carbon-oxygen double bonds, they must have been involved in the carbon-carbon double bond.

$$CH_3CH=CHCH_2CH_3$$

5.12 Acidity of Terminal Alkynes

$$CH_3CH_2C\equiv CH + NaNH_2 \longrightarrow CH_3CH_2C\equiv CNa + NH_3$$

5.13 Terpenes

    a) monocyclic monoterpene   b) acyclic monoterpene

    c) bicyclic sesquiterpene   d) acyclic tetraterpene

    e) acyclic sesquiterpene   f) tricyclic diterpene

    g) monocyclic monoterpene   h) acyclic monoterpene

5.14 Addition Reactions of Alkenes: Section 5.1

a) $CH_3(CH_2)_3\underset{Br}{CH}-\underset{Br}{CH_2}$ b) $CH_3CH_2\underset{Cl}{CH}-\underset{Cl}{CHCH_3}$ c)

d)   e) $CH_3CH_2\underset{Cl}{\overset{CH_3}{C}}CH_3$ f) $CH_3\underset{|}{\overset{CH_3}{C}}-CH_2CH_3$ g) $CH_3\overset{CH_3}{CH}CH_2\underset{OH}{\overset{CH_3}{C}}CH_3$

h) $CH_3\underset{OH}{CH}CH_3$ i) $\underset{Cl}{CH_2}-\underset{Cl}{CH}-\underset{Cl}{CH}-\underset{Cl}{CH_2}$ j) $(CH_3)_2\underset{Cl}{C}CH_2CH_2CH_3$

k) $CH_3(CH_2)_4CH_2CH_2CH_3$ l)

5.15 Addition Reactions of Alkynes: Section 5.2

a) $CH_3CH_2\underset{Cl}{\overset{}{C}}=\underset{Cl}{CH}$ b) $CH_3CH_2\underset{Br}{\overset{Br}{C}}-\underset{Br}{\overset{Br}{C}}CH_3$ c) $CH_3CH_2CH_2CH_2CH=CH_2$

d) $CH_3CH_2CH_2CH_2CH_2CH_3$

e) $CH_3\overset{\overset{\displaystyle CH_3}{|}}{C}HC\overset{\overset{\displaystyle CH_3}{|}}{=}C\overset{}{}HCH_3$ $\underset{Cl\ \ \ H}{}$

f) $CH_3CH_2CH_2\overset{\overset{\displaystyle Br}{|}}{\underset{\underset{\displaystyle Br}{|}}{C}}CH_3$

g) $CH_3CH_2CH_2\overset{}{\underset{\underset{\displaystyle Br}{|}}{C}}{=}CH_2$

h) [cyclohexyl]—$\overset{\overset{\displaystyle Cl}{|}}{\underset{\underset{\displaystyle Cl}{|}}{C}}$—$\overset{\overset{\displaystyle H}{|}}{\underset{\underset{\displaystyle H}{|}}{C}}$—[cyclohexyl]

5.16 Bromination: Section 5.1B2

Bromination involves cis addition due to an intermediate bromonium ion.

[cyclohexene]   + Br_2 ⟶   [cyclohexane with Br and Br]

5.17 Hydrogenation: Section 5.2B

(a) $CH_3\overset{\overset{\displaystyle CH_3}{|}}{C}HC{\equiv}CCH_3$ + $1H_2$ $\xrightarrow{\ Pd\ }$ [alkene product]

(b)   [cyclopentene with two CH₃]   + $1H_2$ $\xrightarrow{\ Pd\ }$ [cyclopentane product]

5.18 Hydration of Alkynes: Section 5.2D

(a) $CH_3CH_2C{\equiv}CH$ + H_2O $\xrightarrow[HgSO_4]{H_2SO_4}$ $CH_3CH_2\overset{\overset{\displaystyle O}{\|}}{C}CH_3$

(b) $CH_3C{\equiv}CCH_3$ + H_2O $\xrightarrow[HgSO_4]{H_2SO_4}$ $CH_3CH_2\overset{\overset{\displaystyle O}{\|}}{C}CH_3$

5.19 Oxidation of Alkenes: Section 5.7A

(a) $CH_3CH{=}CH_2$ $\xrightarrow{\ KMnO_4\ }$ $CH_3\overset{}{\underset{\underset{\displaystyle OH}{|}}{C}}H\overset{}{\underset{\underset{\displaystyle OH}{|}}{C}}H_2$

(b)

5.20 Acidity of Terminal Alkynes: Section 5.8

(a) $CH_3C \equiv CH$ + $NaNH_2$ $\longrightarrow$ $CH_3C \equiv CNa$ + NH_3

(b) $CH_3CH_2CH_2C \equiv CH$ + $NaNH_2$ $\longrightarrow$ $CH_3CH_2CH_2C \equiv CNa$ + NH_3

(c) $CH_3CH_2C \equiv CCH_3$ + $NaNH_2$ $\longrightarrow$ No Reaction
                                                  Not a terminal alkyne

5.21 Ozonolysis: Section 5.7B

Each place there is a carbon-carbon double bond it cleaves and each carbon becomes a carbon-oxygen double bond.

5.22 Ozonolysis: Section 5.7B

Since all of the examples are hydrocarbons, each place you see a carbon-oxygen double bond, you are looking at a carbon that originally was involved in a carbon-carbon double bond.

c) $CH_3C{=}CHCH_2CH_2CH{=}CCH_3$
       CH_3 CH_3

d) $CH_3CCH_2CCH_2CCH_3$ (with CH_2, CH_2, CH_2 substituents)

e) (structure: cyclohexene with two CH_3 groups)

f) (structure: bicyclic with two CH_3 groups)

5.23 Reaction Mechanisms - Electrophilic Addition to Alkenes:
Section 5.1B

(a) $CH_3CH{=}CH_2 \xrightarrow{Br^+} CH_3CH{-}CH_2 \xrightarrow{Br^-} CH_3CH{-}CH_2$
       with $+$ and Br; final with Br Br

(b) (cyclopentene) $\xrightarrow{Cl^+}$ (chloronium ion $+$) $\xrightarrow{Cl^-}$ (dichloride product)

(c) $CH_3C{=}CHCH_3$ (with CH_3) $\xrightarrow{H^+}$ $CH_3C{-}CHCH_3$ (with CH_3, $+$, H) $\xrightarrow{Cl^-}$ $CH_3C{-}CHCH_3$ (with CH_3, Cl H)

(d) (cyclohexene with CH_3) $\xrightarrow{H^+}$ (cyclohexyl cation $+CH_3$, H) $\xrightarrow{Br^-}$ (product Br CH_3, H)

(e) $CH_3C{=}CH_2$ (with CH_3) $\xrightarrow{H^+}$ $CH_3C{-}CH_2$ (with CH_3, $+$, H) $\xrightarrow{H_2O}$ $CH_3C{-}CH_2$ (with CH_3, $+OH$ H, H) $\xrightarrow{-H^+}$ $CH_3C{-}CH_2$ (with CH_3, OH H)

86

5.24 Reaction Mechanisms - Electrophilic Addition to Alkynes

Section 5.2C

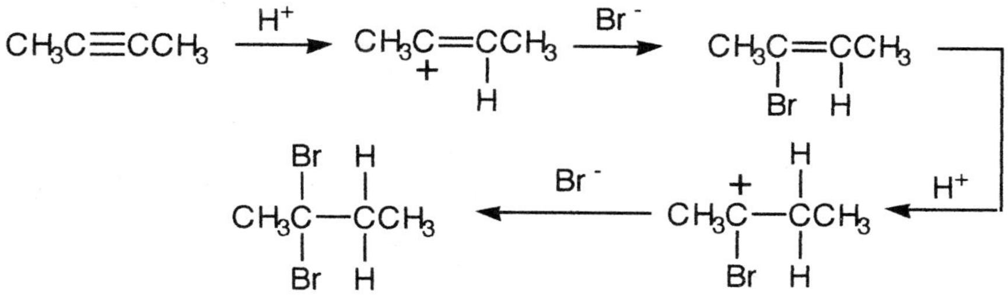

5.25 Electrophilic Addition to Conjugated Dienes: Section 5.4

(a) See problem 5.7a answer.

(b)

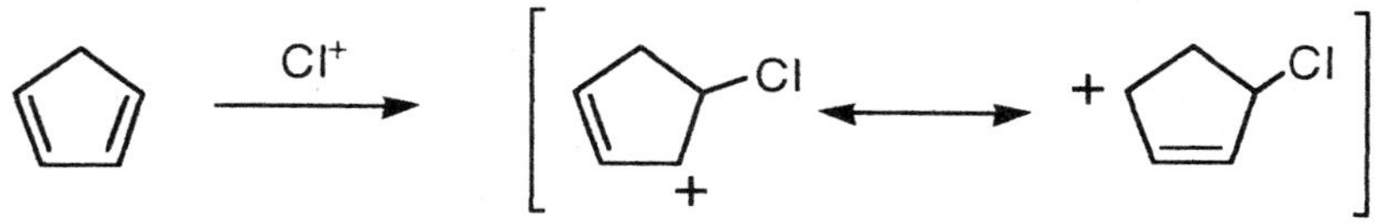

resonance forms

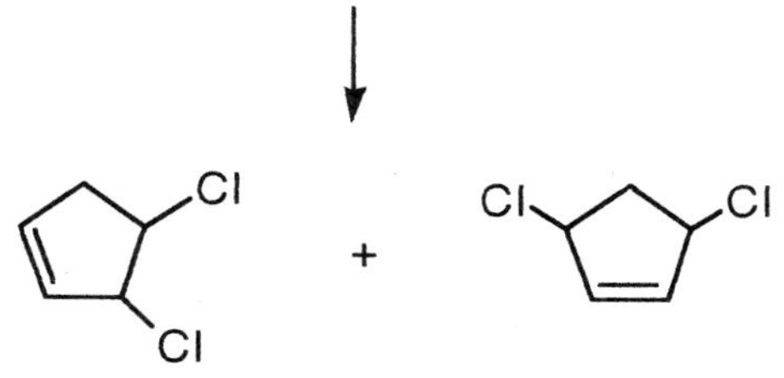

The electrophile attacks one of the double bonds to form an allylic carbocation that is described by two resonance forms. Neutralization forms two products.

1,2 addition   **1,4 addition**

(c)

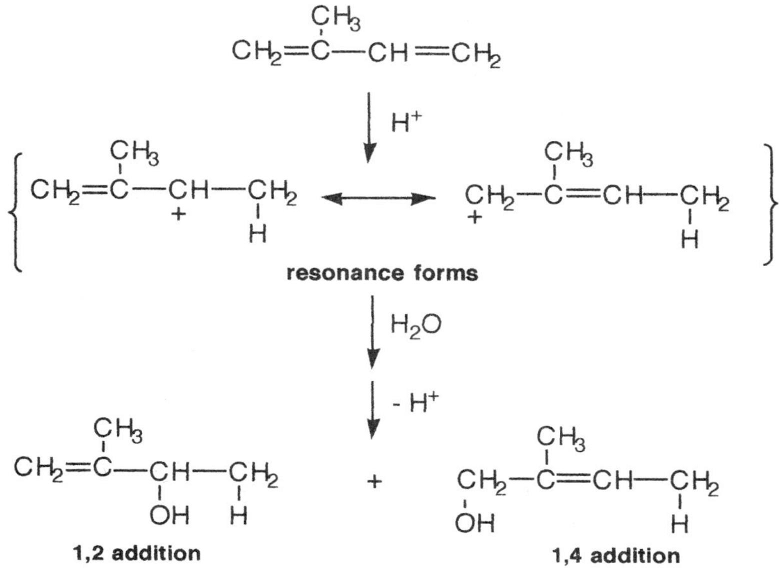

$$CH_2=\overset{\overset{\displaystyle CH_3}{|}}{C}-CH=CH_2$$

$\downarrow H^+$

$$\left\{ CH_2=\overset{\overset{\displaystyle CH_3}{|}}{C}-\underset{+}{CH}-\overset{\overset{\displaystyle }{}}{\underset{H}{CH_2}} \quad\longleftrightarrow\quad \underset{+}{CH_2}-\overset{\overset{\displaystyle CH_3}{|}}{C}=CH-\underset{H}{CH_2} \right\}$$

resonance forms

$\downarrow H_2O$

$\downarrow -H^+$

$$CH_2=\overset{\overset{\displaystyle CH_3}{|}}{C}-\underset{OH}{CH}-\underset{H}{CH_2} \qquad + \qquad \underset{OH}{CH_2}-\overset{\overset{\displaystyle CH_3}{|}}{C}=CH-\underset{H}{CH_2}$$

1,2 addition   **1,4 addition**

1,2 Addition   **1,4 Addition**

5.26 Resonance Forms and Resonance Hybrids: Section 5.5

Resonance forms   **Resonance hybrid**

(a) 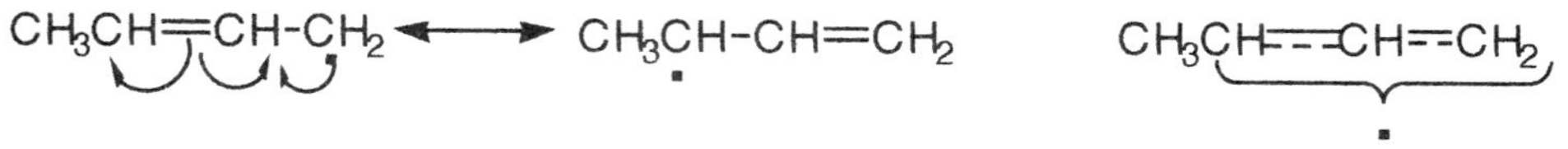 $CH_3CH=CH-CH_2 \longleftrightarrow CH_3CH-CH=CH_2$ $CH_3CH--CH==CH_2$

(b) 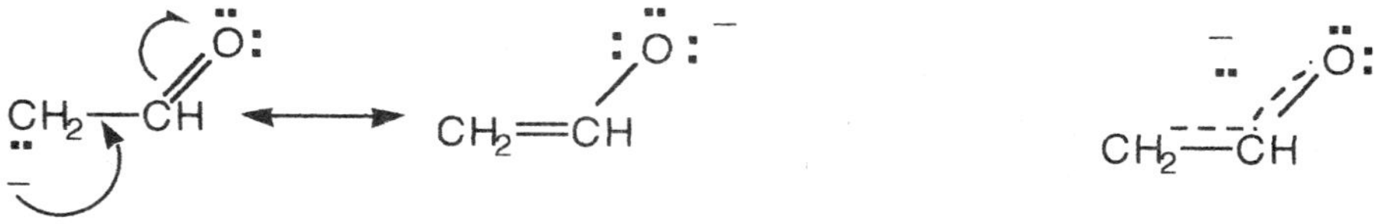 $CH_2-CH \longleftrightarrow CH_2=CH$ $CH_2=CH$

(c)

88

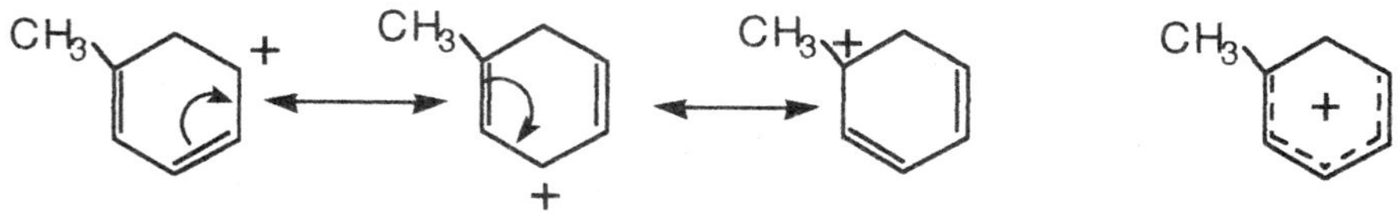

5.27 Resonance Forms and Resonance Hybrids: Section 5.5

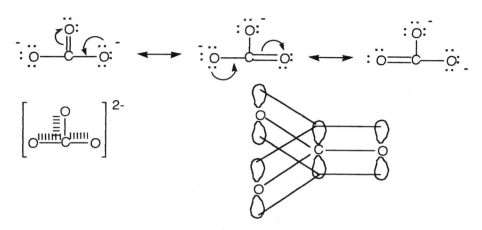

5.28 Addition Polymers: Section 5.3

a) $\sim\!\!\left(\!CH_2\!-\!CF_2\!\right)_{\!n}\!\!\sim$

b) $\sim\!\!\left(\!CH_2\!-\!CH\!\right)_{\!n}\!\!\sim$
   |
   Br

5.29 Reaction Mechanism - Addition Polymers: Section 5.3A-B

(a)   Cationic Polymerization

$H+\quad CH_2\!\!=\!\!CH \longrightarrow H\!:\!CH_2\!-\!CH+\quad CH_2\!\!=\!\!CH \longrightarrow$
         | | |
         CH_3 CH_3 CH_3

$H\!:\!CH_2\!-\!CH\!-\!CH_2\!-\!CH+\quad$ etc. etc. $\xrightarrow{\ :A-\ }\ H\!\sim\!\!\left(\!CH_2\!-\!CH\!\right)_{\!n}\!\!\sim\!A$
         | | |
         CH_3 CH_3 CH_3

(b)   Free-radical Polymerization

$$ROOR \longrightarrow RO\cdot\ +\ RO\cdot$$

                     Cl   Cl   Cl
                     | | |
$RO\cdot\quad CH_2\!\!=\!\!C\cdot \longrightarrow RO\!:\!CH_2\!-\!C\cdot\quad CH_2\!\!=\!\!C \longrightarrow$
                     | | |
                     Cl   Cl   Cl

         Cl   Cl   Cl
         | | |
$RO\!:\!CH_2\!-\!C\!:\!CH_2\!-\!C\cdot\quad$ etc. etc. $\longrightarrow\ RO\!\sim\!\!\left(\!CH_2\!-\!C\!\right)_{\!n}\!\!OR$
         | | |
         Cl   Cl   Cl

(c)   Polymerization by 1,4-Addition: Free-radical, Chain

$$RO:OR \longrightarrow RO\cdot \; + \; RO\cdot$$

$$
RO\cdot \underset{CH_2}{\frown} \underset{C}{\overset{Cl}{|}} \underset{CH}{\frown} CH_2 \; \frown \; CH_2 \underset{C}{\overset{Cl}{|}} CH \frown CH_2 \; \frown \; CH_2 \underset{C}{\overset{Cl}{|}} CH \frown CH_2 \frown
$$

$$
RO: \underset{|}{\overset{Cl}{}} CH_2\overset{Cl}{\underset{|}{C}}{=}CHCH_2 \cdot \; CH_2\overset{Cl}{\underset{|}{C}}{=}CHCH_2 \cdot \; CH_2\overset{Cl}{\underset{|}{C}}{=}CHCH_2 \cdot \xrightarrow{\text{etc}}
$$

$$
RO \Bigl(CH_2\overset{Cl}{\underset{|}{C}}{=}CHCH_2 \Bigr)_n OR
$$

5.30 Synthesis: Sections 4.5, 5.1, 5.2

a) $A = CH_3CH_2CH_2CH{=}CH_2$ or $CH_3CH_2CH{=}CHCH_3$

b) $B = CH_3CH_2CHX_2$ (X = Cl, Br, I)   c) $C = CH_3\overset{CH_3}{\underset{|}{C}}HCH_2\overset{}{\underset{|}{C}}HCH_3$

$\overset{}{\underset{OH}{}}$

d) $D = CH_3\overset{CH_3}{\underset{|}{C}}HCH{=}CH_2$ e) $E =$ ⬡–OH $F =$ ⬡

f) $G = CH_3\overset{}{\underset{X}{|}}CHCH_3$ or $CH_3CH_2CH_2X$ $H = CH_3CH{=}CH_2$

$X = Cl, Br, I$

g) $I = CH_3CH_2CH{=}CH_2 / CH_3CH{=}CHCH_3$ $J = CH_3CH_2\overset{}{\underset{Br}{|}}CHCH_3$

$K = CH_3CH{=}CHCH_3$

h) $L = CH_3CH_2OH$ $M = CH_2{=}CH_2$ $N = \overset{}{\underset{Br}{|}}CH_2{-}\overset{}{\underset{Br}{|}}CH_2$ $O = HC{\equiv}CH$

5.31 Industrial Reactions

a) $CH_2{=}CH_2 + H_2O \xrightarrow[\text{catalyst}]{\text{acid}} CH_3CH_2OH$

b) $CH_3CH{=}CH_2$ + H_2O $\xrightarrow[\text{catalyst}]{\text{acid}}$ $CH_3\underset{\underset{OH}{|}}{C}HCH_3$

c) $CH_2{=}CH_2$ + Cl_2 $\longrightarrow$ $\underset{\underset{Cl}{|}}{C}H_2{-}\underset{\underset{Cl}{|}}{C}H_2$ $\xrightarrow[\text{KOH}]{\text{base}}$ $CH_2{=}\underset{\underset{Cl}{|}}{C}H$

d) $H_2C{=}CH{-}C{\equiv}CH$ + HCl $\longrightarrow$ $CH_2{=}CH{-}\underset{\underset{Cl}{|}}{C}{=}CH_2$

e) $HC{\equiv}CH$ + $2\,Cl_2$ $\longrightarrow$ $H{-}\underset{\underset{Cl}{|}}{\overset{\overset{Cl}{|}}{C}}{-}\underset{\underset{Cl}{|}}{\overset{\overset{Cl}{|}}{C}}{-}H$ $\xrightarrow{\text{1 KOH}}$ $H{-}\underset{\underset{Cl}{|}}{C}{=}\underset{\underset{Cl}{|}}{C}{-}Cl$

f) $HC{\equiv}CH$ + HCl $\longrightarrow$ $CH_2{=}\underset{\underset{Cl}{|}}{C}H$ $\xrightarrow{Cl_2}$ $\underset{\underset{Cl}{|}}{C}H_2{-}\underset{\underset{Cl}{|}}{C}H$ $\xrightarrow{\text{1 KOH}}$ $CH_2{=}\underset{\underset{Cl}{|}}{\overset{\overset{Cl}{|}}{C}}$

5.32 Gasoline Octane Boosters

TBA $CH_3\underset{\underset{|}{}}{\overset{\overset{CH_3}{|}}{C}}{=}CH_2$ + $H_2\ddot{O}{:}$ $\xrightarrow{H_2SO_4}$ $CH_3\overset{\overset{CH_3}{|}}{\underset{\underset{:\ddot{O}H}{|}}{C}}CH_3$

$\downarrow$ H+

$CH_3\overset{\overset{CH_3}{|}}{\underset{\underset{H}{|}}{\overset{}{C}}}{-}\overset{}{C}H_2$ $\xrightarrow{H_2\ddot{O}{:}}$ $CH_3\overset{\overset{CH_3}{|}}{\underset{\underset{+:O{-}H}{|}\;\underset{H}{|}}{C}}{-}CH_3$ $\xrightarrow{-H+}$

MTBE $CH_3\overset{\overset{CH_3}{|}}{C}{=}CH_2$ + $CH_3\ddot{O}H$ $\xrightarrow{H_2SO_4}$ $CH_3\overset{\overset{CH_3}{|}}{\underset{\underset{:\underset{}{O}CH_3}{|}}{C}}CH_3$

$\downarrow$ H+

$CH_3\overset{\overset{CH_3}{|}}{\underset{\underset{H}{|}}{C}}{-}CH_2$ $\xrightarrow{CH_3\ddot{O}H}$ $CH_3\overset{\overset{CH_3}{|}}{\underset{\underset{+:OCH_3}{|}\;\underset{H}{|}}{C}}CH_3$ $\xrightarrow{-H+}$

5.33 Reaction Mechanisms: Section 5.1B.2-3

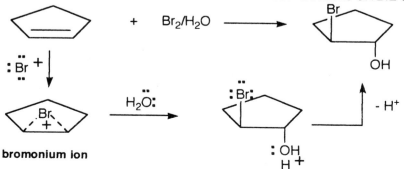

5.34 IUPAC Nomenclature: Section 3.2

a) 2-methyl-2-butene;   b) 2-chloro-1,3-butadiene;

c) 6-ethyl-3-methyl-4-nonyne;   d) 2,2,5,5-tetramethylhexane;

e) 7-ethyl-2-isopropyl-1,3,5-cycloheptatriene;   f) 5-ethyl-3, 6-decadiyne;

g) 2-hexen-4-yne

5.35 IUPAC Nomenclature: Section 3.2

a) [structure: chlorocyclohexane with CH₃CCH₃ / CH₃ substituent]

b) $CH_3CHCH_2CH_2CH_3$ with CH_3 substituent

c) $CH_3CH_2CH\!-\!CHCH_2CH_3$ with CH_2CH_3 and CH_2CH_3 substituents

d) $CH_3C\!\equiv\!CCH_2CH_2CH_3$

e) $CH_3CH_2C\!\equiv\!CCHCH_2CH_2CH_3$ with $CH_2CH_2CH_3$ substituent

f) $CH_3CHCH\!=\!CHCH_2CH_2CH_3$ with Br substituent

g) $CH_3CH\!=\!CHC\!=\!CHC\!=\!CHCHCH_3$ with CH_3, CH_3, CH_2CH_3 substituents

h) $H_2C\!=\!CHC\!\equiv\!CH$

5.36 Reaction of Bromine Water with Alkenes: Sections 5.1B.2-3

[reaction mechanism scheme: cyclopentene + Br₂/H₂O → bromo-hydroxycyclopentane; showing Br⁺ addition to form bromonium ion, H₂O: attack, then loss of H⁺]

bromonium ion

5.37 Hydration: Section 5.1

Pay attention to orientation of addition as explained in Section 5.1C.

(a) $CH_3CH=CHCH_2CH_3$ (b) $CH_3CH_2CH_2CH=CH_2$ (c)

5.38 Reaction Mechanism: Section 5.1

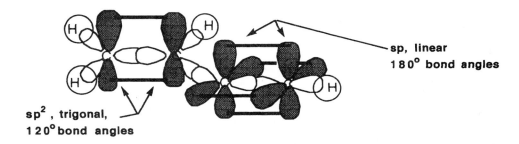

5.39 Molecular Orbitals and Bonding: Section 1.5E-F

sp, linear
180° bond angles

sp², trigonal,
120° bond angles

5.40 Reaction Mechanism: Section 5.1B.3

$CH_3CH=CHCH_3$ + $CH_3\ddot{O}H$ $\xrightarrow{HCl}$ $CH_3\underset{H}{C}H-\underset{:\ddot{O}CH_3}{C}HCH_3$

$H^+ \downarrow$

$CH_3\underset{H}{C}H-\overset{+}{C}HCH_3$ $\xrightarrow{CH_3\ddot{O}H}$ $CH_3\underset{H}{C}H-\underset{\underset{H\,+}{:\ddot{O}CH_3}}{C}HCH_3$ $\xrightarrow{-H^+}$

5.41 Hydrogenation: Section 5.2B

Cis addition occurs.

(a) $CH_3CH_2C≡CCH_3$ + $1H_2$ $\xrightarrow{Pt}$ [cis-alkene structure: CH_3CH_2 and CH_3 on same side]

(b)   [structure with CH_3 branch: $CH_3CH_2CHC≡CCH_3$]   + $1H_2$ $\xrightarrow{Pt}$ [product cis-alkene structure]

5.42 Hydrogenation: Section 5.2B

[cyclohexene ring with two CH_3 groups]   + H_2 $\xrightarrow{Pt}$ [cyclohexane ring with CH_3 CH_3 and H H]

5.43 Carbocations: Section 5.1C.1

$CH_3CH_2CH_2CH_2^+$ $CH_3CHCH_2CH_3^+$ [structure: $CH_3CCH_3^+$ with CH_3]
      $1°$ $2°$ $3°$

5.44 Reactions of Alkynes: Section 5.2

(a) $CH_3CH_2CH_2C≡CH$ +   2HBr $\longrightarrow$ [product: $CH_3CH_2CH_2CCH_3$ with two Br]

(b) $CH_3CH_2C≡CCH_3$ +   2HCl $\longrightarrow$ [product: $CH_3CH_2CCH_2CH_3$ with two Cl]

(c) $CH_3CH_2C≡CCH_3$ + $2Cl_2$ $\longrightarrow$ [product: $CH_3CH_2C-CCH_3$ with Cl Cl / Cl Cl]

5.45 Units of Unsaturation: Sections 3.6, 5.1A.2, 5.2A

1-Buten-3-yne has one triple bond and one double bond. This represents three units of unsaturation. One mole of the compound will add three moles of bromine, one mole to the double bond and two to the triple bond.

5.46 Units of Unsaturation: Sections 3.6, 5.1A.4, 5.2A-B

Since the compound is non-cyclic all the units of unsaturation must be in the form of carbon-carbon double bonds or triple bonds. Four mole-equivalents of hydrogen are consumed so there must be four units of unsaturation: four double bonds, two triple bonds, or one triple and two double bonds.

starting material C_8H_{10} + $4H_2$ $\longrightarrow$ C_8H_{18} **hydrogenation product**

5.47 Hydration of Alkynes: Section 5.2D

5.48 Reaction Mechanisms

$$CH_3\overset{|}{\underset{:\overset{..}{O}CH_3}{C}}HCH_3 \quad + \quad H\ddot{B}r: \quad \longrightarrow \quad CH_3\overset{|}{\underset{:\overset{..}{B}r:}{C}}HCH_3 \quad + \quad CH_3\overset{..}{O}H$$

$$H^+ \downarrow$$

$$CH_3\overset{|}{\underset{\underset{\overset{..}{H} \ +}{:\overset{..}{O}CH_3}}{C}}HCH_3 \quad \xrightarrow{- CH_3\overset{..}{O}H} \quad CH_3\overset{+}{C}HCH_3 \qquad \qquad :\overset{..}{\overset{..}{B}}r: \ ^-$$

5.49 1,4 Addition: Section 5.4

5.50 Allylic Carbocations: Section 5.4-5.5

The three resonance forms show where this resonance stabilized carbocation can be neutralized.

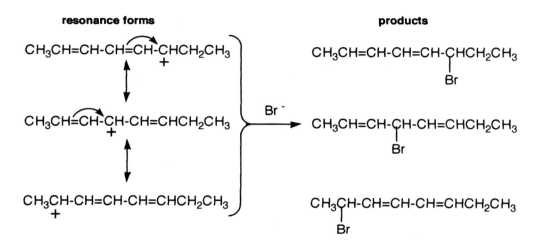

6

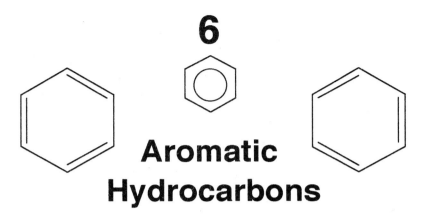

Aromatic
Hydrocarbons

CHAPTER SUMMARY

Aromatic compounds are compounds that are similar to **benzene** in structure and chemical behavior. Benzene, C_6H_6, is a cyclic compound commonly written as a hexagon with alternating double and single bonds. Actually it is a **resonance hybrid** of the two **resonance forms** written with alternating double and single bonds. All six carbons are equivalent, all six hydrogens are equivalent, and all the carbon-carbon bonds are equivalent and intermediate in length between a single bond and double bond. Each carbon is trigonal, sp^2 hybridized, has 120° bond angles. There is a p-orbital on each carbon and the six overlap continuously around the ring.

Benzene and aromatic compounds in general are characterized by **unusual stability**. This is seen in their characteristic reactions, **substitution** where the electronic character of the system is preserved and in heats of hydrogenation from which a **resonance energy** of 36 kcal/mole can be calculated.

Naphthalene, anthracene, and phenanthrene are simple fused ring aromatic systems. **Monosubstituted benzenes** are named as derivatives of benzene or by common names such as **toluene, benzaldehyde, benzoic acid, benzensulfonic acid, phenol, and aniline. Disubstituted benzenes** can be named using **ortho** (1,2), **meta**

(1,3), or **para** (1,4). Substituents on the nitrogen of aniline are located by capital **N**. The prefix for benzene is **phenyl**.

The characteristic reaction of benzene and its derivatives is **electrophilic aromatic substitution.** In these reactions, a hydrogen on the benzene ring can be replaced by a chlorine (**chlorination**), a bromine (**bromination**), an alkyl or acyl group (**Friedel-Crafts alkylation or acylation**), a nitro group (**nitration**), or a sulfonic acid group (**sulfonation**). These groups attack the electron-rich benzene ring as positive species, **electrophiles**; a **carbocation** results. Hydrogen ion is lost from the ring as the carbocation is neutralized and the benzene ring is regenerated.

Groups already present on a benzene ring direct the orientation of substitution of incoming groups. **Electron-donating groups** stabilize the intermediate carbocation and direct the incoming electrophile to the **ortho and para positions.** **Electron-withdrawing groups** destabilize the carbocation and the incoming electrophile is directed to the **meta position.** Electron-donating groups increase the negative character of the ring and its attractiveness to electrophile. As a result they increase reactivity and are called **activating groups.** Electron-withdrawing groups decrease the negative character of the ring and are **deactivating groups.** The directing and activating or deactivating effects of substituents must be taken into account in devising synthesis schemes.

Alkyl side chains on benzene can be **oxidized** to **carboxylic acids** using potassium permanganate.

Connections 1 is about cancer and carcinogens, many of which are aromatic compounds.

Connections 2 describes the structure of gasoline molecules and refinery methods for converting crude oil into gasoline.

Connections 3 is about herbicides that are phenoxyacetic acid derivatives.

SOLUTIONS TO PROBLEMS

6.1 Bonding in Aromatic Compounds

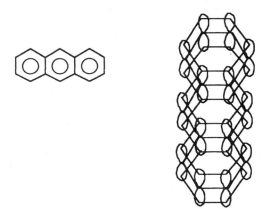

6.2 Positional Isomers

a) $C_{10}H_7Br$

b) $C_{14}H_9Br$

c) $C_{14}H_9Br$

6.3 Nomenclature

a) bromobenzene;   b) isopropylbenzene;   c) butylbenzene

d) ethylbenzene

6.4 Nomenclature

a) m-ethylbenzaldehyde;   b) p-dibromobenzene;

c) o-chlorobenzenesulfonic acid;   d) N-propyl-m-nitroaniline

6.5 Nomenclature

a) 1-chloro-3-isopropyl-5-nitrobenzene   b)   2,4-dibromobenzoic acid;

c) 2,3,4,5,6-pentachlorophenol

d) 5-bromo-2-chloro-N,N-dimethylaniline

6.6 Nomenclature

a) 4-methyl-2-phenylhexane;   b) 3-benzyl-2-pentene

c) 1-(m-bromophenyl)-2-(p-chlorophenyl) ethyne

6.7 Electrophilic Aromatic Substitution

p-xylene

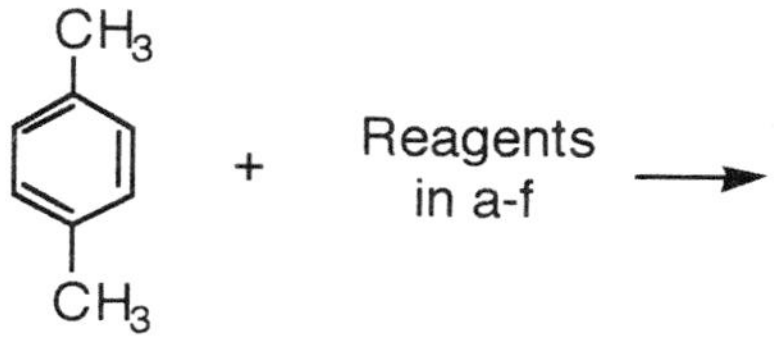

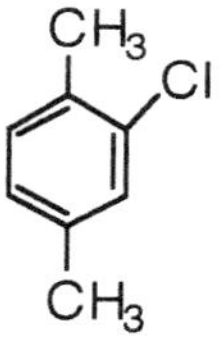

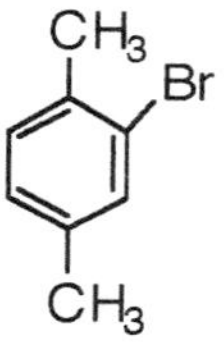

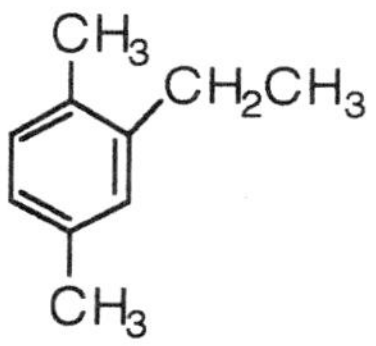

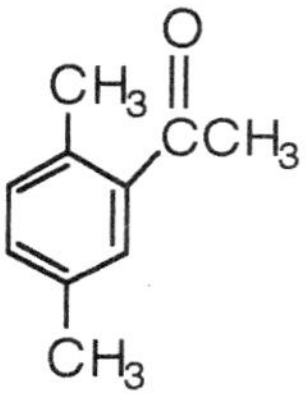

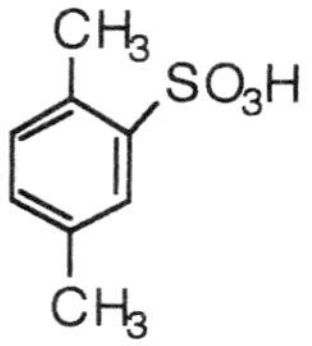

6.8 Electrophilic Aromatic Substitution - Bromination

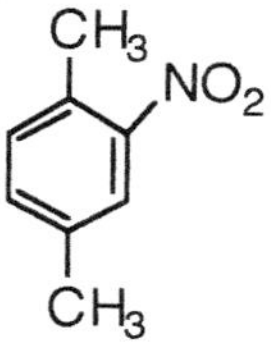

Generation of the Electrophile $Br_2 \; + \; FeBr_3 \longrightarrow Br^+ \; + \; FeBr_4^-$

Two-step Substitution

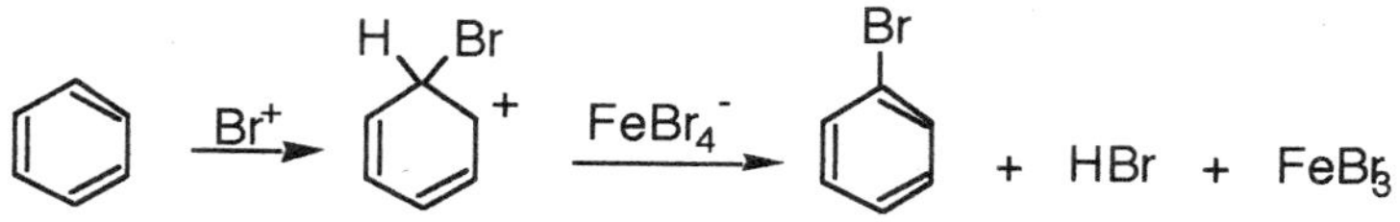

6.9 Electrophilic Aromatic Substitution - Alkylation and Acylation

(a) Acylation

Generation of the Electrophile

$$CH_3\overset{O}{\underset{\|}{C}}Cl + AlCl_3 \longrightarrow CH_3\overset{O}{\underset{\|}{C}}+ + AlCl_4^-$$

Two-step Substitution

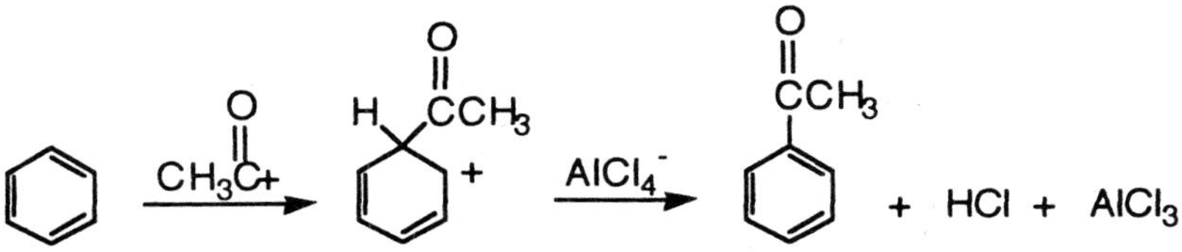

(b) Alkylation

Generation of the Electrophile

$$CH_3CH_2Cl + AlCl_3 \longrightarrow CH_3CH_2+ + AlCl_4^-$$

Two-step Substitution

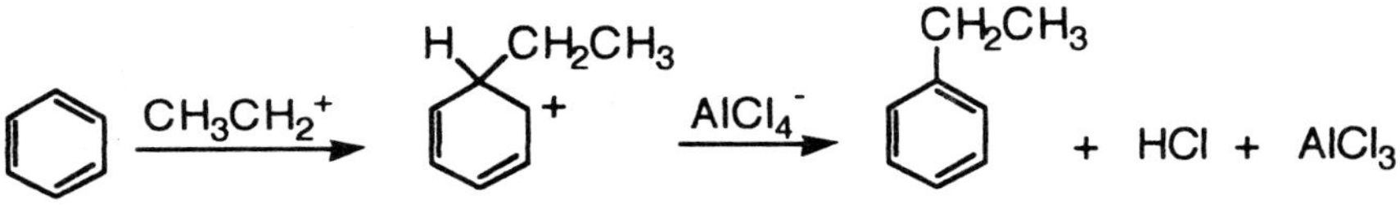

6.10 Electrophilic Aromatic Substitution - Nitration

Generation of the Electrophile

$$HNO_3 + H_2SO_4 \longrightarrow NO_2+ + HSO_4^- + H_2O$$

Two-step Substitution

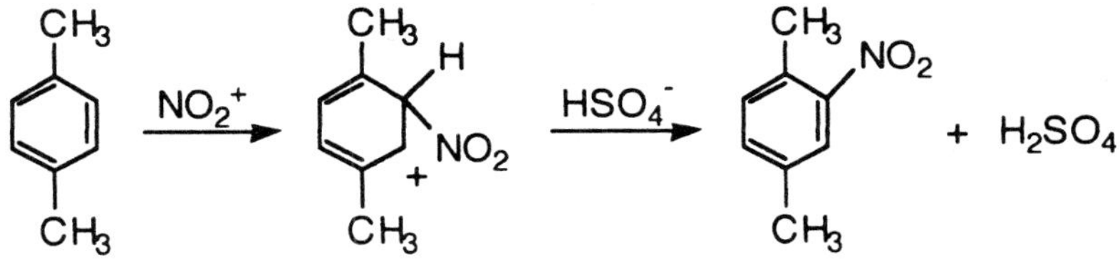

6.11 Electrophilic Aromatic Substitution - Sulfonation

Generation of the Electrophile

$$2 H_2SO_4 \longrightarrow SO_3H^+ \quad HSO_4^- + H_2O$$

Two-step Substitution

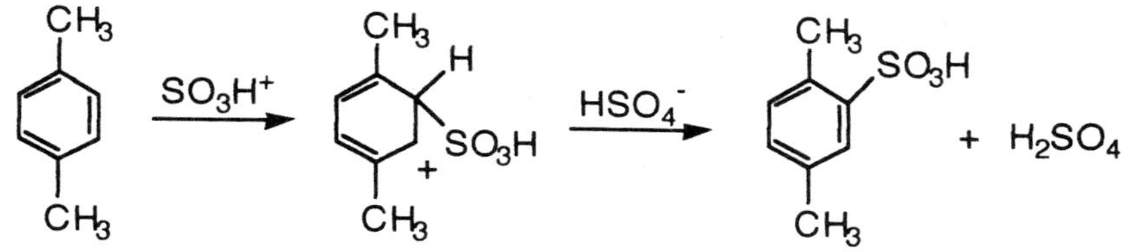

6.12 Electrophilic Aromatic Substitution Reactions

a)

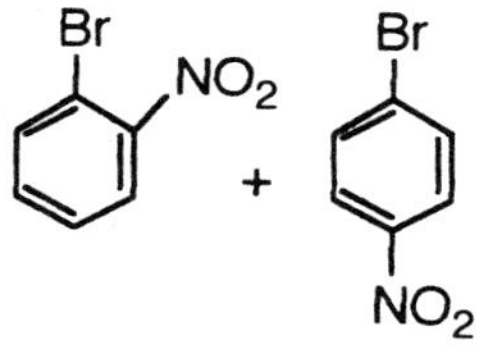

+

;

b)

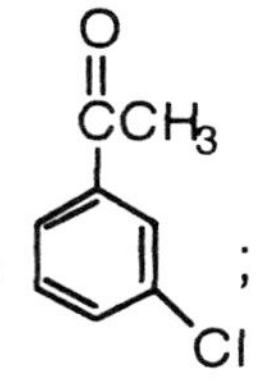

;

c)

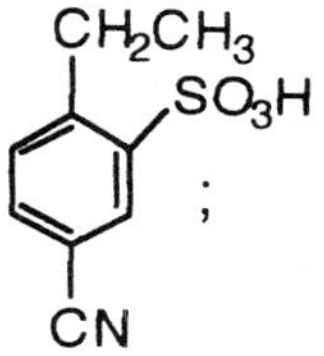

;

d)

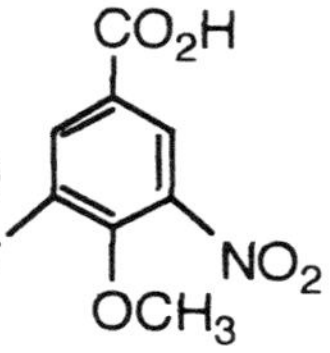

6.13 Synthesis Problems

a)

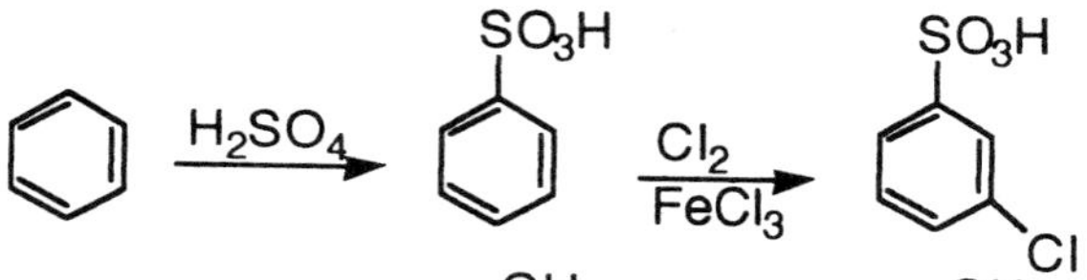

b)

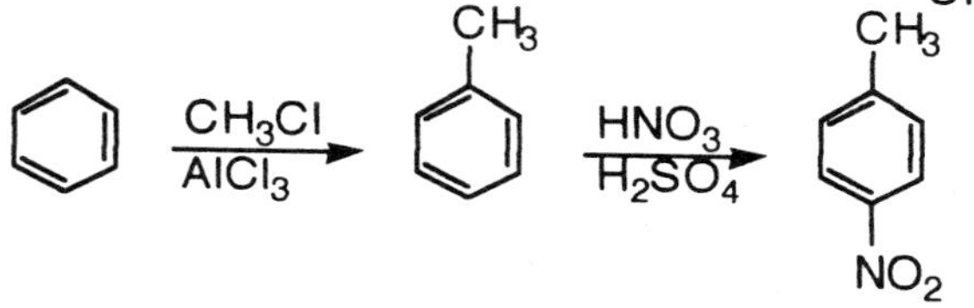

+ ortho isomer

6.14 Activating and Deactivating Groups

(a) methoxybenzene > benzene > chlorobenzene

(b) phenol > p-nitrophenol > nitrobenzene

(c) p-methylaniline > toluene > m-chlorotoluene

6.15 Oxidation of Alkylbenzenes

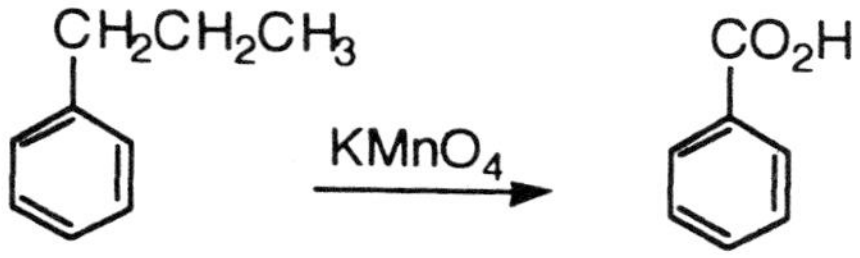

6.16 Bonding Pictures: Section 6.2B

 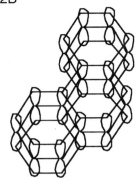

6.17 Molecular Formulas: Section 6.2
   (a) $C_{16}H_{10}$ (b) $C_{20}H_{12}$

6.18 Nomenclature: Section 6.3B-C
   a) fluorobenzene;   b) hexylbenzene;   c) t-butylbenzene;
   d) iodobenzene;   e) o-dichlorobenzene;   f) m-dibromobenzene;
   g) p-bromochlorobenzene;   h) m-iodobenzoic acid;
   i) p-ethylbenzaldehyde;   j) o-nitrophenol;

6.19 Nomenclature: Section 6.3D
   a) 2,4-dichlorotoluene;   b) 3-bromo-5-methylaniline;
   c) 2-ethyl-5-methylbenzenesulfonic acid;   d) 2,4,6-tribromophenol;
   e) 1-bromo-2-chloro-3-ethyl-5-nitrobenzene;   (f) 1,3,5-trinitrobenzene

6.20 Nomenclature of Substituted Anilines: Section 6.3C
   a) m-ethylaniline;   b) N-ethylaniline;
   c) m-nitro-N,N-diethylaniline;   d) 2,4-dichloro-N-ethyl-N-methylaniline

6.21 Nomenclature Using Benzene as a Prefix: Section 6.3E
   a) 2,4-dimethyl-2-phenylpentane;
   b) 2,4-dimethyl-6-(m-nitrophenyl)heptane;
   c) 2-(2,4-dibromophenyl)-5-ethyl-2-heptene;
   d) 3-benzyl-1-bromobutane

6.22 Nomenclature of Polynuclear Aromatic Compounds:
Section 6.3A

a) 1-bromo-5-fluoronaphthalene;   b) 1,4-dinitronaphthalene;

c) 9-methylanthracene;   d) 9-ethylphenanthrene

6.23 Nomenclature: Section 6.3

6.24 Positional Isomers: Section 6.3A

a) 1,2,3-tribromobenzene;   1,2,4-tribromobenzene;
1,3,5-tribromobenzene

b) 1,2-dibromo-3-chlorobenzene;   2,4-dibromo-1-chlorobenzene;
1,4-dibromo-2-chlorobenzene;   1,3-dibromo-2-chlorobenzene;
1,2-dibromo-4-chlorobenzene;   1,3-dibromo-5-chlorobenzene

c) 1-bromo-2-chloro-3-fluorobenzene;
1-bromo-2-chloro-4-fluorobenzene;
2-bromo-1-chloro-4-fluorobenzene;
2-bromo-1-chloro-3-fluorobenzene;
4-bromo-2-chloro-1-fluorobenzene;
1-bromo-3-chloro-5-fluorobenzene;
2-bromo-4-chloro-1-fluorobenzene;

1-bromo-3-chloro-2-fluorobenzene;

1-bromo-4-chloro-2-fluorobenzene;

4-bromo-1-chloro-2-fluorobenzene;

d) 1,2; 1,3; 1,4; 1,5; 1,6; 1,7; 1,8; 2,3; 2,6; and 2,7-dibromonaphthalenes

e) 1,2; 1,3; 1,4; 1,10; 1,5; 1,6; 1,7; 1,8; 1,9; 2,3; 2,10; 2,6; 2,7; 2,9; and 9,10-dinitroanthracenes

f) 1,2; 1,3; 1,4; 1,5; 1,6; 1,7; 1,8; 1,9; 1,10; 2,3; 2,4; 2,5;   2,6; 2,7; 2,9; 2,10; 3,4; 3,5; 3,6; 3,9; 3,10; 4,5; 4,9; 4,10; and 9,10-dinitrophenanthrenes

6.25 Positional Isomers: Section 6.3

6.26 Positional Isomers: Section 6.3

6.27 Reactions of Aromatic Compounds: Section 6.4

First look at the reagent and decide what is going to substitute for a hydrogen on the benzene ring - a halogen, alkyl, acyl, nitro, or sulfonic acid group. Then look at the groups on the ring and determine where they direct - ortho/para or meta. Place the incoming group where it is directed by the existing groups. See Example 6.5 in the text.

a) 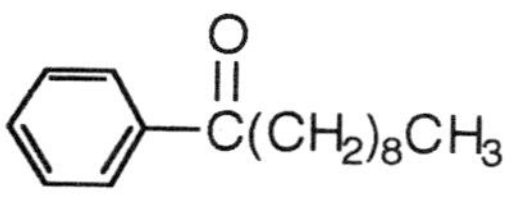 ... b) 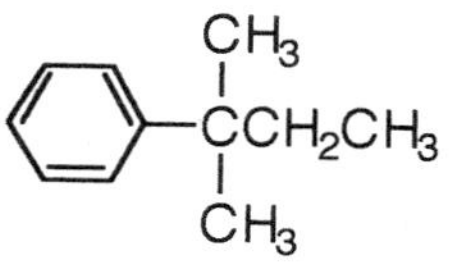 ... c) 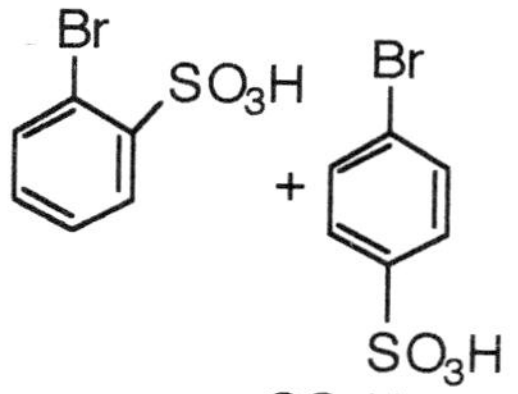 ...

d) 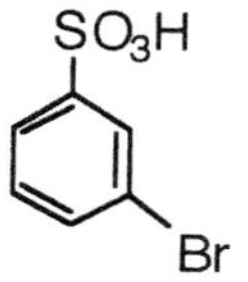 ... e) 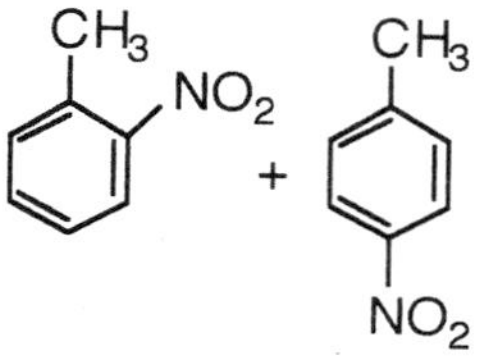 ... f) 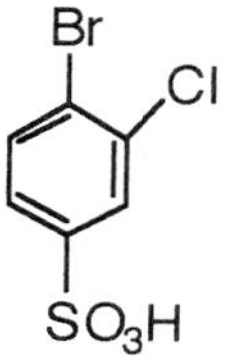 ... g)

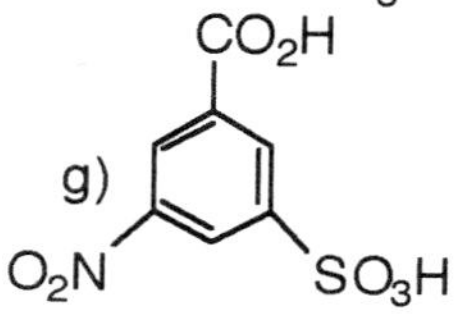

h) 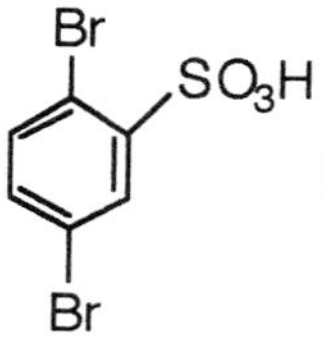 ... i) 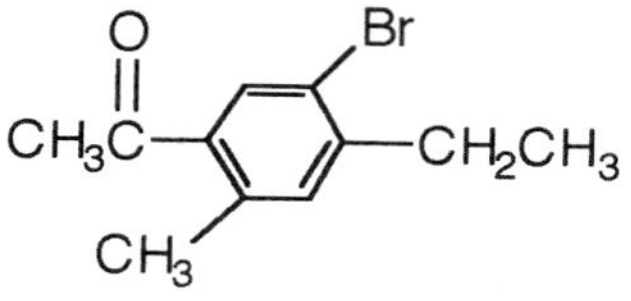 ... j)

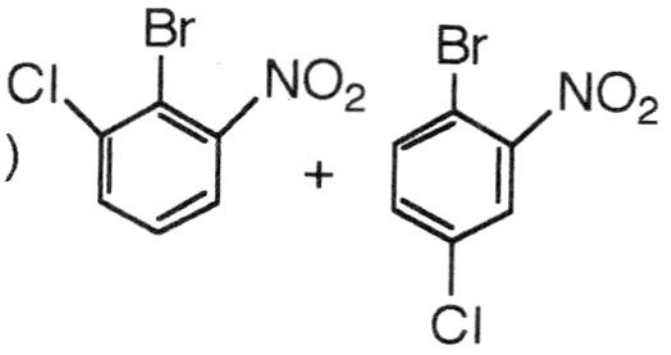

k)

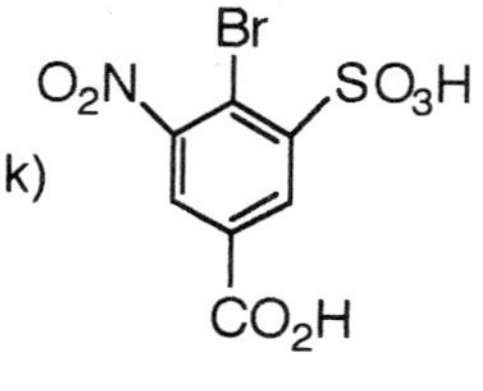

6.28 Reactions of Aromatic Compounds: Section 6.4

See explanation on problem 6.27.

a) 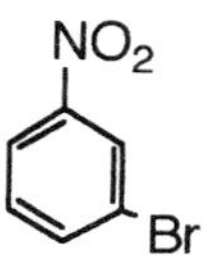 ... b) 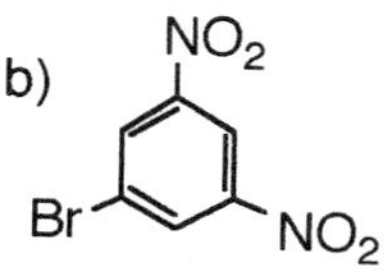 ... c) 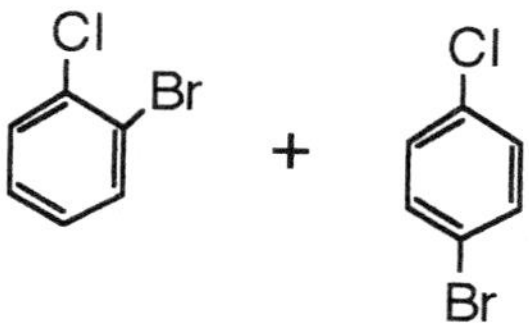 ...

d)

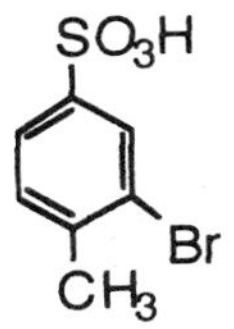

e)

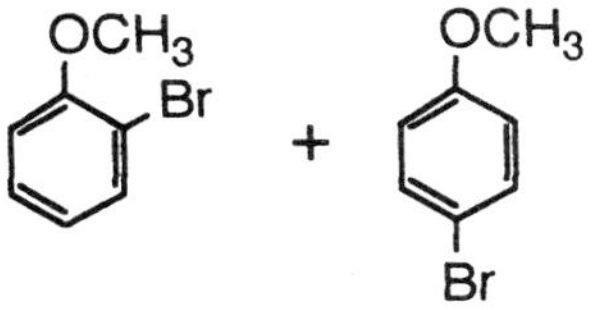

6.29 Reaction Mechanisms: Section 6.4B

Following is the general mechanism for electrophilic aromatic substitution. First the electrophile is generated. The two-step substitution occurs: the electrophile bonds to the ring forming a carbocation followed by elimination of a hydrogen ion to regenerate the benzene ring.

Generation of the Electrophile

$$E\text{-}A \quad + \quad catalyst/reagent \longrightarrow E^+$$

Two-step Substitution

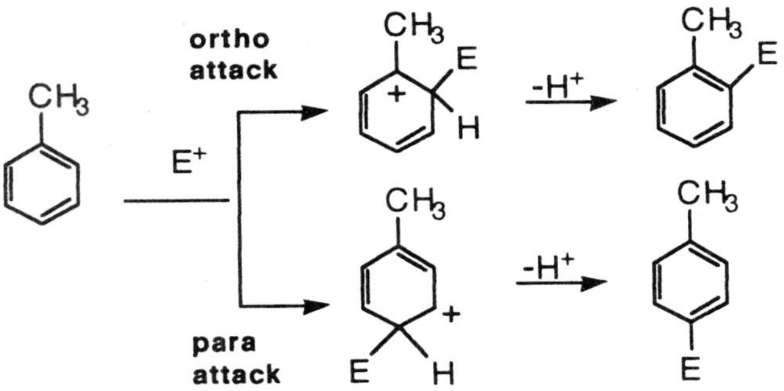

The mechanism is the same for all cases, only the electrophile differs. Following are the equations for generation of the electrophiles.

a) $E^+ = Cl^+$ $Cl_2 + FeCl_3 \longrightarrow Cl^+ + FeCl_4^-$

b) $E^+ = CH_3CH_2\overset{O}{\overset{\|}{C}}+$ $CH_3CH_2\overset{O}{\overset{\|}{C}}Cl + AlCl_3 \longrightarrow CH_3CH_2\overset{O}{\overset{\|}{C}}+ \ +AlCl_4^-$

c) $E^+ = CH_3\overset{+}{C}HCH_3$ $CH_3\underset{Cl}{\overset{|}{C}}HCH_3 + AlCl_3 \longrightarrow CH_3\overset{+}{C}HCH_3 + AlCl_4^-$

d) $E^+ = NO_2^+$ $HNO_3 + H_2SO_4 \longrightarrow NO_2^+ + HSO_4^- + H_2O$

e) $E^+ = \overset{+}{S}O_3H$ $2\,H_2SO_4 \longrightarrow \overset{+}{S}O_3H + HS\bar{O}_4 + H_2O$

Below is a specific example using acylation, part (b).

Generation of the Electrophile

Two-step Substitution

In this reaction, the catalyst is regenerated when hydrogen ion reacts with the aluminum tetrachloride anion.

$$AlCl_4^- \quad + \quad H^+ \quad \longrightarrow \quad AlCl_3 \quad + \quad HCl$$

6.30 Reactions of Aromatic Compounds: Section 6.4

To predict each product, first determine what group will be introduced on the benzene ring. If one or more groups are already on the ring, determine where they direct (o,p or m) and bond the incoming group accordingly.

c) E = 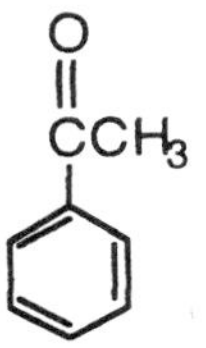   F = 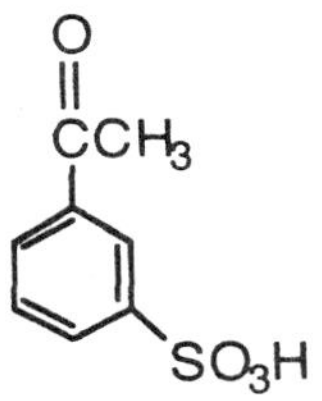   G =

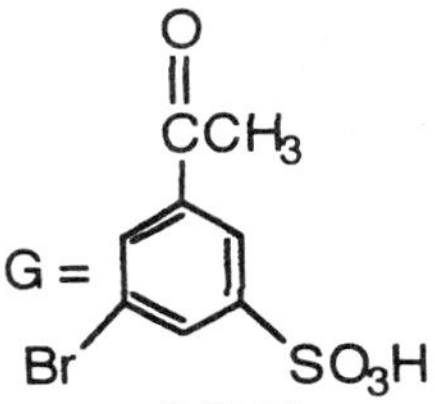

d) H = 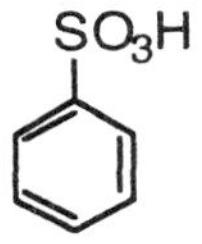   I = 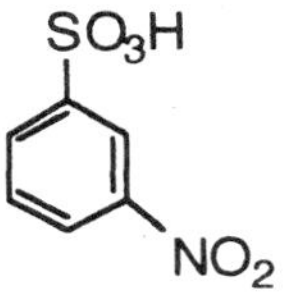   J =

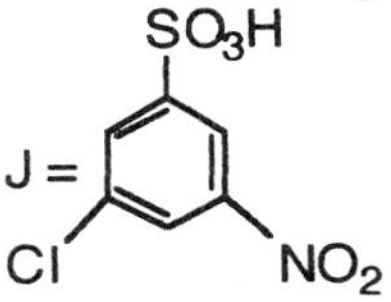

6.31 Oxidation of Alkylbenzenes: Section 6.5

a) 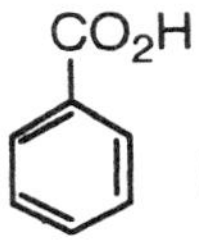   b) 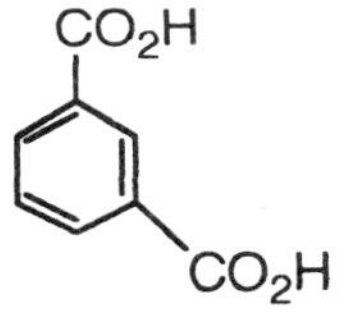   c) A = 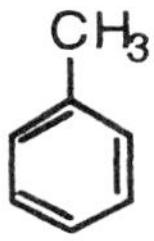   B = 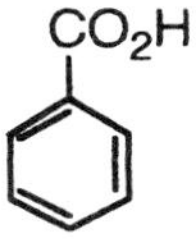   C =

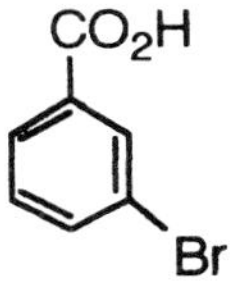

6.32 Synthesis: Section 6.4 and 6.5

To make m-bromobenzoic acid from toluene one should first oxidize the methyl group to a carboxylic acid, which is a meta director, and then introduce the bromine. If the bromine is introduced first, it will go ortho and para since the methyl group is an ortho/para director.

6.33 Synthesis: Section 6.4

The chlorine is an ortho/para director and the sulfonic acid group is a meta director. Since the desired product is p-chlorobenzenesulfonic acid, the chlorine should be introduced first so it can direct the sulfonic acid group para.

6.34 Synthesis: Sections 6.4-6.5

Draw the compound you are trying to synthesize. Determine the reagents needed to introduce each group. Then determine the order in which to introduce groups. For example, in a disubstituted benzene, cover one group with your finger. Does the remaining group direct so that the group you have covered would go where you want it?

a) $\underset{FeBr_3}{\xrightarrow{Br_2}}$ $\underset{FeCl_3}{\xrightarrow{Cl_2}}$

b) + CH_3CHCH_3 with Cl $\xrightarrow{AlCl_3}$ $\xrightarrow{H_2SO_4}$

c) $\xrightarrow{H_2SO_4}$ $\underset{FeBr_3}{\xrightarrow{Br_2}}$

d) $\underset{H_2SO_4}{\xrightarrow{HNO_3}}$ $\underset{FeCl_3}{\xrightarrow{Cl_2}}$

e) $\underset{FeCl_3}{\xrightarrow{Cl_2}}$ $\underset{H_2SO_4}{\xrightarrow{HNO_3}}$

f) $\underset{AlCl_3}{\xrightarrow{CH_3CH_2Cl}}$ $\underset{H_2SO_4}{\xrightarrow{HNO_3}}$ $\underset{FeBr_3}{\xrightarrow{Br_2}}$

g) $\underset{AlCl_3}{\xrightarrow{CH_3Cl}}$ $\xrightarrow{KMnO_4}$ $\underset{H_2SO_4}{\xrightarrow{HNO_3}}$

h) $\underset{AlCl_3}{\xrightarrow{CH_3Cl}}$ $\underset{H_2SO_4}{\xrightarrow{HNO_3}}$ $\xrightarrow{KMnO_4}$

6.35 Synthesis: Sections 6.4-6.5

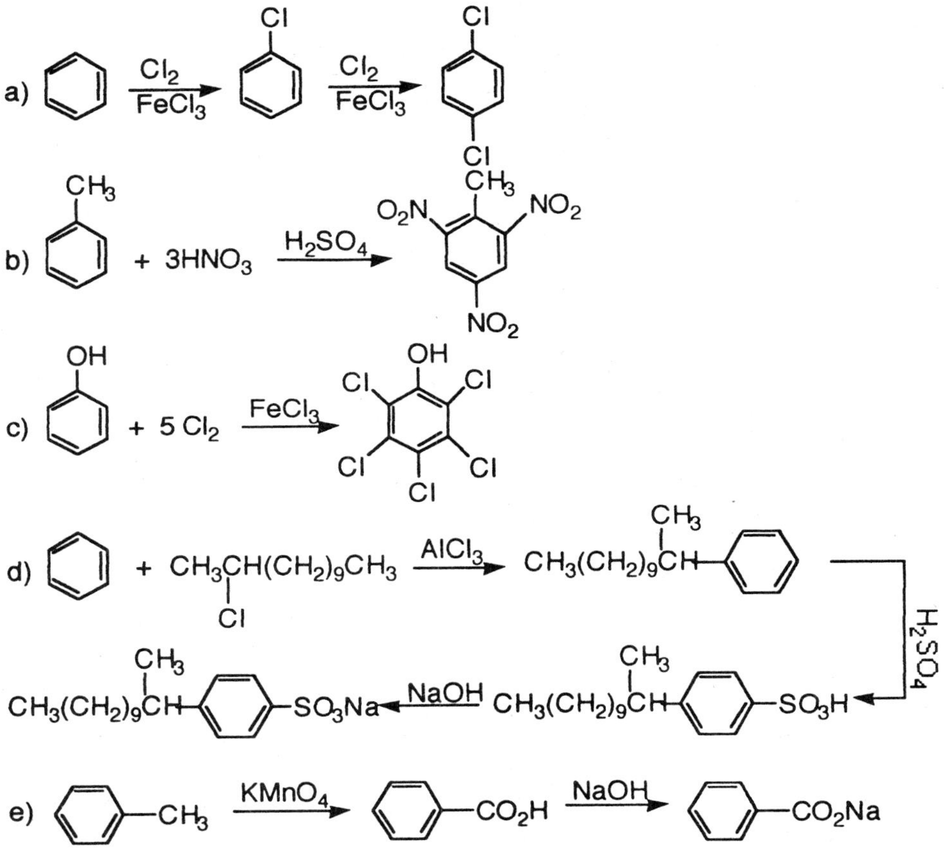

a)

b)

c)

d)

e)

6.36 Reaction Mechanisms: Sections 5.1B and 6.4B

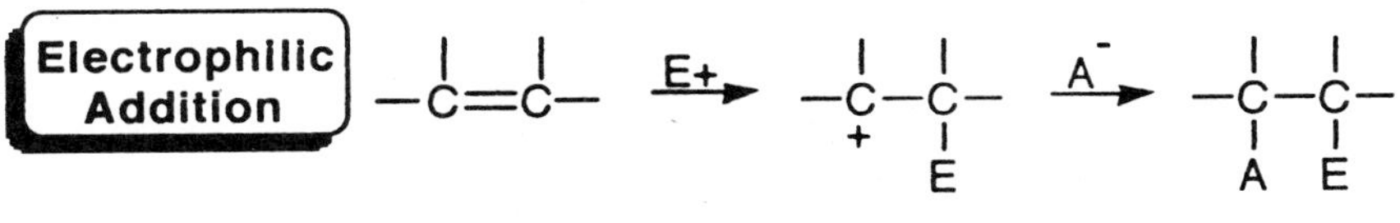

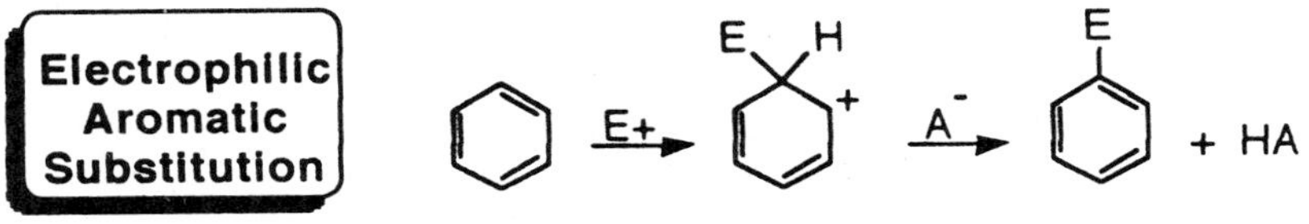

6.37 Reaction Mechanisms: Sections 4.4B and 6.4B.1

Electrophilic Aromatic Substitution

Generation
of the
Electrophile

$$Br_2 + FeBr_3 \longrightarrow Br^+ + FeBr_4^-$$

Two-step
Substitution

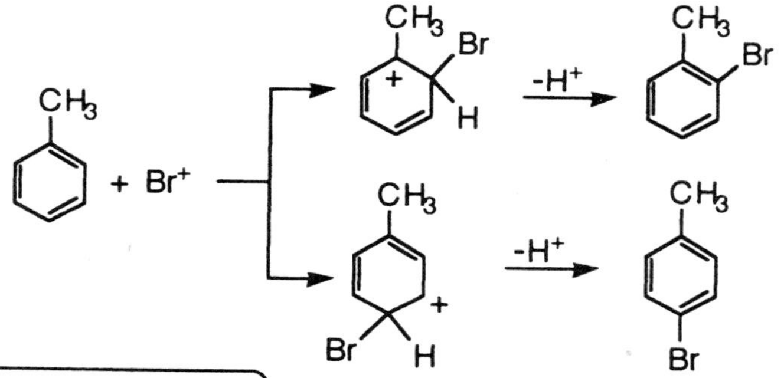

Free-radical Chain Bromination

Initiation $Br_2 \xrightarrow{\text{light}} 2\ Br\cdot$

Propagation

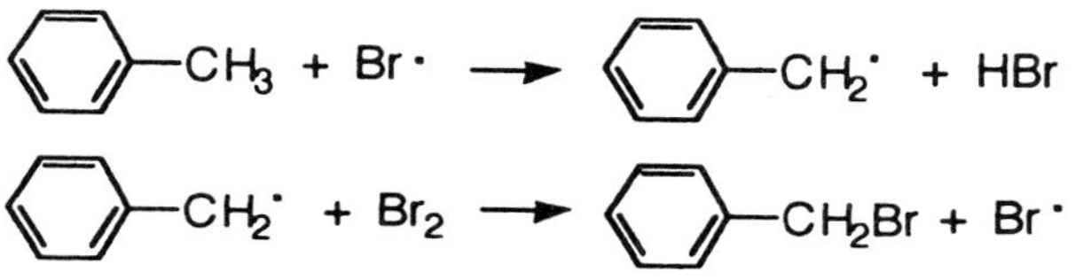

6.38 Reaction Mechanisms: Sections 4.5C, 5.1B, and 6.4B.2

The mechanism of electrophilic substitution for the synthesis of ethylbenzene is one of two-step substitution: the electrophile bonds and forms a carbocation which is neutralized upon elimination of hydrogen ion.

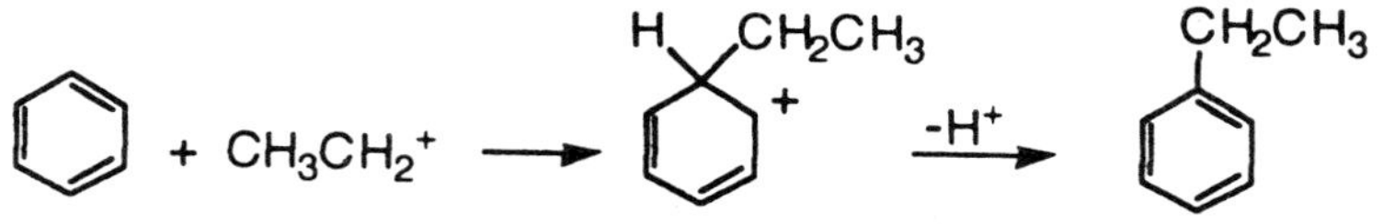

The difference in the three procedures described is in the way the electrophile is generated.

112

$$CH_3CH_2Cl + AlCl_3 \longrightarrow \underline{CH_3CH_2^+} + AlCl_4^-$$

$$CH_3CH_2OH \xrightarrow{H_2SO_4} CH_3CH_2\overset{H}{\overset{+}{O}}H \longrightarrow \underline{CH_3CH_2^+} + H_2O$$

$$CH_2{=}CH_2 \xrightarrow{H_2SO_4} \underline{CH_3CH_2^+}$$

6.39 Activating and Deactivating Groups

a) nitrobenzene < benzene < phenol

b) chlorobenzene < benzene < aniline

c) benzoic acid < p-methylbenzoic acid < p-xylene

d) p-chloronitrobenzene < p-nitrotoluene < benzene < toluene < p-xylene

6.40 Activating and Deactivating Groups

a)   **The nitro group is deactivating; as a result, substitution occurs on the other ring.**

b)   **The methoxy group is activating and directs substitution to the ring it occupies.**

6.41 Physical Properties: Section 2.9

a) Ethylbenzene has a greater molecular weight.

b-d) The compound with the highest melting point in each case is the most symmetrical and consequently, forms a very strong and stable crystal lattice.

6.42 Gasoline: Connection 6.2

1) Hydrocarbons with 5-10 carbons
2) Branched hydrocarbon chains
3) Unsaturated, cyclic and especially aromatic hydrocarbons

Research Octane Number	100	101	91

6.43 Production of Gasoline: Connection 6.2

a) Isomerization;   b) Cracking;   c) Isomerization or Aromatization;
d) Alkylation or Polymerization;   e) Alkylation;   f) Aromatization

6.44 Basicity of Aniline

Any group that increases the availability of the electron pair of nitrogen will increase basicity and those that decrease availability will decrease basicity. Electron-withdrawing groups like nitro pull electrons from the ring and from the amine group whereas releasing groups do the opposite. Thus electron-withdrawing groups decrease basicity and electron-releasing groups increase basicity. Since resonance effects occur between positions in an ortho or para relationship, these groups will have greater effect if ortho or para rather than meta.

6.45 Aromaticity: Section 6.2

The two electrons on pyrrole's nitrogen exist in a p-orbital and are part of the aromatic sextet of electrons; sp^2 hybridization allows this. Pyrrolidine is not aromatic, there are no adjacent p-orbitals to overlap in resonance, and thus the nitrogen is sp^3 hybridized.

7

$$CO_2H$$

H ━━ C ━━ OH

$$CH_3$$

$$CO_2H$$

HO ━━ C ━━ H

$$CH_3$$

Optical Isomerism

CHAPTER SUMMARY

   Isomers are compounds with identical molecular formulas but different structural formulas. **Structural or constitutional isomers** differ in the bonding arrangement of atoms; different atoms are attached to one another in the isomers. There are three types of structural isomers. **Skeletal isomers** differ in their carbon skeletons or chains. In **positional isomers**, the difference is in the position of a non-carbon group or multiple bond. **Functional isomers** belong to different groups or classes of organic compounds. In **stereoisomerism** the same atoms are bonded to one another but their orientation in space differs; there are three types of stereoisomerism. **Geometric** or **cis-trans isomerism** refers to the orientation of groups around a double bond or on a ring. **Conformational isomers** differ in the extent of rotation around a carbon-carbon single bond. **Optical isomers** are related as mirror images (to some degree).

   A carbon with four different bonded groups is called a **chiral carbon;** the four groups can exist in two arrangements which are mirror images. **Enantiomers** are optical isomers that are non-superimposable mirror images. All physical properties are identical for these two isomers except the direction of rotation of **plane polarized light.** One rotates plane polarized light to the right and is termed **dextrorotatory (d,+);** the other rotates the light an equal amount in the opposite direction, to the left, and is termed **levorotatory (l,-).** The rotation of plane polarized light is measured with an instrument called a **polarimeter.** A compound that rotates plane polarized light is said to be **optically active** or **chiral.** A **chiral compound** or **optically active compound** is not superimposable on its mirror image. A **racemic mixture** is a 50/50 mixture of enantiomers; because the enantiomers cancel each others' rotation of plane polarized light, a racemic mixture is **optically inactive** (does not rotate plane polarized light).

Enantiomers can be drawn using **wedges and dashes** to show the tetrahedral geometry or by using **Fischer projections** in which the tetrahedral nature is assumed. Drawings can be compared for superimposability or non-superimposability by **physically maneuvering** structures in a way to maintain the **configurational** relationships or **interchanging** groups; one interchange gives the mirror image, two maintains the original configuration but from a different perspective.

Optical isomers with **one chiral carbon** can only exist as a pair of enantiomers. More possibilities exist if there are two or more chiral carbons. A **meso compound** has more than one chiral center and is superimposable on its mirror image; meso compounds are optically inactive. **Diastereomers** are optical isomers that are not mirror images; all physical properties of diastereomers are usually different. Drawing optical isomers of a formula should be done in a systematic fashion and in pairs of mirror images. These mirror images can be tested for superimposability. The maximum number of optical isomers possible for a compound is 2^n where n is the number of chiral carbons; this is known as the **van't Hoff rule.**

The configuration of a chiral carbon can be described by the **R,S system.** The groups connected to the chiral carbon are assigned priorities. The molecule is then visualized so that the group of lowest priority is directed away from the observer. The remaining three groups are in a plane and are visualized from highest to lowest priority. If in doing this the eye moves **clockwise**, the configuration is **R**; if the eye moves **counterclockwise**, the configuration is **S**. Priority depends on the atomic number of atoms directly attached to the chiral carbon. If two or more are identical one proceeds along the groups until differences are found. In double and triple bonds the groups are considered to be duplicated or triplicated. The configuration of **geometric isomers** can be expressed using the **E,Z system.** If the two high priority groups are on the same side of the double bond, **E** is assigned; if they are on opposite sides, the configuration is **Z**.

Since enantiomers have identical physical properties they cannot be separated by physical means. They can be separated by **resolution through diastereomers.** In this method, enantiomers are converted to diastereomers by reaction with a pure optically active compound. Diastereomers have different physical properties and can be separated. After separation, the diastereomers are converted back to the original enantiomers.

Chiral carbons can be **generated** during chemical reactions. If a chiral carbon is generated in a compound that previously had no chiral carbons, a pair of enantiomers results; they are formed in equal amounts. If a chiral carbon is generated in a compound that already has a chiral carbon, a pair of diastereomers results; they are formed in unequal amounts.

Connections 7.1 gives a historical perspective of the discovery of optical isomerism.

Connections 7.2 is about optical isomerism in the biological world.

SOLUTIONS TO PROBLEMS

7.1   Chiral Objects

The answers to this question can vary in a few items depending on the type of item being considered or depending on one's concept of the item. Most are fairly straightforward, however.

Chiral Objects: a, c, d, f, h, j, k, m, n, o, r, s

7.2   Chiral Carbons

(a) the carbon with the bromine; (b) carbons 3 and 5 (with methyl groups);
(c) the carbon with the methyl group.

7.3   Chiral Carbons

Two of the nine isomers of C_7H_{16} have chiral carbons.

$$CH_3CH_2\overset{\overset{\displaystyle CH_3}{|}}{C}HCH_2CH_2CH_3 \qquad\qquad CH_3\overset{\overset{\displaystyle CH_3}{|}}{\underset{\underset{\displaystyle CH_3}{|}}{C}}HCH_2CH_3$$

7.5   Enantiomers and Chiral Carbons

The following compounds are the only ones with chiral carbons and thus the only ones that can exist as a pair of enantiomers.

(b) 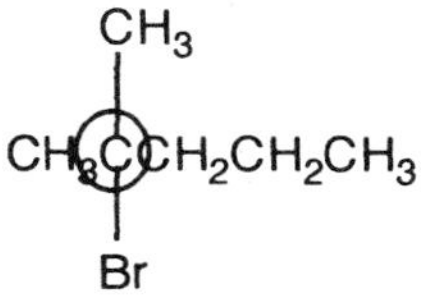 $CH_3CCH_2CH_2CH_3$ with CH_3 at top and Br at bottom

(c)

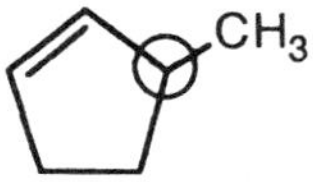

7.6 Drawing Enantiomers

See Example 7.4 in the text for assistance.

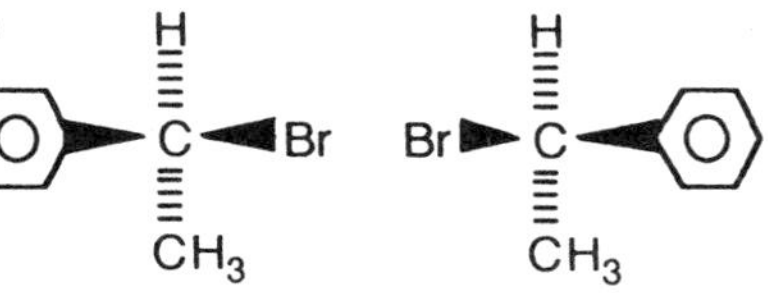

a) 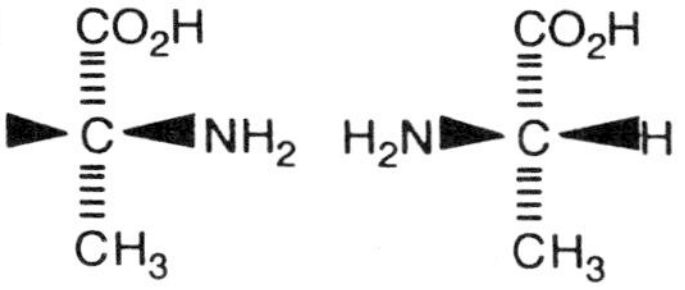
$$H-C-NH_2 \quad H_2N-C-H$$
with CO_2H at top and CH_3 at bottom

b)
$$C_6H_5-C-Br \quad Br-C-C_6H_5$$
with H at top and CH_3 at bottom

c)
$$CH_3(CH_2)_3-C-(CH_2)_4CH_3 \quad CH_3(CH_2)_4-C-(CH_2)_3CH_3$$
with CH_2CH_3 at top and $(CH_2)_2CH_3$ at bottom

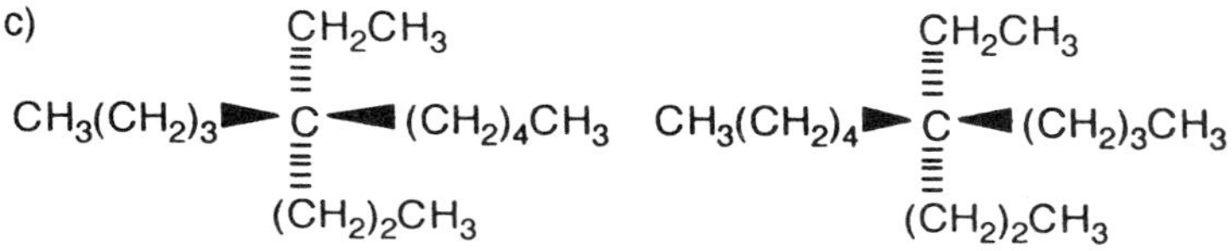

7.7 Maneuvering Optical Isomers

See Example 7.5 for assistance. Either physical maneuvering or interchanging of groups will work though the latter is probably faster and offers less chance of error.

Identical: c, d, e,   Mirror Image: a, b, f, g, h

7.8 Optical Isomers of Threonine

Draw the optical isomers systematically and in pairs of mirror images. Start out drawing the wedge/dash representation you see in the structures below. Pick two groups, the acid and methyl in the example shown, and put one at the top and one at the bottom; they do not change positions. Now put the two hydrogens on one side and the other two groups on the other; draw the mirror image. Finally put the hydrogens on opposite sides; draw the mirror image. Compare the pairs of mirror images for superimposability. In this case, since the top and bottom of the molecule are different, there is no possibility of rotation to superimpose; there are two pairs of enantiomers.

CO₂H structures:

$$\begin{array}{cccc}
\text{CO}_2\text{H} & \text{CO}_2\text{H} & \text{CO}_2\text{H} & \text{CO}_2\text{H} \\
\text{H}\blacktriangleright\text{C}\blacktriangleleft\text{NH}_2 & \text{H}_2\text{N}\blacktriangleright\text{C}\blacktriangleleft\text{H} & \text{H}_2\text{N}\blacktriangleright\text{C}\blacktriangleleft\text{H} & \text{H}\blacktriangleright\text{C}\blacktriangleleft\text{NH}_2 \\
\text{H}\blacktriangleright\text{C}\blacktriangleleft\text{OH} & \text{HO}\blacktriangleright\text{C}\blacktriangleleft\text{H} & \text{H}\blacktriangleright\text{C}\blacktriangleleft\text{OH} & \text{HO}\blacktriangleright\text{C}\blacktriangleleft\text{H} \\
\text{CH}_3 & \text{CH}_3 & \text{CH}_3 & \text{CH}_3 \\
\underline{A} & \underline{B} & \underline{C} & \underline{D}
\end{array}$$

Enantiomers: AB, CD   Diastereomers: AC, AD, BC, BD

7.9   Drawing Optical Isomers

See problem 7.8 for a brief description of the systematic method for drawing optical isomers. Draw the isomers in pairs of mirror images.

(a) The top half of the molecule is different from the bottom and thus there is no possibility of rotation to attempt superimposability of mirror images. There are two pairs of enantiomers.

$$\begin{array}{cccc}
\text{CH}_3 & \text{CH}_3 & \text{CH}_3 & \text{CH}_3 \\
\text{H}\blacktriangleright\text{C}\blacktriangleleft\text{Br} & \text{Br}\blacktriangleright\text{C}\blacktriangleleft\text{H} & \text{Br}\blacktriangleright\text{C}\blacktriangleleft\text{H} & \text{H}\blacktriangleright\text{C}\blacktriangleleft\text{Br} \\
\text{H}\blacktriangleright\text{C}\blacktriangleleft\text{Cl} & \text{Cl}\blacktriangleright\text{C}\blacktriangleleft\text{H} & \text{H}\blacktriangleright\text{C}\blacktriangleleft\text{Cl} & \text{Cl}\blacktriangleright\text{C}\blacktriangleleft\text{H} \\
\text{CH}_3 & \text{CH}_3 & \text{CH}_3 & \text{CH}_3 \\
\underline{A} & \underline{B} & \underline{C} & \underline{D}
\end{array}$$

Enantiomers: AB, CD   Diastereomers: AC, AD, BC, BD

(b) This molecule can be drawn so that the top and bottom halves are identically constituted thus allowing for 180° rotation to test for superimposability. There is one pair of enantiomers and one meso structure. B when rotated 180° is superimposable on A, its mirror image. Thus A is a meso structure. C and D are not superimposable and are enantiomers.

b)
$$\begin{array}{cccc}
\text{CH}_3 & \text{CH}_3 & \text{CH}_3 & \text{CH}_3 \\
\text{H}\blacktriangleright\text{C}\blacktriangleleft\text{Br} & \text{Br}\blacktriangleright\text{C}\blacktriangleleft\text{H} & \text{Br}\blacktriangleright\text{C}\blacktriangleleft\text{H} & \text{H}\blacktriangleright\text{C}\blacktriangleleft\text{Br} \\
\text{H}\blacktriangleright\text{C}\blacktriangleleft\text{Br} & \text{Br}\blacktriangleright\text{C}\blacktriangleleft\text{H} & \text{H}\blacktriangleright\text{C}\blacktriangleleft\text{Br} & \text{Br}\blacktriangleright\text{C}\blacktriangleleft\text{H} \\
\text{CH}_3 & \text{CH}_3 & \text{CH}_3 & \text{CH}_3 \\
\underline{A} & \underline{B} & \underline{C} & \underline{D}
\end{array}$$

A = B; A is a meso structure   C and D are enantiomers
Diastereomers: AC and AD

7.10 Optical Isomerism in Cyclic Compounds

(a) This compound has symmetry and is capable of having meso structures.

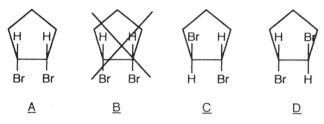

A=B; A is a meso   C and D are enantiomers
Diastereomers: AC, AD

(b) There is no symmetry in this molecule and thus rotations to find superimposable mirror images will be fruitless.

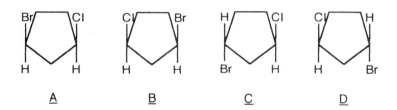

Enantiomers: AB, CD   Diastereomers: AC, AD, BC, BD

7.11 Specification of Configuration: Group Priorities

a) $I > Br > Cl > F$;   b) $Br > OCH_3 > CH_3 > H$;

c) $OH > CH_2OH > CH_2CH_2Br > CH_2CH_2CH_3$.   d) $Cl > SH > CO_2H > CH_2OH$

7.12 Specification of Configuration: R and S

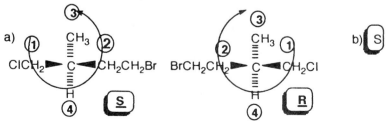

7.13 Z and E Configurations

a) [structure] Z   E   b) [structure] Z   E

7.14 Generation of a Chiral Carbon

Butane has no chiral carbons but a chiral carbon is generated in 2-bromobutane. As a result a pair of enantiomers forms in equal amounts. When one chiral carbon is generated in a compound with none, a pair of enantiomers always forms in equal amounts.

$$CH_3CH_2CH_2CH_3 \;+\; Br_2 \longrightarrow$$ [structures]

7.15 Generation of a Chiral Carbon

A new chiral carbon is generated in a compound with one already. A pair of diastereomers is formed in unequal amounts. Notice that the configuration of the original chiral carbon remains the same (S) but the new chiral carbon appears as mirror images (R and S).

[structure] S + HBr $\longrightarrow$ [structures] R,S   S,S

7.16 Chiral Carbons: Section 7.1 and 7.2A

Chiral carbons have four different bonded groups; they are circled in the following compounds. The maximum number of possible optical isomers is 2^n where n is the number of chiral carbons (van't Hoff rule).

a)

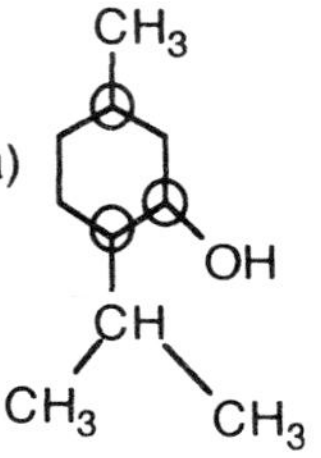

(8)

b)

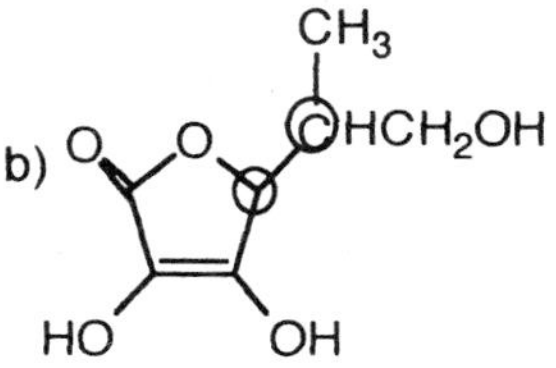

(4)

c) 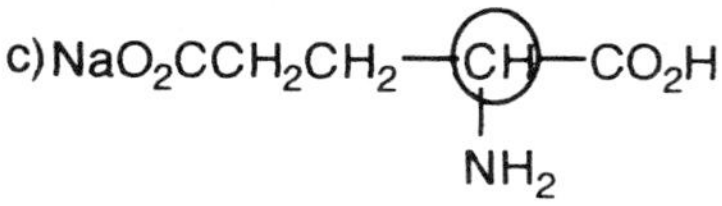 $NaO_2CCH_2CH_2$—CH—CO_2H with NH_2

(2)

d)

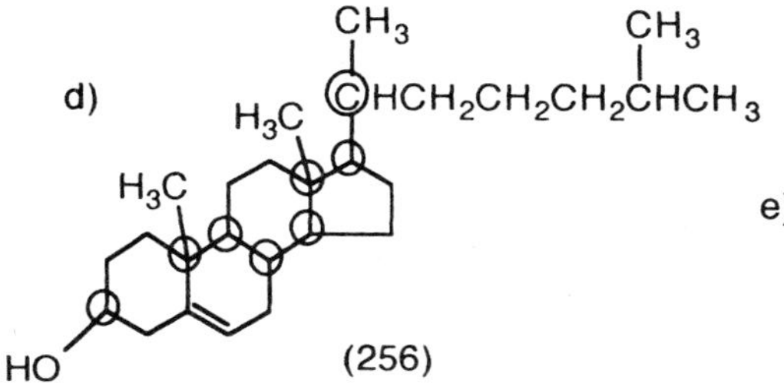

(256)

e) 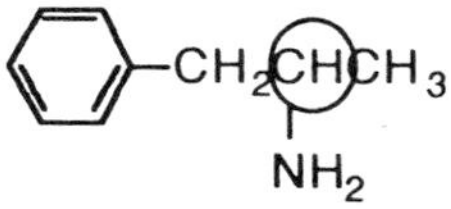 —CH_2CHCH_3 with NH_2

(2)

f)

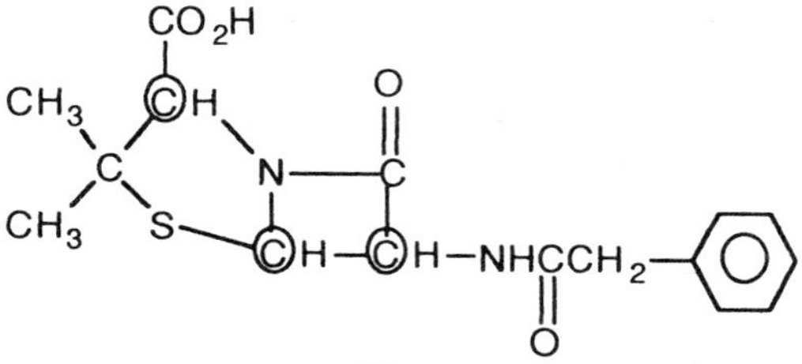

(8)

g)

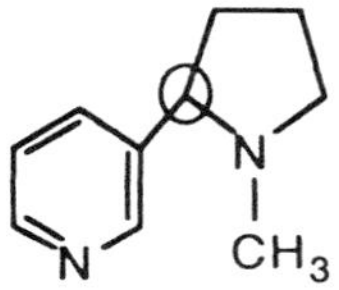

(2)

h) 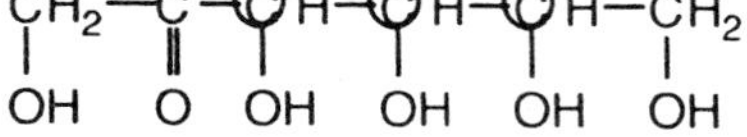 CH_2—C—CH—CH—CH—CH_2 with OH, O, OH, OH, OH, OH

(8)

7.17-7.18 Chiral Carbons and Enantiomers: Sections 7.1 and 7.2

Chiral carbons are circled. Enantiomers shown with wedges/dashes.

(a) 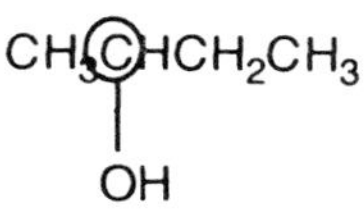 CH_3CHCH$_2$CH$_3$ with OH

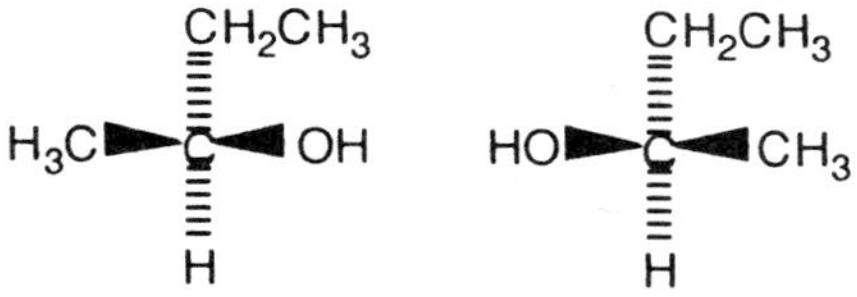

(b) 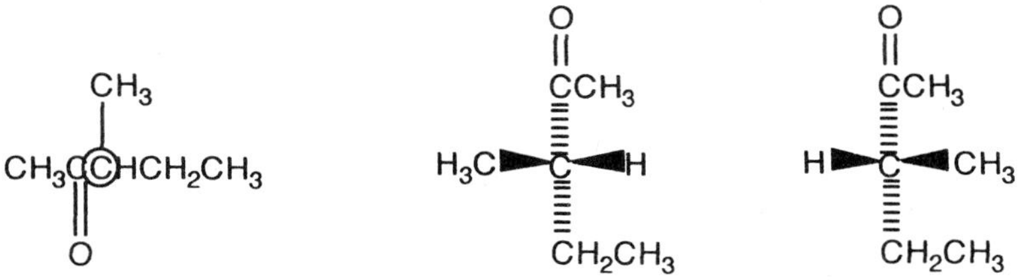

(c) 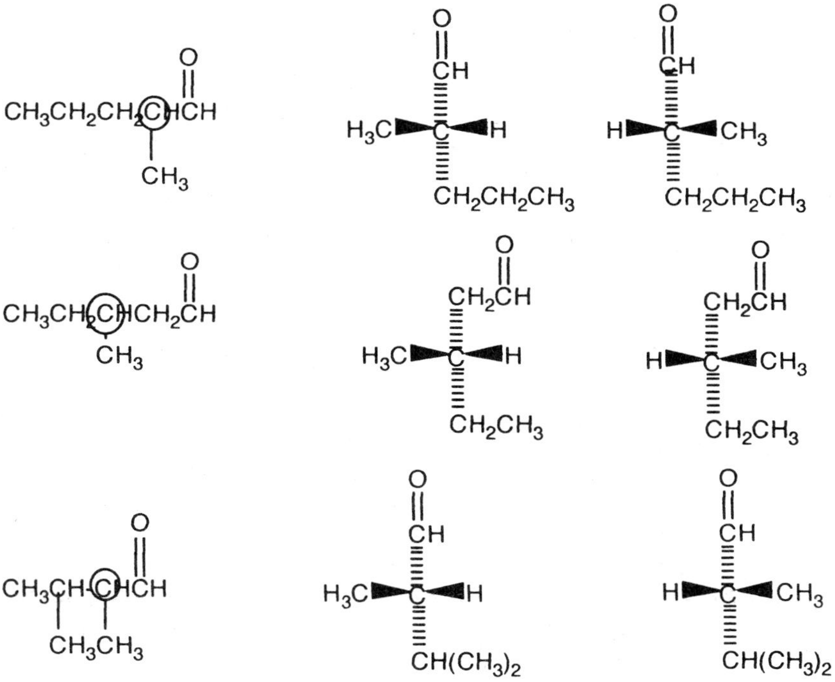

(d) 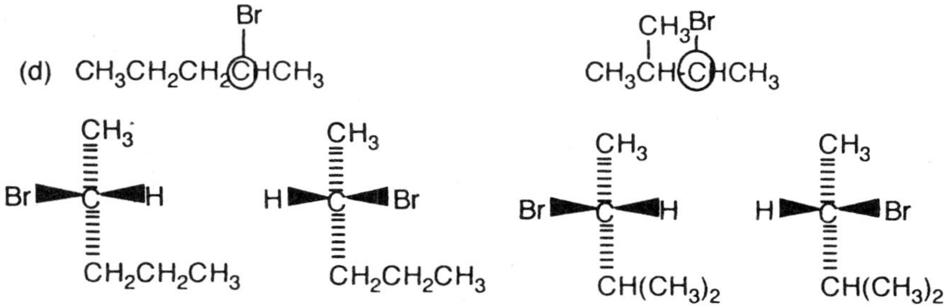

7.18 Enantiomers

Please see problem 7.17.

7.19 Enantiomers and Diastereomers: Section 7.5B

$$
\begin{array}{ccc}
\text{CH}_3 & \text{CH}_3 & \text{CH}_3 \\
\text{H}\!-\!\text{C}\!-\!\text{Br} & \text{Br}\!-\!\text{C}\!-\!\text{H} & \text{Br}\!-\!\text{C}\!-\!\text{H} \\
\text{H}\!-\!\text{C}\!-\!\text{CH}_3 & \text{H}_3\text{C}\!-\!\text{C}\!-\!\text{H} & \text{H}\!-\!\text{C}\!-\!\text{CH}_3 \\
\text{CH}_2\text{CH}_3 & \text{CH}_2\text{CH}_3 & \text{CH}_2\text{CH}_3
\end{array}
$$

.Enantiomers   **Diastereomer**

7.20 Enantiomers: Section 7.2 and 7.4

a)

$$
\text{C}_6\text{H}_5\!-\!\overset{\text{OH}}{\underset{\text{H}}{\text{C}}}\!-\!\text{CH}_2\text{CH}_3 \qquad\qquad
\text{CH}_3\text{CH}_2\!-\!\overset{\text{OH}}{\underset{\text{H}}{\text{C}}}\!-\!\text{C}_6\text{H}_5
$$

b)

$$
\text{CH}_3\overset{}{\underset{\text{O}}{\text{C}}}\!-\!\overset{\text{CH}_3}{\underset{\text{H}}{\text{C}}}\!-\!\text{CH}_2\text{Br} \qquad\qquad
\text{BrCH}_2\!-\!\overset{\text{CH}_3}{\underset{\text{H}}{\text{C}}}\!-\!\overset{}{\underset{\text{O}}{\text{CCH}_3}}
$$

c)

$$
\text{CH}_2\!=\!\text{CH}\!-\!\overset{\text{Cl}}{\underset{\text{H}}{\text{C}}}\!-\!\text{CH}_3 \qquad\qquad
\text{CH}_3\!-\!\overset{\text{Cl}}{\underset{\text{H}}{\text{C}}}\!-\!\text{CH}\!=\!\text{CH}_2
$$

7.21 Optical Isomers: Section 7.5

In working the following problems, it is important to draw the isomers in pairs of mirror images and in an orderly fashion. It will be easiest in this way to identify enantiomers, diastereomers and meso compounds since their definitions involve mirror image relationships. Before working with these examples, be sure you are thoroughly familiar with the definitions in Table 7.1 and the examples explained in sections 7.4 and 7.5.

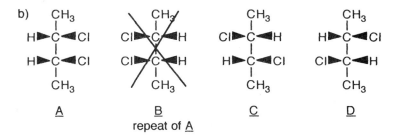

a)

$$\text{A} \qquad \text{B} \qquad \text{C} \qquad \text{D}$$

Enantiomers: AB, CD   Diasteromers: AC, AD, BC, BD

b)

$$\text{A} \qquad \text{B} \qquad \text{C} \qquad \text{D}$$
$$\text{repeat of } \underline{A}$$

Enantiomers: CD   Meso: A   Diastereomers: AC, AD

c)

$$\text{A} \qquad \text{B} \qquad \text{C} \qquad \text{D}$$

Enantiomers: AB, CD   Diastereomers: AC, AD, BC, BD

125

d)

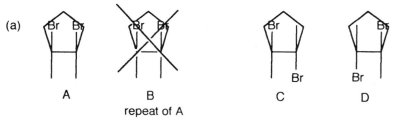

A   B   C   D
            repeat of A

Enantiomers: CD   Meso: A   Diastereomers: AC, AD

7.22   Optical Isomers: Section 7.6

(a)

A   B   C   D
        repeat of A

Enantiomers: CD   Meso: A   Diastereomers: AC, AD

(b)

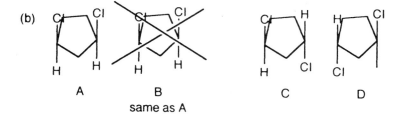

A   B   C   D
      same as A

Enantiomers: CD   Meso: A   Diastereomers: AC, AD

(c)

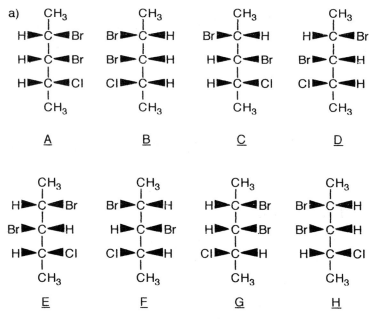

Enantiomers: AB, CD   Diastereomers: AC, AD, BC, BD

7.23 Optical Isomers: Section 7.5

enantiomers: AB, CD, EF, GH   meso: none

diastereomers: AC, AD, AE, AF, AG, AH, BC, BD, BE, BF, BG, BH, CE, CF,

CG, CH, DE, DF, DG, DH, EG, EH, FG, FH

b)

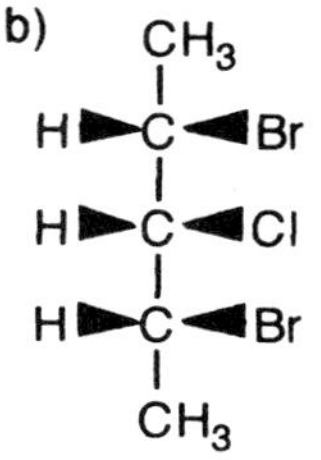

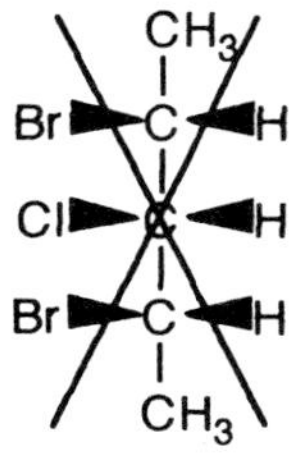

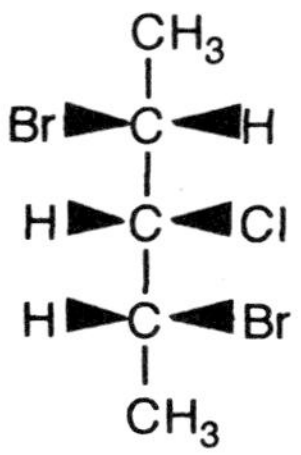

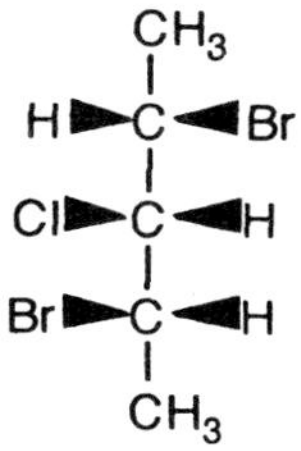

A	B	C	D
	repeat of A		

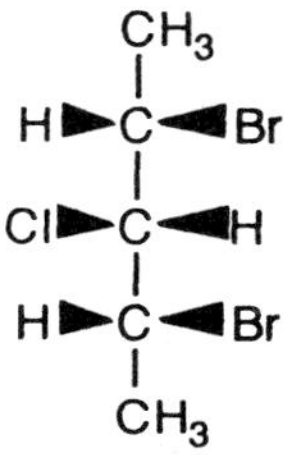

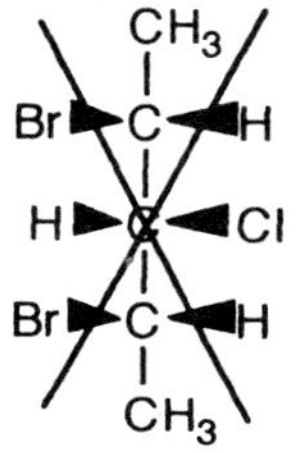

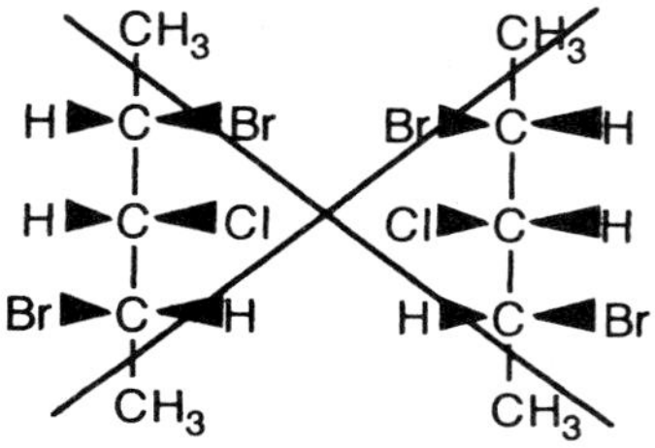

E	F	G	H
	repeat of E	same as C and D (G = D, H = C)	

enantiomers: CD   meso: A, E   diastereomers: AC, AD, AE, CE, DE

7.24 R,S Configurations: Section 7.7

(a) Br > OCH$_3$ > CHO > H   (b) C(CH$_3$)$_3$ > CH(CH$_3$)$_2$ > CH$_2$CH$_3$ > CH$_3$

(c) Br > F > CH$_2$Cl > CH$_2$CH$_2$I   (d) OCH$_3$ > NH$_2$ > CN > H

7.25 Specification of Configuration - R,S: Section 7.7A

For assistance, see Examples 7.6-7.8 for determining group priorities and Example 7.9 for assigning R or S.

a) **S**   Br > Cl > CH$_3$ > H

b) **R**   OH > CH$_2$CH$_3$ > CH$_3$ > H

c) **R**   CH(CH$_3$)$_2$ > CH$_2$CH$_2$Cl > CH$_2$CH$_2$CH$_2$Br > CH$_2$CH$_2$CH$_3$

d) **R**   I > Br > Cl > F

e) **S**   CH$_2$Br > CH$_3$CFCl > (CH$_3$)$_3$C > H

f) **S**   OH > CH$_2$CH$_2$CH$_3$ > CH$_2$CH$_3$ > CH$_3$

g) **R** $NH_2 > CH_2CO_2H > CH_3 > H$

h) **R** $Cl > OCH_3 > (CH_2)_8CH_3 > H$

7.26   Specification of Configuration - Z,E: Section 7.7B

a) **E** $CH_3 > H$ and $Cl > CH_3$

b) **Z** $(CH_3)_2CH > CH_2CH_2CH_3$ and $SCH_3 > OCH_2CH_3$

c) **Z** $CH_2OH > CH_3$ and $Br > CH_2OH$

d) **E** $Br > H$ and $F > CH_2Cl$

e) **E,Z** $Br > H$ and $CH=CHBr > H$ on both double bonds

7.27   Specification of Configuration - R,S,Z,E: Section 7.7

R,Z $Br > CH=CHCH_3 > CH_3 > H$ on chiral carbon

$CH_3CHBr > H$ and $CH_3 > H$ on double bond

7.28   Specification of Configuration: Section 7.7

Lets put each chiral carbon in a form that can easily be read. First on both the top and bottom chiral carbons interchange groups to get the hydrogens, the lowest priority group on each carbon, projecting behind the paper. Now interchange any other two groups to return to the original configuration, but keep the hydrogens back. Since the low priority groups are behind the plane we can read the configuration directly.

2S, 3R   2-bromo-3-chloropentane

7.29 Specification of Configuration - R,S: Section 7.7

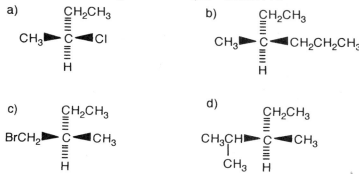

7.30 Specification of Configuration - Z,E: Section 7.7B

a)

$$CH_3 \quad CH_2CH_3$$
$$C=C$$
$$H \qquad H$$

b)

$$I \qquad CH_2CH_3$$
$$C=C$$
$$Br \qquad Br$$

c)

$$H \qquad H$$
$$C=C$$
$$CH_3 \qquad C=C \quad CH_3$$
$$\qquad H \qquad H$$

7.31 Newman and Fischer Projections: Sections 2.7 and 7.5

We are drawing Newman projections of E, G, and H in section 7.5B. The two acid groups, top and bottom, are behind the plane and we would see them as down in the Newman projection. The other groups are coming out of the plane and we display them accordingly in the Newman.

E = meso compound   G and H = pair of enantiomers

7.32 R,S Configurations: Section 7.7A

To determine configuration is not as difficult as it might seem. Look at the Newman; you have three groups sticking out at you and one, the other carbon, projected behind. Exchange the back group, an ethyl, with the front hydrogen, the low priority group. Now you have hydroxy, acid, ethyl (this is the priority) in front. Exchange two other groups in front to restore the original configuration and determine R or S.

7.33 Stereochemistry and Chemical Reactions: Section 7.9

(a) One chiral carbon is formed in a molecule that had no previous chiral carbons. Enantiomers are formed in equal amounts.

(b) A new chiral carbon is formed in an optically active compound with one chiral carbon. A pair of diastereomers are formed in unequal amounts.

7.34 Optical Isomers: Sections 7.4-7.5

e)

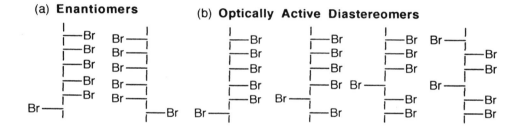

$$
\begin{array}{cc}
\text{CH}_3 & \text{CH}_3 \\
\text{H} \blacktriangleright \text{C} \blacktriangleleft \text{Br} & \text{H} \blacktriangleright \text{C} \blacktriangleleft \text{Br} \\
\text{H} \blacktriangleright \text{C} \blacktriangleleft \text{Br} & \text{Br} \blacktriangleright \text{C} \blacktriangleleft \text{H} \\
\text{H} \blacktriangleright \text{C} \blacktriangleleft \text{Br} & \text{Br} \blacktriangleright \text{C} \blacktriangleleft \text{H} \\
\text{H} \blacktriangleright \text{C} \blacktriangleleft \text{Br} & \text{H} \blacktriangleright \text{C} \blacktriangleleft \text{Br} \\
\text{CH}_3 & \text{CH}_3
\end{array}
$$

7.35 Optical Isomers: Section 7.5

The isomers are shown in simplified stick drawings. Since the top and bottom groups are identical, both methyl, they are not shown. Likewise, the hydrogens, the other atoms on each of the chiral carbons, are not shown.

(a) Enantiomers   **(b) Optically Active Diastereomers**

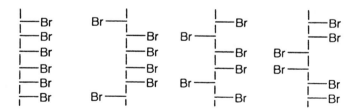

(c) Meso Compounds

7.36 Optical Isomers: Section 7.5

(a) D-galactose is optically active. The top and bottom halves of the molecule are different so there is no possibility of rotating to test for superimposability on a mirror image. D-galactose is a pure enantiomer. However, the product of the reaction is a meso compound and not optically active. The aldehyde group on top was changed to an alcohol, the same as the bottom. If you draw the mirror image and rotate it 180°, you will find it is superimposable on the product shown.

(b)

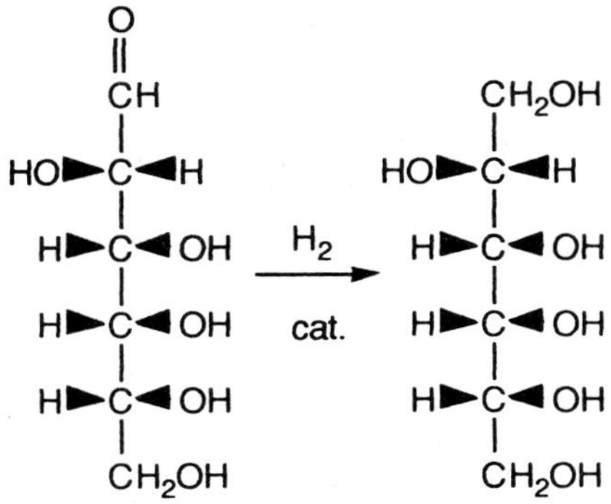

Optically active product; the starting
material is a diastereomer of
galactose.

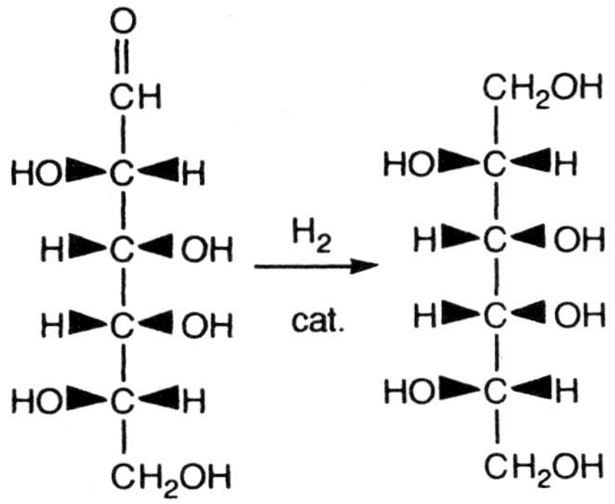

Optically inactive product: the
starting material is the enantiomer of
galactose. The product shown is the
same as that formed from galactose.

7.37 R,S Configurations: Section 7.7A

> **This is
> 2-bromo-3-chlorobutane**

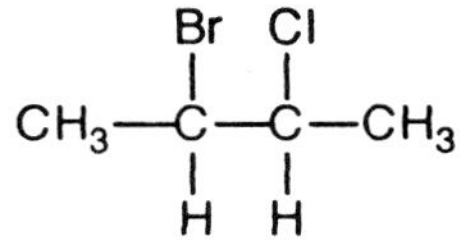

CH_3—C—C—CH_3
(Br, Cl, H, H)

> **First draw the Fischer
> projection. For each
> chiral carbon, put the
> lowest priority group
> (H) behind the paper.**

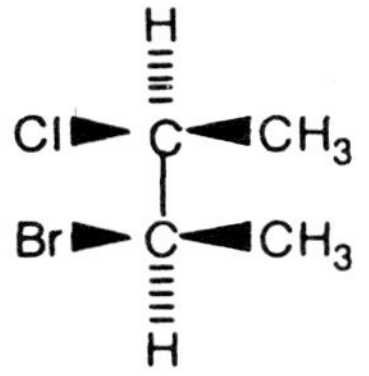

Cl ► C ◄ CH_3 C-3   S
Br ► C ◄ CH_3 C-2   R

> **Now look up the C_2-C_3
> bond of the eclipsed
> Fischer projection and
> translated it to a
> Newman projection.**

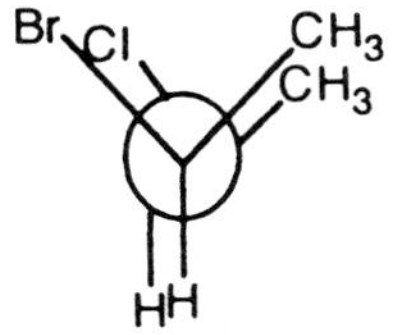

C_2 in front
C_3 behind

(b)

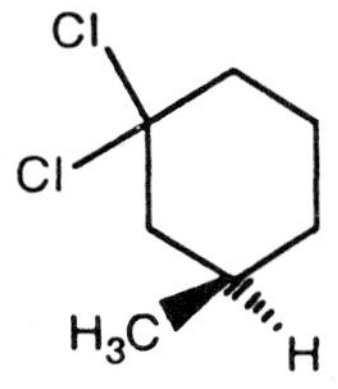

7.38 Stereoisomers: Section 7.7

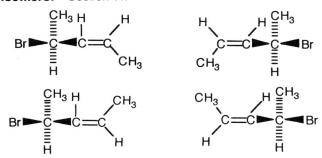

7.39 R and S Designations: Section 7.7A

A way to do this problem is to move the lowest priority group on each carbon to a vertical bond going behind the paper by the interchange method and then interchange two other groups to get back to the original configuration. Then you can read the R,S directly. For example, let's use structure A in the answer to problem 7.8. This compound is 2R,3R.

The configurations in Problem 7.8 are:

A	B	C	D
2R, 3R	2S, 3S	2S, 3R	2R, 3S

7.40 Optical Isomers without Chiral Carbons

7.41   Optical Isomers without Chiral Atoms

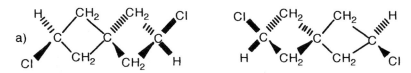

a)

The middle carbon is common to both rings. Since it is tetrahedral, one ring will be in the plane of the paper and other perpendicular (in and out of the paper). The H and Cl on the ring in and out of the paper are above and below the ring and thus in the plane of the paper. Those on the other ring are in front of and behind the paper plane. One cannot superimpose both rings and the Cl's (or H's) simultaneously.

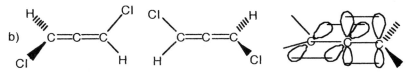

b)

Since the middle carbon is involved in both double bonds, it has two p-orbitals which are perpendicular (90°) to each other. Thus the two pi-bonds are in perpendicular planes. Consequently, the H's and Cl's on each end are in perpendicular planes. No amount of rotating or turning the molecules will allow the simultaneous superimposition of both chlorines.

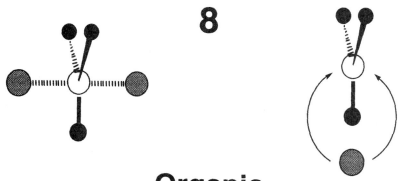

8

Organic
Halogen Compounds

CHAPTER SUMMARY

Alkyl halides are **organic halogen compounds** in which one or more hydrogens of a hydrocarbon have been replaced with a halogen. These compounds can be classified as **primary, secondary,** or **tertiary** depending on whether there are one, two, or three carbons respectively connected to the carbon bearing the halogen. In **aryl halides** the halogen is directly attached to a benzene or other aromatic hydrocarbon ring and in **benzylic halides**, the halogen is on a carbon directly attached to a benzene ring. If the halogen is directly attached to a carbon-carbon double bond, it is termed **vinyl**, and if it is attached to a carbon directly attached to the double bond it is **allylic. IUPAC nomenclature** involves using the prefixes fluoro, chloro, bromo, and iodo to designate halogen in a molecule.

Alkyl halides are generally **water insoluble** and have a **greater density** than water. Their **boiling points increase with molecular weight**; alkyl iodides have higher boiling points than the corresponding alkyl bromides which boil at higher temperatures than the chlorides. Alkyl halides have a variety of uses including as insecticides, dry cleaning agents, and in polymers.

A characteristic reaction of alkyl halides is **nucleophilic substitution**. In this reaction, a **nucleophile (Lewis base)** replaces a halide ion, the **leaving group.**

Chloride, bromide, and iodide are effective leaving groups; common nucleophiles include OH^-, SH^-, NH_2^-, and their derivatives, as well as cyanide and acetylide ions. Neutral nucleophiles include water, alcohols, and amines.

Nucleophilic substitution occurs by two general mechanisms depending on the structures of the alkyl halide and nucleophile. The S_N2 **mechanism** is a one step process involving both the alkyl halide and nucleophile simultaneously. The nucleophile enters as the halide leaves, attacking the carbon from the side opposite to that from which the halide departs. The reaction is **bimolecular;** this means the reaction rate depends on the concentrations of both the alkyl halide and the nucleophile. The reaction involving optically active halides occurs with **inversion of configuration.**

The S_N1 **mechanism** is a two step process. In the first step the negative halide ion departs leaving a **carbocation intermediate.** In the second step the carbocation is neutralized by the nucleophile. S_N1 reactions commonly occur in neutral or acid conditions with neutral nucleophiles. The reaction rate is dependent on the slow step, carbocation formation from the alkyl halide, and is termed **unimolecular.** Reaction of an optically active alkyl halide by S_N1 results in the formation of a pair of enantiomers, an **optically inactive racemic mixture**, since the intermediate carbocation can be attacked from either side by the nucleophile.

Several factors influence whether a reaction will occur by an S_N1 or S_N2 mechanism: **carbocation stability, steric effects, strength of nucleophile, and the solvent. Tertiary halides** tend to react by the S_N1 process because they can form the relatively stable tertiary carbocations and because the presence of three large alkyl groups sterically discourages attack by the nucleophile on the carbon-halogen bond. The S_N2 reaction is favored for **primary halides** because it does not involve a carbocation intermediate (primary carbocations are unstable) and because primary halides do not offer as much steric hindrance to attack by a nucleophile as do the more bulky tertiary halides. **Strong nucleophiles** favor the S_N2 mechanism and **polar solvents** promote S_N1 reactions.

Alkyl halides also undergo **dehydrohalogenation** reactions in which **elimination** of a hydrogen and halogen from adjacent carbons produces a double bond. The reaction mechanisms are analogous to those of nucleophilic substitution. The E_2 mechanism is a concerted one-step process in which a nucleophile abstracts a hydrogen ion from one carbon while the halide is leaving from an adjacent one. The E_1 mechanism is two-steps and involves a **carbocation intermediate** formed upon departure of the halide ion in the first step. E_2 reactions are **bimolecular** and the reaction rate depends on the concentrations of both the alkyl halide and nucleophile. E_1

reaction rates depend on the slowest step, formation of the carbocation, and are influenced only by the concentration of the alkyl halide; the reaction is **unimolecular.** E_2 reactions involve **anti elimination** and produce a specific alkene, either cis or trans. E_1 reactions involve an intermediate carbocation and thus give products of both **syn and anti elimination.**

Nucleophilic substitution and elimination are competitive processes. Which prevails depends on a variety of factors; one important consideration is the stability of the alkene that would result from elimination. Since tertiary halides form the more stable highly substitued alkenes, they are more likely to react by elimination than primary halides.

Connections **8.1** is about drug design.

Connections **8.2** is about the thyroid hormone.

SOLUTIONS TO PROBLEMS

8.1   Nomenclature

a) 2-chloropentane;   b) 1,4-dibromo-2-butene;   c) p-difluorobenzene;

(d) 1,1,1-trichloro-2,2-difluoroethane

8.2   Nomenclature

a) CBr_4;   b) CH_2Br_2;   c) CHI_3;   d) $CH_2=CHBr$;

e) $O_2N-\!\!\left\langle\!\!\bigcirc\!\!\right\rangle\!\!-CH_2Cl$;   f) CH_3CHCH_3
$\qquad\qquad\qquad\qquad\qquad\qquad\quad |$
$\qquad\qquad\qquad\qquad\qquad\qquad\quad I$

8.3   Nucleophilic Substitution

CH_3I + reagents in a-i $\longrightarrow$

a) CH_3OH;   b) $CH_3OCH_2CH_2CH_3$;   c) CH_3SH;   d) CH_3SCH_3;   e) CH_3NH_2;

f) $CH_3NHCH_2CH_3$;   g) $CH_3N(CH_3)_2$;   h) CH_3CN;   i) $CH_3C\!\equiv\!CCH_3$

8.4   Nucleophilic Substitution

(a) $CH_3CH_2CH_2Br$ + $NaCN$ $\longrightarrow$ $CH_3CH_2CH_2CN$ + $NaBr$

(b) $CH_3CH_2CH_2CH_2Cl$ + $NaOH$ $\longrightarrow$ $CH_3CH_2CH_2CH_2OH$ + $NaCl$

(c) CH_3I + $NaSCH_3$ ⟶ CH_3SCH_3 +   NaI

8.5   Nucleophilic Substitution

(a)
$$CH_3\underset{\underset{\displaystyle OH}{|}}{CH}CH_2CH_3$$

(b)
$$CH_3\underset{\underset{\displaystyle OCH_3}{|}}{CH}CH_2CH_3$$

(c)
$$CH_3\underset{\underset{\displaystyle N(CH_3)_2}{|}}{CH}CH_2CH_3$$

8.6   S$_N$2 Mechanism

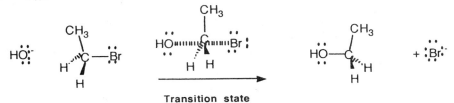

Transition state

8.7   Reaction Rates

(a) 3x   (b) 4x   (c) 6x

8.8   S$_N$2 Mechanism with Stereochemistry

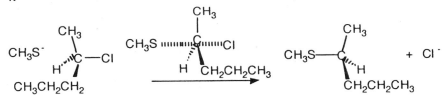

Pure enantiomer; optically active.

Transition state showing nucleophile attacking from opposite side of leaving bromide.

Pure enantiomer; optically active; inverted mirror image configuration.

8.9-8.10   Inversion of Configuration

(a)
S $CH(CH_3)_2$

$$H_3C \!-\! \overset{}{\underset{\underset{\displaystyle H}{\vdots}}{C}} \!-\! I \;\;\; + \;\; NaOH \rightarrow$$

R $CH(CH_3)_2$

$$HO \!-\! \overset{}{\underset{\underset{\displaystyle H}{\vdots}}{C}} \!-\! CH_3$$

(b)
S CH_2CH_3

$$Br \!-\! \overset{}{\underset{\underset{\displaystyle CH_3}{\vdots}}{C}} \!-\! H \;\;\; + \;\; NaOCH_3 \rightarrow$$

R CH_2CH_3

$$H \!-\! \overset{}{\underset{\underset{\displaystyle CH_3}{\vdots}}{C}} \!-\! OCH_3$$

8.11 Nucleophilic Substitution

$$CH_3\overset{\overset{\displaystyle CH_3}{|}}{\underset{\underset{\displaystyle Cl}{|}}{C}}CH_2CH_3 \; + \; CH_3CH_2OH \longrightarrow CH_3\overset{\overset{\displaystyle CH_3}{|}}{\underset{\underset{\displaystyle OCH_2CH_3}{|}}{C}}CH_2CH_3 \; + \; HCl$$

8.12 S_N1 Mechanism

$$CH_3\overset{\overset{\displaystyle CH_3}{|}}{\underset{\underset{\displaystyle Cl}{|}}{C}}CH_2CH_3 \longrightarrow CH_3\overset{\overset{\displaystyle CH_3}{|}}{\underset{\underset{\displaystyle +}{}}{C}}CH_2CH_3 \xrightarrow{\;\overset{..}{C}H_3CH_2\overset{..}{O}H\;}$$

$$CH_3\overset{\overset{\displaystyle CH_3}{|}}{\underset{\underset{\displaystyle H\,+}{\underset{\displaystyle OCH_2CH_3}{|}}}{C}}CH_2CH_3 \longrightarrow CH_3\overset{\overset{\displaystyle CH_3}{|}}{\underset{\underset{\displaystyle OCH_2CH_3}{|}}{C}}CH_2CH_3$$

8.13 Reaction Rates

    (a) no effect   (b) 2x   (c) 4x   (d) 3x

8.14 S_N1 and S_N2 Reaction Mechanisms

    (a) S_N1

Pure enantiomer;
optically active.

Nucleophile attacks
planar carbocation
equally from either side.

Both inversion and retention
of configuration occur equally.
A pair of enantiomers is the
result. This is an optically
inactive racemic mixture.

(b) S$_N$2 Mechanism

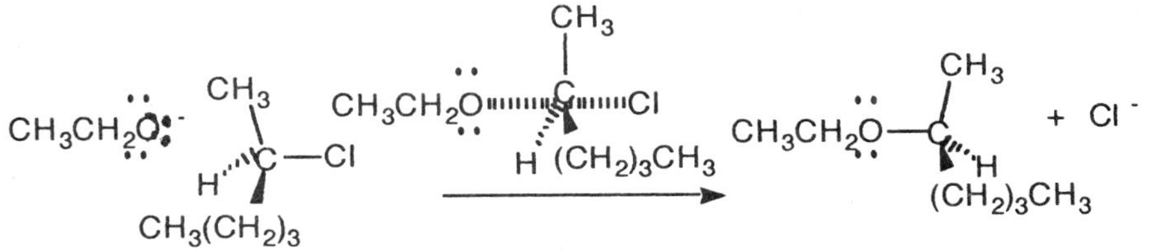

| Pure enantiomer; optically active. | Transition state showing nucleophile attacking from opposite side of leaving chloride | Pure enantiomer: optically active; mirror image configuration |

8.15 -8.16   Racemization in S$_N$1 Reactions

(a)

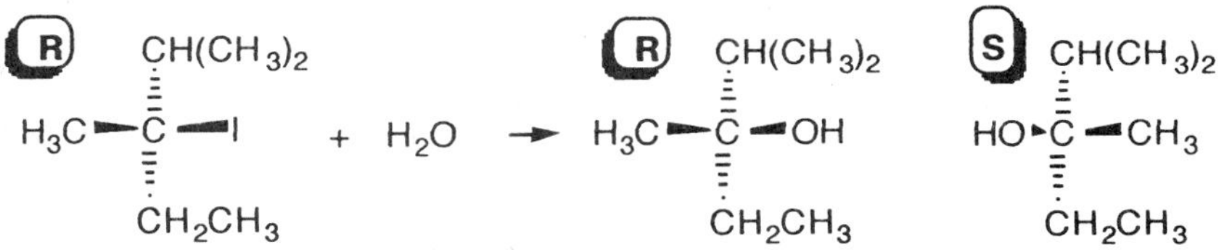

(b)

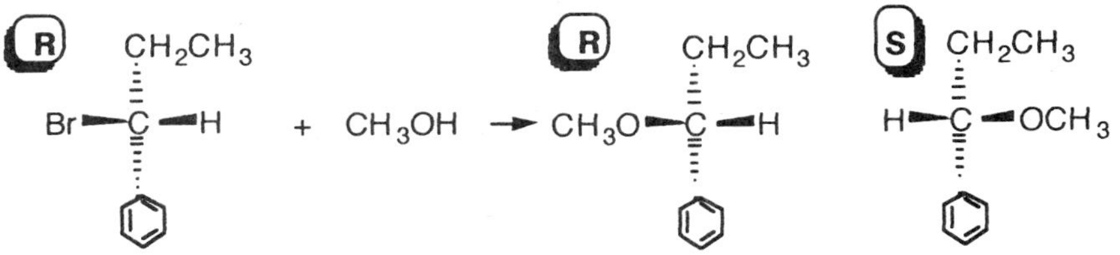

8.17   S$_N$1 Mechanism and Carbocation Stability

Tertiary carbocations are quite stable so tertiary halides tend to react by S$_N$1 mechanisms since the mechanism involves carbocation intermediates. Primary carbocations are unstable and primary halides react by the S$_N$2 mechanism in which there is no carbocation intermediate.

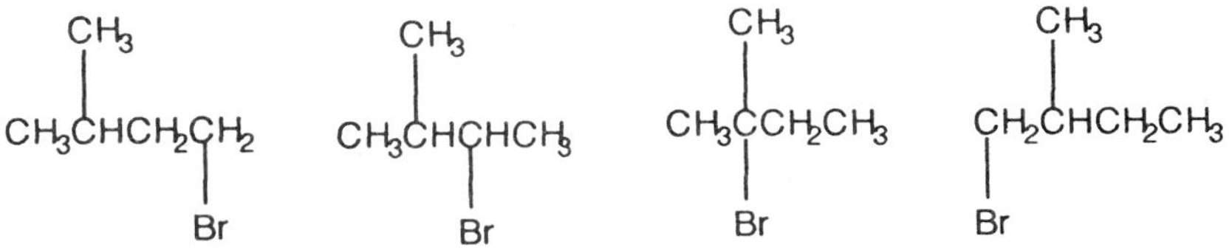

1° halide	2° halide	3° halide	1° halide
S$_N$2	S$_N$2 S$_N$1	S$_N$1	S$_N$2

(a) CH_3O^- because it is negative;   (b) NH_2^- because nitrogen is less electronegative than oxygen;   (c) CH_3NH_2 because nitrogen is less electronegative than oxygen;   (d) CH_3CH_2SH because sulfur is less electronegative than oxygen.

8.20 Predicting Mechanisms

(a) S_N2: the reactants are a primary halide and a strong, negative nucleophile.

(b) S_N1: the reactants are a tertiary halide and a neutral nucleophile.

8.21 E₁ and E₂ Mechanisms

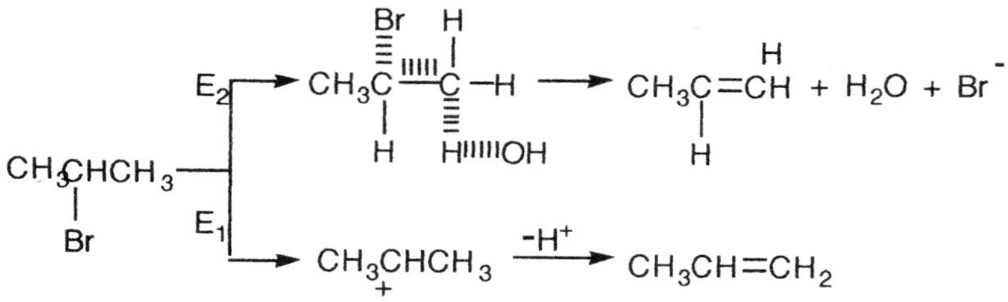

8.22 Syn and Anti Elimination

The E_2 reaction proceeds exclusively via anti elimination whereas the E_1 reaction is capable of both. Rotate the carbon-carbon bond to postion for syn and anit elimination. Eliminate the H and Br as highlighted and then look down the axis from the front to back carbon and imagine a double bond has formed. Translate this into the products shown.

anti elimination

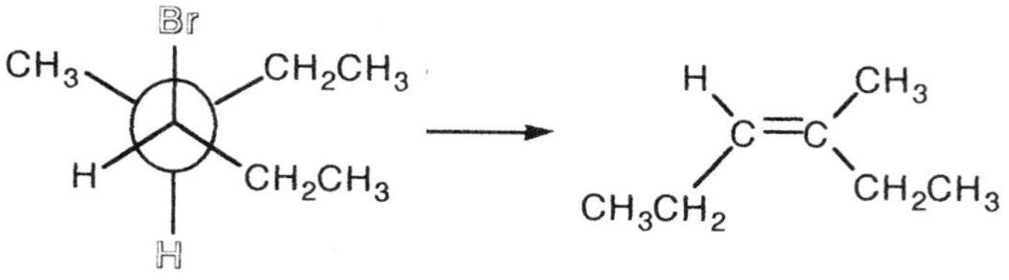

syn elimination

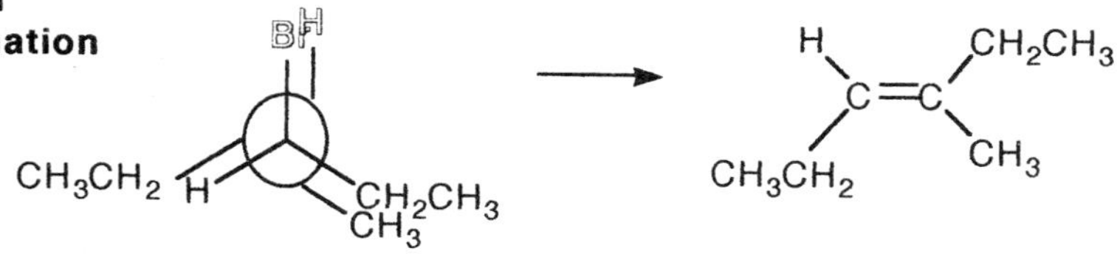

8.23 Williamson Synthesis

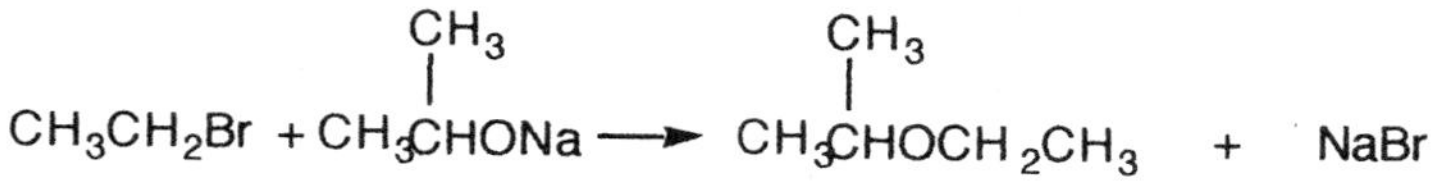

142

syn elimination

8.23 Williamson Synthesis

$$CH_3CH_2Br + CH_3\overset{\underset{\displaystyle CH_3}{|}}{C}HONa \longrightarrow CH_3\overset{\underset{\displaystyle CH_3}{|}}{C}HOCH_2CH_3 \ + \ NaBr$$

8.24 IUPAC Nomenclature: Section 8.1

(a) 3-bromopentane;   (b) chloroethane;   (c) 1-iodopropane;

(d) 1,1,1-tribromo-4,4-dimethylpentane;   (e) 2-iodo-2-methylbutane;

(f) 2,4,6-trichloroheptane;   (g) trans (or E) 5-bromo-6-methyl-2-heptene;

(h) 5-iodo-2-hexyne;   (i) ortho bromochlorobenzene;

(j) 1-(p-fluorophenyl)-3-methyl-1-butyne

8.25 Nomenclature: Section 8.1B

a) $CHCl_3$ b) $\underset{\underset{\displaystyle Br}{|}}{C}H_2 - \underset{\underset{\displaystyle Br}{|}}{C}H - \underset{\underset{\displaystyle Cl}{|}}{C}H_2$ c) $\underset{\underset{\displaystyle Cl}{|}}{C}H_2 - \underset{\underset{\displaystyle Cl}{|}}{C}H_2$ d) CCl_4

e)   f) CF_2Cl_2

8.26 Common Nomenclature: Section 8.1C

a) CH_3Br;   b) CH_2Cl_2;   c) $CHBr_3$;   d) CF_4;   e) $CH_2=CHCH_2I$;

f) $CH_2=CHCl$;   g) $CH_3CHClCH_2CH_3$;   h) $CH_3CHBrCH_3$

8.27 Nucleophilic Substitution: Section 8.4A

a) $CH_3\overset{\underset{\displaystyle OH}{|}}{C}HCH_3$ b) $-CH_2CN$ c) CH_3CH_2SH

d) $CH_3CH_2CH_2\overset{\underset{\displaystyle CH_3}{|}}{N}CH_3$ e) $CH_3\overset{\underset{\displaystyle CH_3}{|}}{C}HOCH_2CH_3$ f) $CH_3\overset{\underset{\displaystyle CH_3}{|}}{C}HCH_2SCH_3$

g) $\langle$O$\rangle$CH$_2$CH$_2$C$\equiv$CCH$_3$ h) CH$_3$NH$_2$

8.28 Nucleophilic Substitution: Section 8.4B

(a) CH$_3$CCH$_2$CH$_3$ with CH$_3$ above and OCH$_3$ below   (b) $\langle$O$\rangle$CH$_2$NCH$_3$ with CH$_3$ above   (c) CH$_3$CH$_2$CH$_2$OCH$-\langle$O$\rangle$ with CH$_3$ above

8.29 Williamson Synthesis of Ethers: Sections 8.4A and 8.6

a) CH$_3$CH$_2$CH$_2$ONa + CH$_3$Cl $\longrightarrow$ CH$_3$CH$_2$CH$_2$OCH$_3$ + NaCl

b) CH$_3$CHONa (with CH$_3$ above) + CH$_3$CH$_2$Cl $\longrightarrow$ CH$_3$CHOCH$_2$CH$_3$ (with CH$_3$ above) + NaCl

8.30 Nucleophilic Substitution in Preparing Alkynes: Sections 5.8 and 8.4A

a) HC$\equiv$CH $\xrightarrow{\text{NaNH}_2}$ HC$\equiv$CNa $\xrightarrow{\text{CH}_3\text{CH}_2\text{Br}}$ HC$\equiv$CCH$_2$CH$_3$

b) HC$\equiv$CH $\xrightarrow{\text{NaNH}_2}$ HC$\equiv$CNa $\xrightarrow{\text{CH}_3\text{CH}_2\text{Cl}}$ HC$\equiv$CCH$_2$CH$_3$

$\Big\rvert$ NaNH$_2$

CH$_3$C$\equiv$CCH$_2$CH$_3$ $\xleftarrow{\text{CH}_3\text{I}}$ NaC$\equiv$CCH$_2$CH$_3$ $\longleftarrow$

c) HC$\equiv$CH $\xrightarrow{\text{NaNH}_2}$ HC$\equiv$CNa $\xrightarrow{\langle O \rangle-\text{CH}_2\text{Br}}$ HC$\equiv$CCH$_2-\langle$O$\rangle$

8.31 Nucleophilic Substitution: Section 8.4A

a)   CH$_2$=CHCH$_2$Br + NaSH $\longrightarrow$ CH$_2$=CHCH$_2$SH + NaBr

b)   CH$_2$=CHCH$_2$Br + NaSCH$_2$CH=CH$_2$ $\longrightarrow$ CH$_2$=CHCH$_2$SCH$_2$CH=CH$_2$ + NaBr

8.32 Synthesis: Section 8.4A

a) $CH_3(CH_2)_8CH_2Cl$ + $NaNH_2$ $\longrightarrow$ $CH_3(CH_2)_8CH_2NH_2$ + $NaCl$

b) CH_3CH_2Cl + $NaSCH_3$ $\longrightarrow$ $CH_3CH_2SCH_3$ + $NaCl$

c) $CH_3CH_2CH_2CH_2Br$ + $NaOH$ $\longrightarrow$ $CH_3CH_2CH_2CH_2OH$ + $NaBr$

8.33 S_N1 and S_N2 Mechanisms: Section 8.4G

Summary of Characteristics

		S_N1	S_N2
a)	Rate Expression	Rate = k (RX)	Rate = k(RX)(Nu)
b)	Reaction Intermediates	carbocation	none
c)	Stereochemistry	racemization	inversion of configuration
d)	Relative Rates of Reaction of 1°, 2°, 3° Halides	3° > 2° > 1°	1° > 2° > 3°
e)	Effect of Increasing Concentration of Nucleophile	none, reaction rate is independent of nucleophile	reaction rate is increased
f)	Effect of Increasing Concentration of Alkyl Halide	reaction rate is increased	reaction rate is increased
g)	Effect of Ionic or Polar Solvent	increases rate and likelihood of S_N1	decreases rate and likelihood of S_N2
h)	Effect of Non-Polar Solvent	decreases rate and likelihood of S_N1	increases rate and likelihood of S_N2
i)	Effect of Bulky Groups Around Reaction Center	S_N1 favored since intermediate has only 3 groups on carbon	S_N2 disfavored since transition state is pentavalent thus increasing steric crowding
j)	Strength of Nucleophile	disfavors S_N1	favors S_N2

8.34 Nucleophilic Substitution Mechanisms

S$_N$2

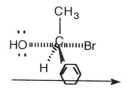

Pure enantiomer;
optically active.

Transition state
showing nucleophile
attacking from opposite
side of leaving bromide.

Pure enantiomer;
optically active;
mirror image
configuration.

S$_N$1

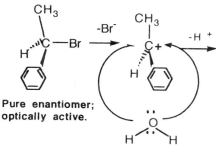

Pure enantiomer;
optically active.

Nucleophile attacks
planar carbocation
equally from either side.

Both inversion and retention
of configuration occur equally.
A pair of enantiomers is the
result. This is an optically
inactive racemic mixture.

8.35 Elimination Reactions: Sections 4.5 and 8.5

a) $CH_3CH=CH_2$

b) $CH_3CH=CHCH_2CH_3$

c) $\begin{array}{c} CH_3 \\ | \\ CH_3C=CHCH_2CH_3 \end{array}$

d) $CH_3C\equiv CH$

8.36 Elimination Reaction Mechanisms: Section 8.5

E₁

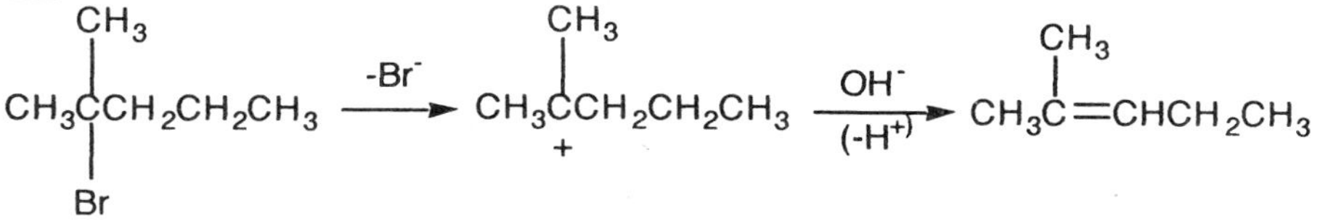

$$CH_3\overset{\displaystyle CH_3}{\underset{\displaystyle Br}{\overset{|}{\underset{|}{C}}}CH_2CH_2CH_3} \xrightarrow{-Br^-} CH_3\overset{\displaystyle CH_3}{\overset{|}{\underset{+}{C}}}CH_2CH_2CH_3 \xrightarrow[(-H^+)]{OH^-} CH_3\overset{\displaystyle CH_3}{\overset{|}{C}}=CHCH_2CH_3$$

STEP 1: Bromide departs and is solvated by the aqueous ethanol solvent. A carbocation intermediate results.

STEP 2: Hydroxide abstracts the hydrogen ion to complete the elimination. The carbocation is neutralized, the double bond forms.

E₂

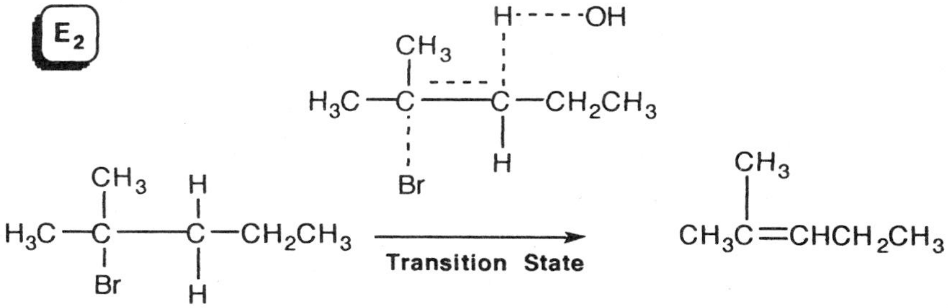

$$CH_3\overset{CH_3}{\overset{|}{C}}=CHCH_2CH_3$$

Transition State

The **E₂** mechanism is a concerted one-step process. The base abstracts the hydrogen ion as the solvent aids in removal of the bromide ion.

8.37 S_N1 and S_N2 Stereochemistry: Section 8.4 D.2 and E.2

a)

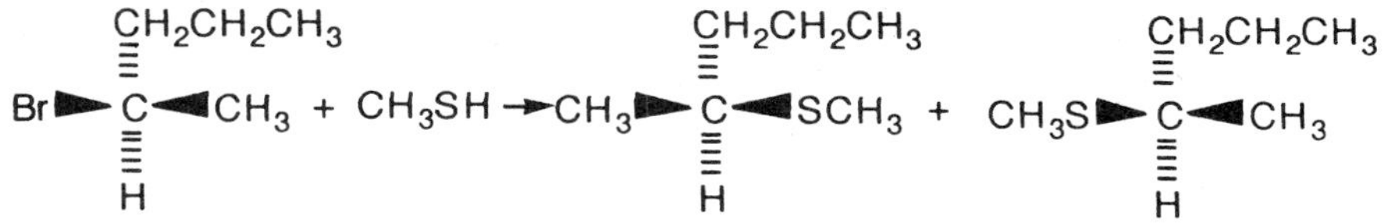

b)

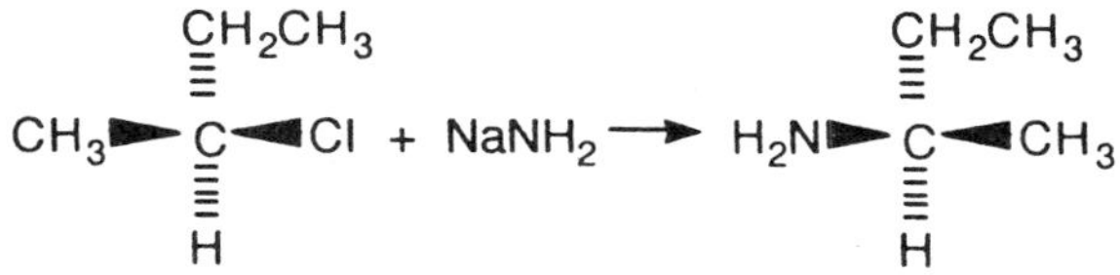

c) $CH_3 \blacktriangleright \underset{H}{\overset{CH(CH_3)_2}{C}} \blacktriangleleft I$ + $NaOCH_2CH_3 \longrightarrow CH_3CH_2O \blacktriangleright \underset{H}{\overset{CH(CH_3)_2}{C}} \blacktriangleleft CH_3$

d) $Cl \blacktriangleright \underset{H}{\overset{(CH_2)_3CH_3}{C}} \blacktriangleleft CH_2CH_3$ + $H_2O \longrightarrow HO \blacktriangleright \underset{H}{\overset{(CH_2)_3CH_3}{C}} \blacktriangleleft CH_2CH_3$ + $CH_3CH_2 \blacktriangleright \underset{H}{\overset{(CH_2)_3CH_3}{C}} \blacktriangleleft OH$

8.38 S_N1 and S_N2 Stereochemistry: Sections 8.4D.1 and E.1

S_N2 reactions proceed with inversion of configuration. Since both starting materials were optically active, the products are also optically active.

(a) $H_3CO \blacktriangleright \underset{H}{\overset{CH_3}{C}} \blacktriangleleft CH_2CH(CH_3)_2$

(b) $H \blacktriangleright \underset{CN}{\overset{C_6H_5}{C}} \blacktriangleleft CH_3$

8.39 E_1 and E_2 Stereochemistry: Section 8.5

$\boxed{E_2}$ $\underset{H}{\overset{CH_3}{}}C = C\overset{CH_3}{\underset{C_6H_5}{}}$ $\boxed{E_1}$ $\underset{H}{\overset{CH_3}{}}C = C\overset{CH_3}{\underset{C_6H_5}{}}$ $\underset{CH_3}{\overset{H}{}}C = C\overset{CH_3}{\underset{C_6H_5}{}}$

8.40 E_1 and E_2 Stereochemistry: Section 8.5

First draw the compound without stereochemistry. Then we will convert it to a Newman projection with the 2R, 3S configuration. This is most easily done by drawing an eclipsed Newman projection and putting the two low priority groups down and placing the others to conform to the stated configuration. Finally, rotate the Newmans to syn and anti elimination and draw the products.

E_2: anti elimination only

E_1: both syn and anti eliminations are possible

8.41 E₁ and E₂ Stereochemistry: Sections 4.5B and 8.5

E₂ reactions proceed by anti elimination. The only anti possibility is shown and it does not give the most stable product (Saytzeff)

E₁ reactions proceed by syn or anti elimination. The syn shown gives the most stable alkene (Saytzeff).

149

8.42 Nucleophilic Substitution Reactions: Section 8.4D.2

This is an S_n2 reaction as a result of the strong nucleophile. S_N2 reactions proceed with inversion of configuration so the product will be trans 1-ethoxy-2-methylcyclopentane.

8.43 Nucleophilic Substitution Reactions: Section 8.4F

$$CH_3CH_2CH_2CH_2CI$$

$$\overset{\overset{\displaystyle CH_3}{|}}{CH_3CHCH_2CI}$$

These are both primary halides and because they do not form stable carbocations and because they are relatively unhindered sterically, they react by S_N2.

$$\overset{\overset{\displaystyle CH_3}{|}}{\underset{\underset{\displaystyle CI}{|}}{CH_3CCH_3}}$$

$$\underset{\underset{\displaystyle CI}{|}}{CH_3CHCH_2CH_3}$$

The tertiary halide on the left is hindered to attack by a nucleophile but forms a stable carbocation. Consequently it reacts by S_N1. The other halide is secondary and can react by either mechanism.

8.44 Nucleophilic Substitution Reactions: Section 8.4F

$$CH_3CH_2CH_2CH_2CH_2Br$$

least

$$\overset{\overset{\displaystyle CH_3}{|}}{\underset{\underset{\displaystyle Br}{|}}{CH_3CCH_2CH_3}}$$

most

8.45 Substitution versus Elimination: Section 8.6

III > IV > II > I

III is a tertiary halide and forms a highly substituted alkene. IV is secondary but forms as highly substituted an alkene as does III. II forms a disubstituted alkene and I forms only a monosubstituted alkene.   Elimination is favored when highly substituted stable alkenes are possible.

9

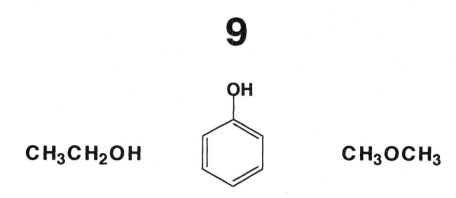

CH_3CH_2OH CH_3OCH_3

Alcohols, Phenols, and Ethers

CHAPTER SUMMARY

Alcohols, phenols, and ethers can be thought of a derivatives of water. Replacement of one hydrogen on water results in an **alcohol**, and replacement of both gives and **ether**. In **phenols**, one hydrogen of water is replaced by an aromatic ring. A **primary alcohol** has only one alkyl group attached to the carbon bearing the OH; a **secondary alcohol** has two and a **tertiary alcohol** has three.

The base name of an alcohol is derived from the Greek for the longest continuous carbon chain followed by the suffix **-ol**. If the alcohol is unsaturated, the double or triple bonds are designated with the suffixes **-en** and **-yn** respectively. The carbon chain is numbered to give the lowest number to the alcohol group. The name of an ether is based on the longest carbon chain connected to the ether oxygen. The other alkyl group is named as an **alkoxy group**. In common nomenclature, alcohols are often name by the alkyl group followed by alcohol (such as ethyl alcohol) and ethers are named using the names of the two alkyl groups followed by ether (such as diethyl ether).

Hydrogen bonding causes the boiling points of alcohols to be higher than those of compounds of similar molecular weight in other functional groups. Hydrogen bonding is an electrostatic attraction between the partially positive OH hydrogen of one molecule and a non-bonding electron-pair on the oxygen of another molecule. Hydrogen bonding occurs in molecules where hydrogen is bonded to a strongly electronegative element such as nitrogen, oxygen, or fluorine.

Alcohols, ethers, and phenols have a variety of important uses. **Methyl alcohol** is used in industrial synthesis, as a solvent, and as a clean burning fuel. **Ethyl alcohol** is beverage alcohol; it is also used as a solvent and antiseptic. **Isopropyl alcohol** is rubbing alcohol. **Ethylene glycol** is antifreeze and **glycerol** is a humectant. **Diethyl ether** is an important solvent and was once widely used as a general anesthetic. **Phenol and many of its derivatives** are used in over-the-counter medications as disinfectants and local anesthetics. They are also used as antioxidants and preservatives.

The **reaction sites** in alcohols, phenols, and ethers are the **polar bonds** (carbon-oxygen and oxygen-hydrogen) and the **lone pairs of electrons** on the oxygen. The unshared electron-pairs on alcohols and ethers make these compounds **Lewis bases.** Most reactions of alcohols involve the **O-H bond, C-O bond, or both.**

The polar O-H bond of alcohols makes them weak acids. By the **Bronsted-Lowry** definition, acids are hydrogen ion donors and bases are hydrogen ion acceptors in chemical reactions. **Strong acids** are 100% ionized in water and **weak acids** are only partially ionized. Weak acids establish an equilibrium in water between their ionized and un-ionized forms. This equilibrium and the strength of an acid is described by the **acidity constant, K_a**. K_a is defined as the concentrations of the ionized forms of the acids (H_3O^+ and A^-) divided by the un-ionized form (HA). The stronger the acid, the greater will be the value of the acidity constant. Acid strengths are also expressed by **pK_a**, which is defined as the negative logarithm of K_a. Numerically smaller pK_a's signify stronger acids and larger pK_a's, weaker acids. Approximate pK_a's include 50 for alkanes, 25 for terminal alkynes, 16 for alcohols, 10 for phenols, 5 for carboxylic acids, and -2 or so for strong inorganic acids. The ion or molecule formed by the loss of a proton from an acid is the **conjugate base.** Strong acids form weak conjugate bases and weak acids form strong conjugate bases.

Phenols are one million to one billion times more acidic than alcohols and this is the characteristic property that distinguishes them. Phenols will react with the base sodium hydroxide but alcohols will not. The acidity of phenols is explained by **resonance stabilization** of the **phenoxide ion**; the negative charge is dispersed throughout the benzene ring as opposed to being concentrated on the oxygen as it is in the **alkoxide ion**.   **Electron-withdrawing groups** on the benzene ring increase the acidity of phenols. Although alcohols will not react with sodium hydroxide as do phenols, they will react with sodium metal to form alkoxide ions and hydrogen gas. Alcohols will also react with organic and inorganic acids to form **esters.**

Alcohols and ethers react with hydrogen halides by nucleophilic substitution.   With alcohols, the OH group is replaced by a halogen; water is the by-product.   In the reaction mechanism, the first step involves formation of an **oxonium ion** by the Lewis acid-base reaction of the hydrogen ion of the hydrogen halide and alcohol oxygen.   In the S_N2 reaction the next step involves displacement of the water molecule by halide ion to form the final products.   In the S_N1 reaction the water molecule departs leaving a carbocation that is neutralized by halide ion.   The S_N2 reaction with an optically active alcohol proceeds with **inversion of configuration** whereas the S_N1 reaction produces **racemization.** Tertiary and secondary alcohols react by the S_N1 mechanism because they can form relatively stable intermediate carbocations; primary alcohols react by the S_N2 mechanism that does not require a carbocation.   The relative rates of reaction are **$3^o > 2^o > 1^o$**. Alcohols can also be converted to alkyl halides using **thionyl chloride** or **phosphorus trihalides.**

Ethers react with hydrogen halides to form an alkyl halide and an alcohol.   The alcohol in turn can react to form a second molecule of alkyl halide and water.   Thus in the presence of two mole-equivalents of hydrogen halide, an ether produces two moles of alkyl halide and one of water.   The reaction mechanism is analogous to that of alcohols and hydrogen halides.   Tertiary and secondary ethers react by the S_N1 mechanism and primary and methyl ether carbons react by S_N2.

Alcohols dehydrate in the presence of strong acids such as sulfuric acid. The reaction proceeds via an **E_1 mechanism.**   The alcohol oxygen is first protonated to give an oxonium ion which loses water to form a carbocation; subsequent loss of hydrogen ion forms the double bond.

Primary alcohols oxidize to carboxylic acids; secondary alcohols oxidize to ketones with chromium trioxide or sodium dichromate. Tertiary alcohols do not oxidize under mild conditions. With

pyridinium chlorochromate (PCC) the oxidation of primary alcohols can be stopped at **aldehydes.**

Epoxides are three-membered cyclic ethers.   Their characteristic chemical property is ring-opening reactions initiated by acid or base. **Thiols** or alkyl hydrogen sulfides are sulfur analogues of alcohols and sulfides are sulfur analogues of ethers.

Connections 9.1 describes the important uses of methyl, ethyl, and isopropyl alcohols.

Connections 9.2 is about neurotransmitters.

Connections 9.3 is about insecticides and nerve gases.

Connections 9.4 describes the chemical basis of methanol and ethylene glycol poisoning.

SOLUTIONS TO PROBLEMS

9.1   Nomenclature of Alcohols
a) 1-butanol; b) 2-butanol;   c) 3,5,5-trimethyl-3-hexanol;
d) 4-methyl-2-cyclohexenol;   e) 5-bromo-3-hexynol

9.2   Nomenclature of Ethers
a) 1-propoxyheptane; b) dimethoxymethane; c)   2-ethoxy-1-ethanol

9.3   Nomenclature of Phenols
(a) meta nitrophenol;   (b) para butoxyphenol

9.4   Nomenclature of Alcohols, Phenols, and Ethers

a) $CH_3\overset{\overset{\displaystyle CH_3}{|}}{\underset{\underset{\displaystyle OH}{|}}{C}}CH_3$ b) $CH_3(CH_2)_3CH_2OH$ c) $CH_3CH_2OCH_2CH_3$

d) CH_3CH_2O-⬠   e)

9.5   Physical Properties and Hydrogen-Bonding

Even though their molecular weights are similar, these three compounds have very different boiling points. Butane has the lowest boiling point because it is a non-polar compound and thus has weak intermolecular attractions. Propanol has an O-H bond and is capable of hydrogen bonding, a phenomenon that causes strong intermolecular attractions and elevated boiling points.   1,2-Ethandiol has two O-H groups and thus greater opportunity for hydrogen bonding; as a result it has a drastically higher boiling point.

9.6   Physical Properties

Ethylbenzene, the third compound and gasoline component, has the lowest boiling point ($136^{\circ}C$) because it has only carbon-carbon and carbon-hydrogen bonds and is a non-polar compound. The first compound, rose oil, has the highest boiling point ($221^{\circ}C$) because it has an O-H bond and is capable of hydrogen-bonding. The middle compound is an ether; though it is polar because of the C-O-C bonds, it is not capable of hydrogen-bonding and thus has an intermediate boiling point ($171^{\circ}C$).

9.7   Hydrogen-Bonding

9.8   Physical Properties and Hydrogen-Bonding

The two compounds are isomers and have the same molecular weight. Butanoic acid has an OH group and is thus capable of hydrogen bonding. Ethyl acetate cannot hydrogen bond. Butanoic acid has the higher boiling point because of its ability to hydrogen bond.

9.9   Reaction Sites

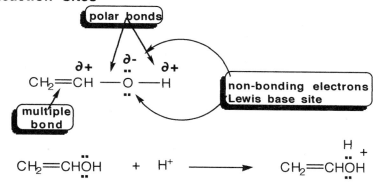

$$CH_2\!=\!CH\overset{..}{\underset{..}{O}}H \quad + \quad H^+ \quad \longrightarrow \quad CH_2\!=\!CH\overset{\overset{H}{\underset{..}{\overset{..}{O}}}\,^+}{O}H$$

9.10 Acids and Bases

$$CH_3OH \quad + \quad HCl \quad \longrightarrow \quad CH_3\overset{H\,+}{O}H \quad + \quad Cl^-$$

$\quad$ **base** $\qquad\qquad$ **acid** $\qquad\qquad\qquad\qquad\qquad$ **conjugate** $\qquad$ **conjugate**
$\qquad\qquad\qquad\qquad\qquad\qquad\qquad\qquad\qquad\qquad$ **acid** $\qquad\qquad$ **base**

9.11 Relative Acidities

(a) pK_a's: $11.8 < 6.2 < 3.4$

(b) K_a's: $9.8 \times 10^{-12} < 6.7 \times 10^{-5} < 3.4 \times 10^{-3}$

(c) $CH_3CH_2CH_3 < CH_3CH_2CH_2OH < CH_3CH_2CO_2H$

9.12 Relative Acidities

(a)   No reaction: the conjugate acid that would be produced, CH_3CH_2OH, is stronger than the original acid, CH_4.   The conjugate base is likewise stronger than the original base. Thus the products of the theoretical neutralization would be more acidic and basic than the original compounds and immediately react to reform them.

(b)   Yes, neutralization would occur: HCl is a stronger acid than the conjugate acid, CH_3CO_2H. The conjugate base is weaker than the original base. Thus the two reactants shown are more reactive, i.e. more acidic and basic, than the theoretical products of the reaction.

9.13 Acidity of Phenols

9.14 Reactions of the O-H Bond of Alcohols: Acidity
   a) No reaction with NaOH
   b) $2\ CH_3CH_2OH + 2\ Na \longrightarrow 2\ CH_3CH_2ONa + H_2$

9.15 Reactions of the O-H Bond of Alcohols: Ester Formation
$CH_3CH_2CH_2CH_2OH + HONO \longrightarrow CH_3CH_2CH_2CH_2ONO + H_2O$

9.16 Reactions of Alcohols with Hydrogen Halides

(b) $CH_3(CH_2)_3CH_2OH + HBr \longrightarrow CH_3(CH_2)_3CH_2Br + H_2O$

9.17 Reactions of Alcohols with Hydrogen Halides: Relative Rates
     The compounds in Problem 9.16 show relative reactions rates in the following order: **a > c > b** since the relative rates of reaction of alcohols with hydrogen halides is **3° > 2° > 1°**.

9.18 Lucas Reagent

Rate of Reaction with the Lucas Reagent

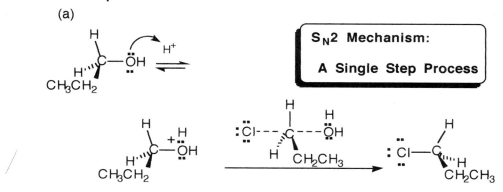

CH_3	CH_3	CH_3	CH_3
$CH_3CHCH_2CH_2$	$CH_3CHCHCH_3$	$CH_3CCH_2CH_3$	$CH_2CHCH_2CH_3$
OH	OH	OH	OH
1⁰ Alcohol one hour with heat	2° Alcohol 5-15 minutes	3° Alcohol instantaneous	1⁰ Alcohol one hour with heat

9.19 Nucleophilic Substitution Mechanisms

(a)

S_N2 **Mechanism:**

A Single Step Process

Primary alcohol protonated to form primary oxonium ion. Oxonium ion is attacked by bromide.

Transition state showing bromide displacing water molecule from the opposite side to form final product.

(b)

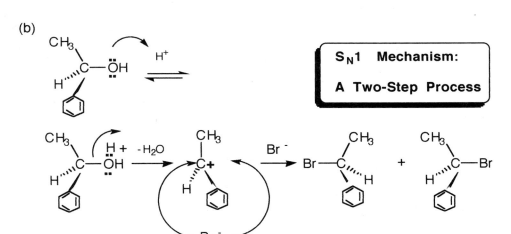

S_N1 Mechanism:

A Two-Step Process

Pure enantiomer; optically active alcohol is protonated to optically active oxonium ion.

Nucleophile, Br⁻ attacks planar carbocation from either side.

Both inversion and retention of configuration occur equally. A pair of enantiomers results. This is an optically inactive racemic mixture.

9.20 Preparation of Alkyl Halides

(a) CH_3CH_2OH + $SOCl_2$ ⟶ CH_3CH_2Cl + SO_2 + HCl

(b) CH_3CH_2OH PBr_3 ⟶ CH_3CH_2Br + $P(OH)_3$

9.21 Preparation of Alkyl Halides

$CH_3CH_2CH_2CH_2OH$ can be converted to   **$CH_3CH_2CH_2CH_2Cl$**

using the following reagents: **(1)   HCl with $ZnCl_2$**
                                (2) $SOCl_2$
                                (3) PCl_3

9.22 Reaction of Ethers with Hydrogen Halides

$$CH_3\underset{\underset{CH_3}{|}}{\overset{\overset{CH_3}{|}}{C}}OCH_3 \quad + \quad HBr \quad \longrightarrow \quad CH_3\underset{\underset{CH_3}{|}}{\overset{\overset{CH_3}{|}}{C}}Br \quad + \quad CH_3OH$$

9.23 S_N1 Mechanism: Ethers and Hydrogen Halides

oxonium ion   carbocation   alkyl bromide product

9.24 Reactions of Ethers with Hydrogen Halides

9.25 S_N2 Mechanism: Ethers and Hydrogen halides

Primary alcohol protonated
to form primary oxonium ion.
Oxonium ion is attacked by
bromide.

Transition state
showing bromide displacing
water molecule from the opposite
side to form the final product.

9.26 Reactions of Ethers with Hydrogen Halides
$ICH_2CH_2CH_2CH_2I$ + H_2O

9.27 Dehydration Reactions

Focus your attention on the OH; remove it and a hydrogen from an adjacent carbon. The double bond forms between these two carbons. When more than one elimination product is possible, the most substituted alkene forms predominantly (Section 4.5B).

(a) $CH_3\overset{\displaystyle |}{\underset{\displaystyle OH}{C}}HCH_3 \xrightarrow{\quad H_2SO_4 \quad} CH_3CH{=}CH_2 \quad + \quad H_2O$

(b) $CH_3\overset{\displaystyle CH_3}{\underset{\displaystyle OH}{\overset{\displaystyle |}{\underset{\displaystyle |}{C}}}}CH_2CH_3 \xrightarrow{\quad H_2SO_4 \quad} CH_3\overset{\displaystyle CH_3}{\overset{\displaystyle |}{C}}{=}CHCH_3 \quad + \quad H_2O$

(c) $CH_3CH\overset{\displaystyle CH_3}{\underset{\displaystyle OH}{\overset{\displaystyle |}{\underset{\displaystyle |}{C}}}}HCH_2CH_3 \xrightarrow{\quad H_2SO_4 \quad} CH_3CH{=}\overset{\displaystyle CH_3}{\overset{\displaystyle |}{C}}CH_2CH_3 \quad + \quad H_2O$

9.28 E$_1$ Elimination Reaction Mechanism

E$_1$ Mechanism for Dehydration of Alcohols

Step 1: Oxygen (Lewis base) protonated by H$^+$ (Lewis acid)

Step 2: Oxonium ion loses water molecule to form carbocation.

Step 3: Carbocation neutralized by elimination of hydrogen ion. C=C results.

9.29 E$_1$ and S$_N$1 Reactions

Whether S$_N$1 or E$_1$, the first thing that happens is oxonium ion formation. Water is lost to generate the carbocation. Whether elimination or substitution occurs is not determined until the carbocation is formed.

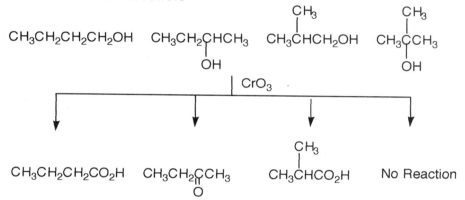

Step 1: Oxygen
(Lewis base)
protonated by H⁺
(Lewis acid).

Step 2: Oxonium
ion loses water
molecule to form
carbocation.

Step 3: Carbocation
neutralized by elimination
of hydrogen ion or by
bonding of chloride ion.

9.30 Oxidation of Alcohols

CH₃CH₂CH₂CH₂OH   CH₃CH₂CHCH₃   CH₃CHCH₂OH   CH₃CCH₃

CrO₃

CH₃CH₂CH₂CO₂H   CH₃CH₂CCH₃   CH₃CHCO₂H   No Reaction

9.31 Oxidation of Alcohols

(a) ⬠ OH and CrO₃   (b) ⬠ CH₂OH and CrO₃   (c) ⬠ CH₂OH and PCC

9.32 Reactions of Epoxides

(a) CH₃CH -CHCH₃   (b) CH₃CH -CHCH₃   (c) CH₃CH -CHCH₃
       OH Br   OH OH   OH OCH₂CH₃

9.33 Reactions of Epoxides

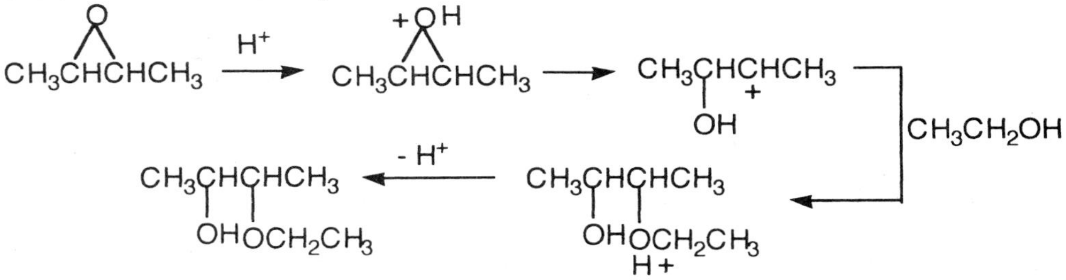

$$CH_3\overset{O}{\overset{\triangle}{CHCHCH_3}} \xrightarrow{H^+} CH_3\overset{+OH}{\overset{\triangle}{CHCHCH_3}} \longrightarrow CH_3\overset{}{\underset{OH}{CHCHCH_3}}\,+$$

$$CH_3CH_2OH$$

$$CH_3\overset{}{\underset{OHOCH_2CH_3}{CHCHCH_3}} \xleftarrow{-H^+} CH_3\overset{}{\underset{\underset{H+}{OHOCH_2CH_3}}{CHCHCH_3}} \longleftarrow$$

9.34 Isomerism and Nomenclature: Section 9.1

(a-c) Alcohols

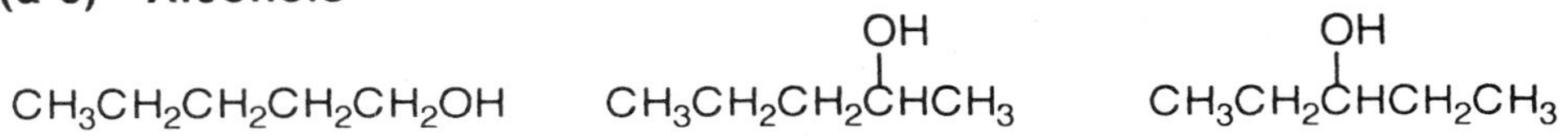

$CH_3CH_2CH_2CH_2CH_2OH$

1°   1-pentanol

$CH_3CH_2CH_2\overset{OH}{\overset{|}{C}HCH_3}$

2°   2-pentanol

$CH_3CH_2\overset{OH}{\overset{|}{C}HCH_2CH_3}$

2°   3-pentanol

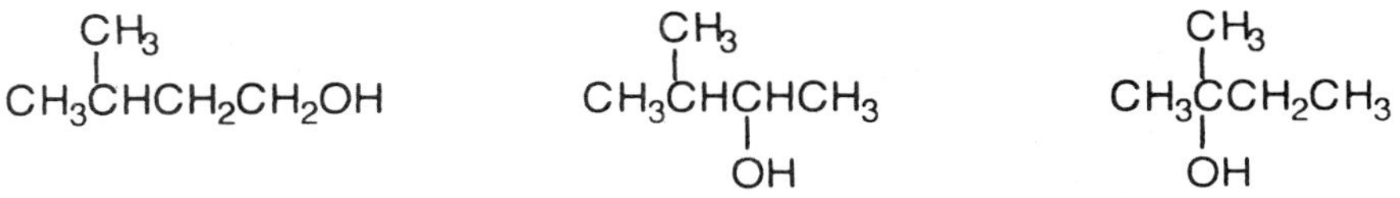

$CH_3\overset{CH_3}{\overset{|}{C}HCH_2CH_2OH}$

1°   3-methyl-1-butanol

$CH_3\overset{CH_3}{\overset{|}{C}HCHCH_3}$
$\underset{OH}{}$

2°   3-methyl-2-butanol

$CH_3\overset{CH_3}{\overset{|}{C}CH_2CH_3}$
$\underset{OH}{}$

3°   2-methyl-2-butanol

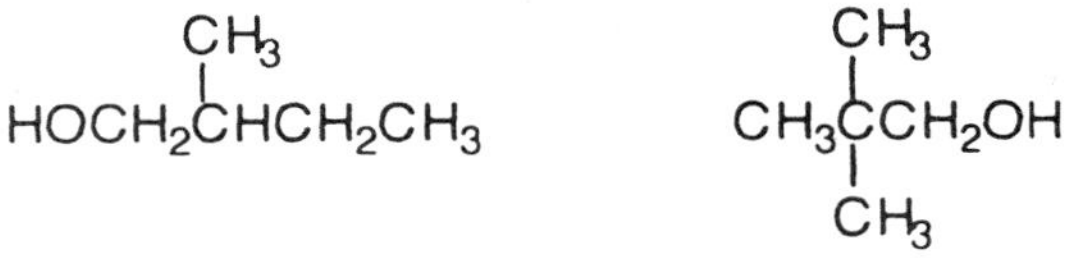

$HOCH_2\overset{CH_3}{\overset{|}{C}HCH_2CH_3}$

1°   2-methyl-1-butanol

$CH_3\overset{CH_3}{\overset{|}{\underset{\underset{CH_3}{|}}{C}}CH_2OH}$

1°   2,2-dimethyl-1-propanol

(d-e) Ethers

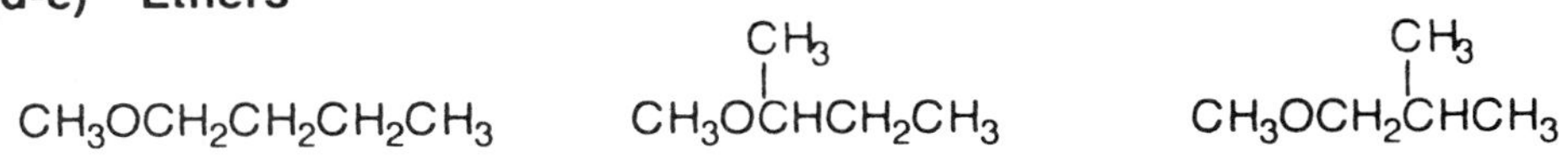

$CH_3OCH_2CH_2CH_2CH_3$

1-methoxybutane

$CH_3O\overset{CH_3}{\overset{|}{C}HCH_2CH_3}$

2-methoxybutane

$CH_3OCH_2\overset{CH_3}{\overset{|}{C}HCH_3}$

1-methoxy-2-methylpropane

$$CH_3OCCH_3 \quad \overset{\overset{CH_3}{|}}{\underset{\underset{CH_3}{|}}{}} \qquad CH_3CH_2OCH_2CH_2CH_3 \qquad CH_3CH_2OCHCH_3 \overset{CH_3}{|}$$

2-methoxy-2-methylpropane   1-ethoxypropane   2-ethoxypropane

9.35 IUPAC Nomenclature of Alcohols: Section 9.1A

(a) 1-nonanol;   (b) 2-hexanol;   (c) 2-methyl-2-butanol;   (d) cyclopentanol;
(e) 3,3-dimethyl-1-butanol;   (f) 4-ethyl-4-methyl-1-cyclohexanol;
(g) 2,2,3-trimethyl-3-pentanol;   (h) 2,5-dimethyl-2-hexanol

9.36 IUPAC Nomenclature of Alcohols: Section 9.1A

(a) 4,5-dibromo-3-hexanol;   (b) 5-methyl-3-heptanol;   (c) 1,5-pentandiol;
(d) 1,3,5-cyclohexantriol

9.37 IUPAC Nomenclature of Alcohols: Section 9.1A

(a) 3-buten-2-ol;   (b) 4-ethyl-2-hexyn-1-ol;   (c) 2,4-hexadien-1,6-diol;
(d) 3-cyclopenten-1-ol;   (e) 2-phenyl-1-ethanol;   (f) 1-hexyn-4-en-3-ol

9.38 IUPAC Nomenclature of Ethers: Section 9.1B

(a) methoxyethane;   (b) ethoxyethane;   (c) 1-ethoxy-6-methoxyhexane;
(d) propoxycyclopentane;   (e) methoxycyclopropane

9.39 IUPAC Nomenclature of Ethers: Section 9.1B

(a) tetramethoxymethane;   (b) 3-methoxy-1-propanol;   (c) 1-ethoxypropene;
(d) 4-methoxy-2-buten-1-ol

9.40 IUPAC Nomenclature of Phenols: Section 9.1C

(a) 2-methylphenol;   (b) 3-bromophenol;   (c) 4-ethylphenol;
d) 4-methoxy-2-nitrophenol
      ortho, meta, and para in a,b,c respectively is correct also.

9.41 Common Nomenclature: Section 9.1D

a) $CH_3\underset{\underset{OH}{|}}{C}HCH_2CH_3$ b) $CH_3\overset{\overset{CH_3}{|}}{\underset{\underset{CH_3}{|}}{C}}CH_2OH$ c) $CH_3CH_2OC\overset{CH_3}{|}HCH_3$

d) ⬡—OCH$_3$ e) CH$_2$=CHCH$_2$OH   f) ⬡—OCH=CH$_2$

9.42 IUPAC Nomenclature: Section 9.1

a)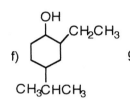

b) CH$_3$CHCH$_2$CH$_2$CH$_2$CH$_3$
       |
       SH

c) CH$_3$CH$_2$SSCH$_2$CH$_2$CH$_3$

d) CH$_3$CH$_2$SCH$_2$CH$_2$CH$_3$

e)

f)

g) CH$_3$CH$_2$CHCH$_2$OH
            |
            OCH$_3$

9.43 Physical Properties: Sections 2.9 and 9.2

(a) **I < II < III**   increasing molecular weight in homologous series

(b) **III < II < I**   more OH groups and thus more hydrogen bonding

(c) **I < II < III**   increasing number OH groups, increasing hydrogen bonding

(d) **III < II < I**   increasing number OH groups, increasing hydrogen bonding

(e) **III < II < I**   increasing number N-H bonds, increasing hydrogen bonding

(f) **II < I**   no hydrogen bonding in II; I has OH and hydrogen bonding

(g) **III < II < I**   OH bond more polar than NH; OH further polarized by C=O

(h) **II < I < III**   II is non-polar; I is polar but no hydrogen bonding; III has OH
                              and thus hydrogen bonding

(i) **III < I < II**   II has strongest hydrogen bonding due to C=O next to OH;
                              III has no hydrogen bonding.

(j) **I < II < III < IV < V < VI**   increasing molecular weight

9.44 Physical Properties: Section 9.2

The ortho isomer in each case undergoes <u>intramolecular</u> hydrogen bonding.

Because of this, the attractions between molecules are diminished and boiling points are lower than might be expected. The relationship between the

two substituents is not favorable for intramolecular hydrogen bonding in the other compounds. Thus, intermolecular hydrogen bonding occurs (as shown with p-nitrophenol) increasing attractions between molecules and thus the boiling points.

9.45 Water Solubility: Section 9.2

Sucrose has 12 carbons; there are OH groups on eight of them. This allows for tremendous hydrogen bonding with water and thus high water solubility.

9.46 Water Solubility: Section 9.2

(a) **hexanol < pentanol < ethanol:**   increasing ratio of OH to hydrocarbon as boiling points increase.

(b) **pentane < heptanol < propanol:**   pentane has no OH and cannot hydrogen bond to water; propanol has a higher ratio of OH to hydrocarbon than heptanol and thus is more like water.

(c) **hexane < hexanol < 1,2-ethanediol:**   hexane has no OH and no hydrogen bonding; hexanol can hydrogen bond with water but has only one OH for all six carbons and has only slight water solubility; ethanediol has an OH on each carbon and is infinitely soluble in water.

(d) **pentane < ethoxyethane < butanol:**   these compounds have similar molecular weights but the first two have no OH and thus no hydrogen bonding; the second is polar and has some slight water solubility and butanol has an OH and thus can hydrogen bond with water.

9.47 Reactive Sites: Section 9.5

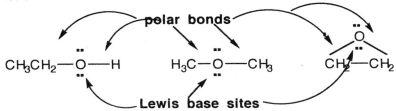

9.48 Lewis Base Character of Alcohols and Ethers: Section 9.6A

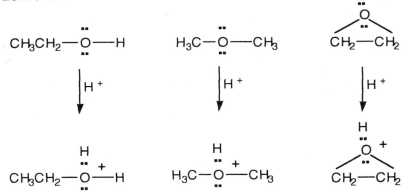

9.49 Acidity: Section 9.6A

(a) **No**: the conjugate acid and base are stronger than the original acid and base

(b) **Yes**: the original acid and base are stronger than the conjugate forms

(c) **Yes**; (d) **Yes**; (e) **Yes**; (f) **No**; (g) **Yes**; (h) **No**

9.50 Acid Base Neutralization: Section 9.6A

Acid		Base		Conjugate Base		Conjugate Acid
(b) $CH_3CH_2SO_3H$	+	CH_3CO_2Na	⟶	$CH_3CH_2SO_3Na$	+	CH_3CO_2H
(c) H_2SO_4	+	$2\ NaOH$	⟶	Na_2SO_4	+	$2\ H_2O$
(d) CH_3OH	+	CH_3Na	⟶	CH_3ONa	+	CH_4
(e) CH_3CO_2H	+	CH_3ONa	⟶	CH_3CO_2Na	+	CH_3OH
(g) ⬡—OH	+	CH_3ONa	⟶	⬡—ONa	+	CH_3OH

9.51 Acidity Constants: Section 9.6A

(I) $CH_3CH_2CH_2OH$ (II) $CH_3CH_2CH_3$ (III) $CH_3CH_2CO_2H$ (IV) H_3C⟨⟩OH

K_a's	10^{-16}	10^{-49}	10^{-5}	10^{-11}
pK_a's	16	49	5	11

Relative Acidities: II < I < IV < III

9.52 Acidity of Phenols: Section 9.6A.2

(a) **III < I < IV < II** : The methyl group is electron-releasing and decreases basicity; it is most effective ortho and para and probably a little more effective ortho due to proximity.

(b) **II < IV < I < III:** The acetyl group is electron-withdrawing and increases acidity; it is most effective ortho and para and probably a little more effective ortho due to proximity.

(c) **II < IV < I < III**: The methyl group decreases acidity and thus IV is more acidic than II. The nitro group increases acidity and III has more of them than does I.

9.53 Acidity of Phenols: Section 9.6A.2

(a) CH_3—⟨⟩—OH + $NaOH$ ⟶ CH_3—⟨⟩—ONa + H_2O

(b) Cl—⟨Cl⟩—OH + $NaOH$ ⟶ Cl—⟨Cl⟩—ONa + H_2O

(c) ⟨NO_2⟩—OH + $NaOH$ ⟶ ⟨NO_2⟩—ONa + H_2O

9.54 Reactions of Alcohols: Sections 9.5-9.9

Reagent	I	II	III
	$CH_3CH_2CH_2CH_2OH$	$CH_3CHCH_2CH_3$ $\quad\ \ $OH	CH_3CCH_3 $\quad$CH_3 / OH

a) Na $CH_3CH_2CH_2CH_2ONa$ $CH_3\underset{ONa}{CHCH_2CH_3}$ $CH_3\underset{\underset{CH_3}{|}}{\overset{\overset{CH_3}{|}}{C}}CH_3$
                                                             ONa

b) H_2SO_4 $CH_3CH_2CH=CH_2$ $CH_3CH=CHCH_3$ $CH_3\underset{\underset{CH_3}{|}}{C}=CH_2$

c) $HCl/ZnCl_2$ $CH_3CH_2CH_2CH_2Cl$ $CH_3\underset{Cl}{CHCH_2CH_3}$ $CH_3\underset{Cl}{\overset{\overset{CH_3}{|}}{C}}CH_3$

d) CrO_3/H^+ $CH_3CH_2CH_2CO_2H$ $CH_3\underset{\underset{O}{\|}}{C}CH_2CH_3$ No Reaction

e) HNO_3 $CH_3CH_2CH_2ONO_2$ $CH_3\underset{ONO_2}{CHCH_2CH_3}$ $CH_3\underset{ONO_2}{\overset{\overset{CH_3}{|}}{C}}CH_3$

9.55 Salts of Alcohols and Phenols: Section 9.6A-B

a) (3-chlorophenol ONa structure)   b) (naphthalen-2-ol ONa structure)   c) $CH_3\underset{}{\overset{\overset{CH_3}{|}}{C}}HCH_2ONa$

9.56 Reactions of Alcohols to Form Alkyl Halides: Section 9.7A-B

a) $CH_3\underset{Br}{CHCH_3}$ b) $CH_3\underset{I}{\overset{\overset{CH_3}{|}}{C}}CH_3$ c) $CH_3\underset{Cl}{CHCH_3}$ d) $CH_3CH_2CH_2Cl$ e) $CH_3\underset{Br}{CHCH_2CH_3}$

9.57 Oxidation of Alcohols: Section 9.9

a) $CH_3\overset{\overset{O}{\|}}{C}OH$ b) $CH_3\underset{\underset{O}{\|}}{\overset{\overset{CH_3}{|}}{C}}HCH_3$ c) $CH_3(CH_2)_{10}\overset{\overset{O}{\|}}{C}H$

9.58 Reactions of Ethers: Section 9.7C

a) $CH_3OH + CH_3Br$ b) $CH_3Br + BrCH_2\underset{\underset{CH_3}{|}}{CHCH_3}$ c) $CH_3CH_2OH + CH_3CH_2Cl$

d) $CH_3\overset{\overset{\displaystyle CH_3}{|}}{C}HI + ICH_2CH_3$ e) $BrCH_2CH_2CH_2CH_2Br$ f) $-CH_2Br + CH_3Br$

9.59 Dehydration of Alcohols: Sections 4.5B and 9.8

The predominant product is shown for each dehydration. Direct your attention to the OH group. Remove it and a hydrogen from an adjacent carbon. Draw a double bond between the two carbons. In cases where there is more than one adjacent carbon with hydrogens, remove the hydrogen from the one with the greatest number of alkyl groups (fewest number of hydrogens) to produce the most substituted alkene (the most stable).

a) $CH_3CH=CHCH_2CH_3$ b) $CH_3\overset{\overset{\displaystyle CH_3}{|}}{C}=CHCH_3$ c) $CH_3\overset{\overset{\displaystyle CH_3}{|}}{C}=\overset{\overset{\displaystyle CH_3}{|}}{C}CH_2CH_2CH_3$

d) $CH=CHCH_3$ e)

9.60 Reactions of Alcohols with Hydrogen Halides: Section 9.7A

Refer to the structures in Problem 9.59. **Replace the OH groups with Br.**

9.61 Reactions of Epoxides: Section 9.10A

a) $CH_3\overset{\overset{\displaystyle \ }{}}{\underset{\underset{\displaystyle OH}{|}}{C}}H-\underset{\underset{\displaystyle OH}{|}}{C}HCH_3$ b) $HOCH_2CH_2OCH_2CH_3$ c) $HOCH_2CH_2\underset{\underset{\displaystyle CH_3}{|}}{N}CH_3$

d) $BrCH_2CH_2OH$ e) $BrCH_2CH_2Br$

9.62 Reaction Mechanisms: Section 9.7A and C

Look at the carbon(s) directly bonded to the oxygens. If a carbon is primary, the mechanism of displacement is S_N2. If it is secondary or tertiary, the mechanism is S_N1.

(a) S_N2: primary alcohol;   (b) S_N2 for both carbons: ether where one carbon is methyl, one is primary;   (c) S_N1: secondary alcohol;   (d) S_N2 for the CH_3 carbon and S_N1 for the secondary carbon;   (e) S_N2 for both carbons since both are primary.

9.63 Nucleophilic Substitution Mechanisms: Sections 9.7A and C

(a)

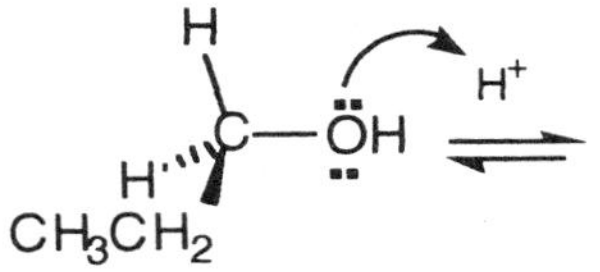

S_N2 Mechanism:

A Single Step Process

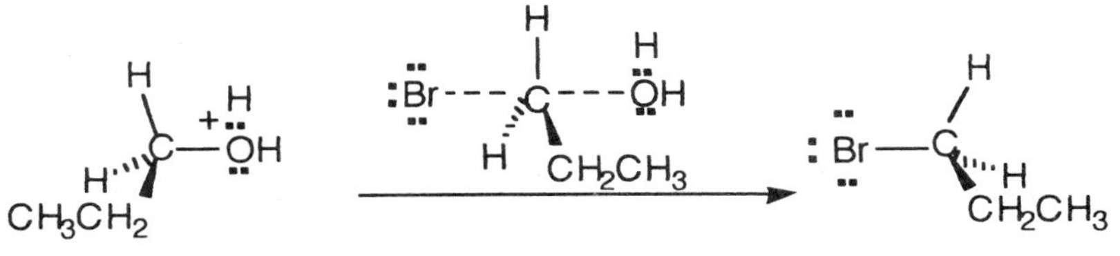

Primary alcohol protonated to form primary oxonium ion. Oxonium ion is attacked by bromide.

Transition state showing bromide displacing water molecule from the opposite side to form final product.

(b)

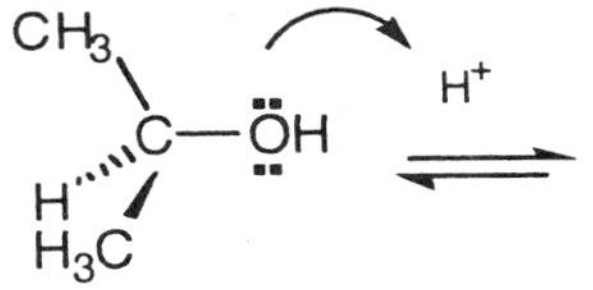

S_N1 Mechanism:

A Two-Step Process

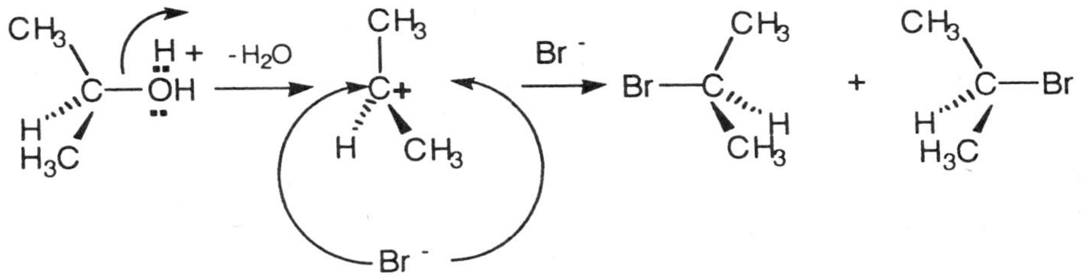

(c)

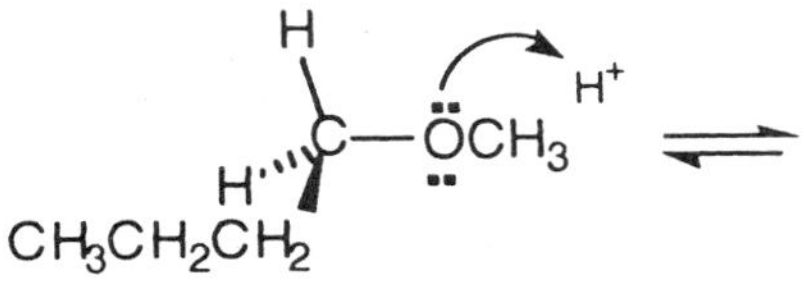

S_N2 Mechanism:

A Single Step Process

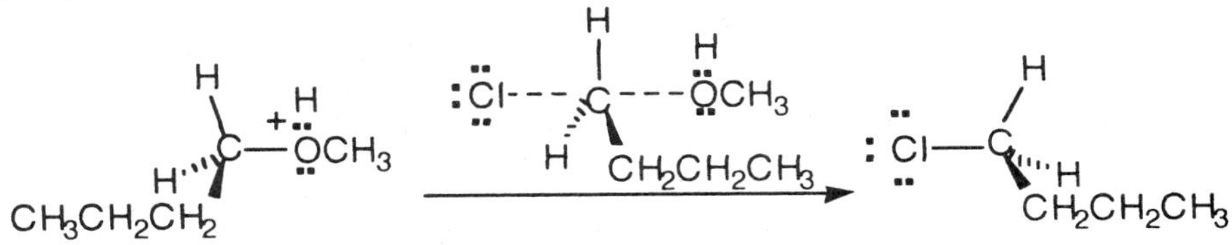

FOLLOWED BY

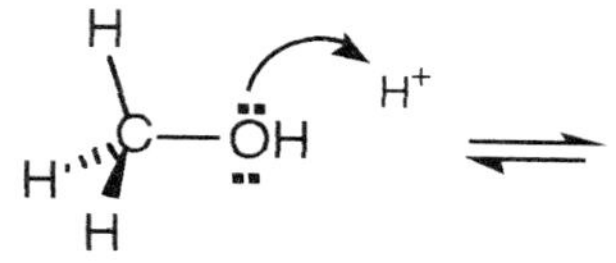

S_N2 Mechanism:

A Single Step Process

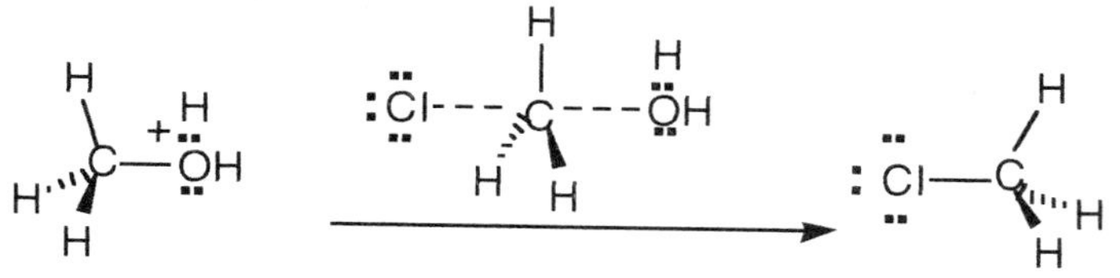

(d)

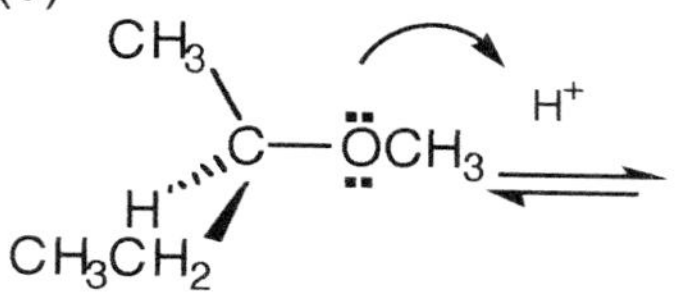

S_N1 Mechanism:

A Two-Step Process

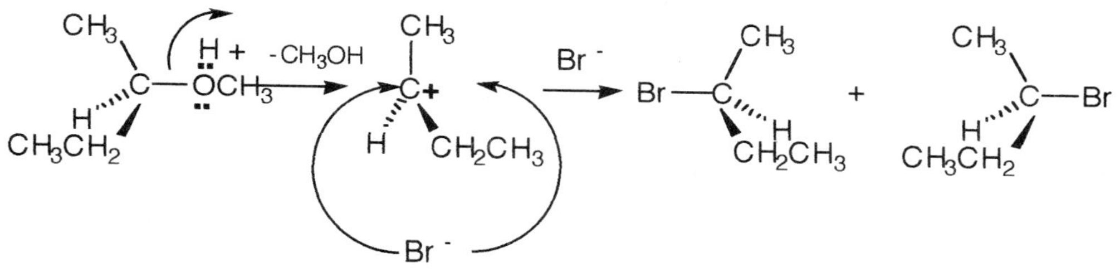

FOLLOWED BY

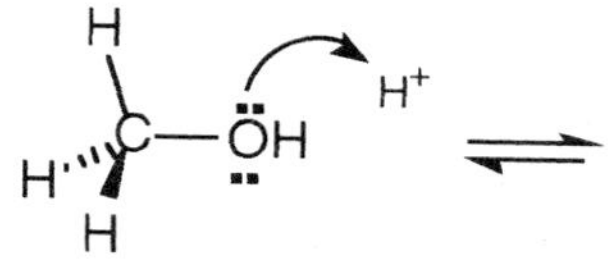

S_N2 Mechanism:

A Single Step Process

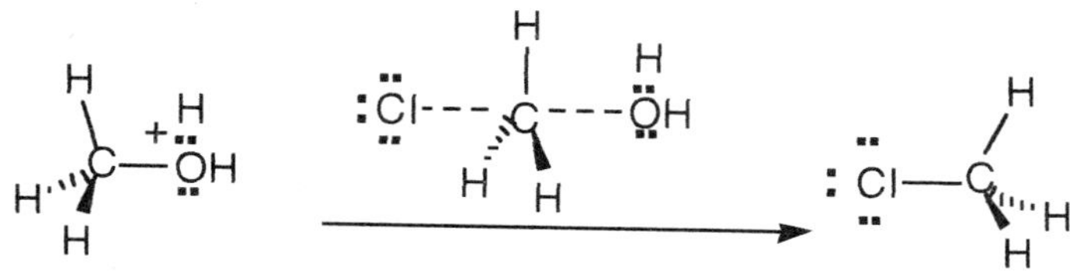

9.64 Nucleophilic Substitution Mechanisms: Section 9.7A and C

(a)

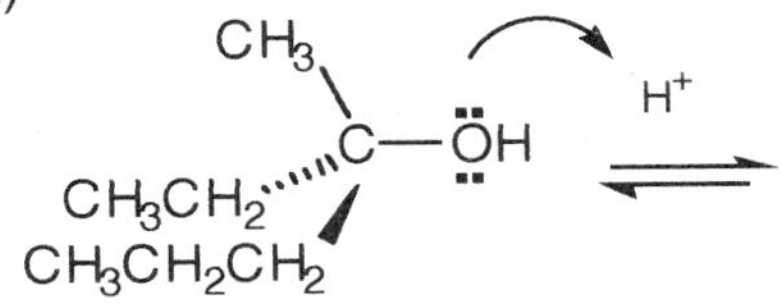

S_N1 Mechanism:

A Two-Step Process

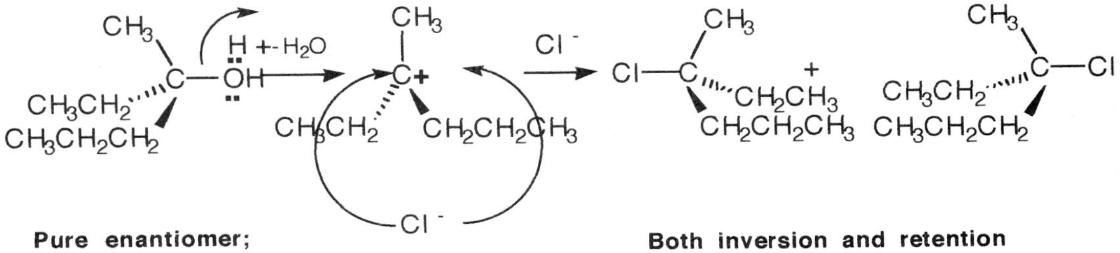

Pure enantiomer; optically active alcohol is protonated to optically active oxonium ion.

Nucleophile, Cl^- attacks planar carbocation from either side.

Both inversion and retention of configuration occur equally. A pair of enantiomers results. This is an optically inactive racemic mixture.

(b)

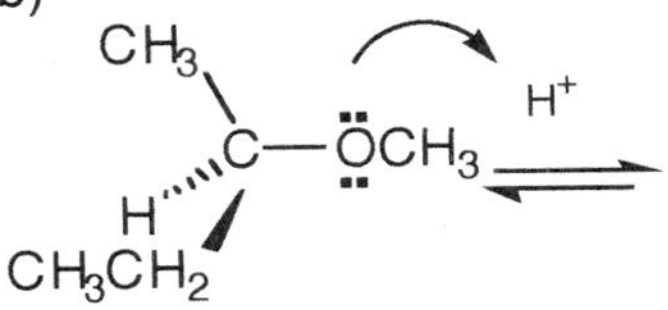

S_N1 Mechanism:

A Two-Step Process

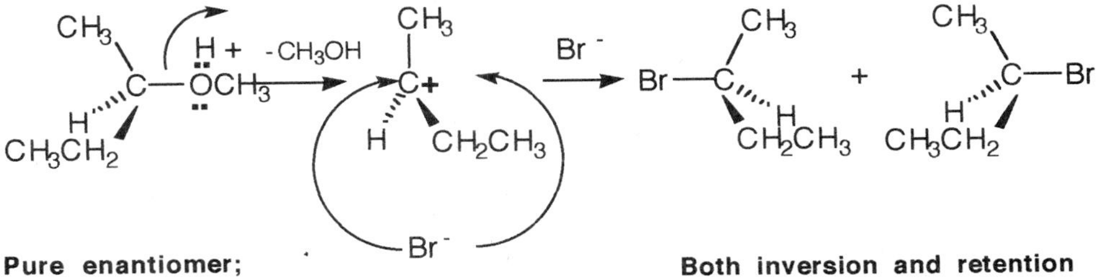

Pure enantiomer; optically active alcohol is protonated to optically active oxonium ion.

Nucleophile, Br^- attacks planar carbocation from either side.

Both inversion and retention of configuration occur equally. A pair of enantiomers results. This is an optically inactive racemic mixture.

9.65 Dehydration Mechanism: Sections 4.5C and 9.8

(a)

E_1 **Mechanism for Dehydration of Alcohols**

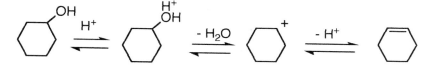

Step 1: Oxygen
(Lewis base)
protonated by H^+
(Lewis acid).

Step 2: Oxonium
ion loses water
molecule to form
carbocation.

Step 3: Carbocation
neutralized by elimination
of hydrogen ion. C=C
results.

(b)

E_1 **Mechanism for Dehydration of Alcohols**

Step 1: Oxygen
(Lewis base)
protonated by H^+
(Lewis acid).

Step 2: Oxonium
ion loses water
molecule to form
carbocation.

Step 3: Carbocation
neutralized by elimination
of hydrogen ion. C=C
results.

9.66 Reaction Mechanisms: Sections 4.5C, 9.7, and 9.8

a) $CH_3CHCH_3 \xrightarrow{H^+} CH_3CHCH_3 \xrightarrow{-H_2O} \left[CH_3CHCH_3 \right] \xrightarrow{-H^+} CH_3CH=CH_2$
     OH OH $+$
                               H^+

b) $CH_3CHCH_3 \xrightarrow{H^+} CH_3CHCH_3 \xrightarrow{-H_2O} \left[CH_3CHCH_3 \right] \xrightarrow{Br^-} CH_3CHCH_3$
     OH OH $+$ Br
                               H^+

c) $CH_3CHCH_3 \xrightarrow{H^+} CH_3CHCH_3 \xrightarrow{-CH_3OH} \left[CH_3CHCH_3 \right] \xrightarrow{Cl^-} CH_3CHCH_3$
     OCH_3 OCH_3 $+$ Cl
                               H^+

9.67 Williamson Synthesis of Ethers: Sections 8.4A, 8.6, 9.4B, 9.6B

(a) $CH_3CH_2CH_2OH$ + Na $\longrightarrow$ $CH_3CH_2CH_2ONa$ + H_2

$CH_3CH_2CH_2OCH_3$ $\longleftarrow$ CH_3Br

(b) $(CH_3)_3COH$ + Na $\longrightarrow$ $(CH_3)_3CONa$ + H_2

$(CH_3)_3COCH_2CH_2CH_3$ $\longleftarrow$ $CH_3CH_2CH_2Cl$

9.68 Qualitative Analysis: Sections 9.6A.2 and 9.7A

(a)   p-Ethylphenol being a phenol is acidic and reacts with sodium hydroxide. Alcohols are not so acidic and do not react with sodium hydroxide. p-Ethylphenol will dissolve in a sodium hydroxide solution and the other compound will not.

$CH_3CH_2-\langle\ \rangle-OH$ + NaOH $\longrightarrow$ $CH_3CH_2-\langle\ \rangle-ONa$ + H_2O

(b)   Treatment of each of these alcohols with the Lucas reagent will produce a turbid mixture as the alkyl halide is formed. However the reaction proceeds at different rates depending on the structure of the alcohol.

3° $CH_3\overset{CH_3}{\underset{OH}{C}}CH_2CH_3$ + HCl $\xrightarrow{ZnCl_2}$ $CH_3\overset{CH_3}{\underset{Cl}{C}}CH_2CH_3$ + H_2O Instantaneous reaction at room temperature

2° $CH_3\overset{}{\underset{OH}{CH}}CHCH_3$ + HCl $\xrightarrow{ZnCl_2}$ $CH_3\overset{}{\underset{Cl}{CH}}CHCH_3$ + H_2O Instantaneous reaction only if heated

1° $CH_3\overset{CH_3}{\underset{}{CH}}CH_2CH_2OH$ + HCl $\xrightarrow{ZnCl_2}$ $CH_3\overset{CH_3}{\underset{}{CH}}CH_2Cl$ + H_2O Slow reaction even if heated

9.69 Epoxide Chemistry: Section 9.10

There are three N-H bonds to add across the epoxide ring.

3 CH_2-CH_2 + NH_3 $\longrightarrow$ $\left(HOCH_2-CH_2\right)_3-N$

$\overset{}{\underset{O}{}}$

9.70 Preparations of Alcohols: Sections 5.1A.3, B.3, C and 9.4A

(a) $CH_3CH=CHCH_3$ + H_2O $\xrightarrow{H_2SO_4}$ $CH_3CH_2\underset{\underset{OH}{|}}{C}HCH_3$

(b) $CH_3CH=\overset{\overset{CH_3}{|}}{C}CH_2CH_3$ + H_2O $\xrightarrow{H_2SO_4}$ $CH_3CH_2\overset{\overset{CH_3}{|}}{\underset{\underset{OH}{|}}{C}}CH_2CH_3$

(c) $CH_3CH_2CH_2CH_2CH=CH_2$ + H_2O $\xrightarrow{H_2SO_4}$ $CH_3CH_2CH_2CH_2\underset{\underset{OH}{|}}{C}HCH_3$

9.71 Williamson Synthesis of Ethers: Sections 8.4A, 8.6, and 9.4B

$CH_3CH_2CH_2ONa$ + CH_3CH_2Br $\longrightarrow$ $CH_3CH_2CH_2OCH_2CH_3$ + $NaBr$

CH_3CH_2ONa + $CH_3CH_2CH_2Br$ $\longrightarrow$ $CH_3CH_2OCH_2CH_2CH_3$ + $NaBr$

9.72 Williamson Synthesis of Ethers: Sections 8.4A, 8.6, and 9.4B

The nucleophilic substitution is on the methyl bromide where competing elimination is not possible. If one used sodium methoxide and isopropyl bromide, elimination would be a competing process.

$(CH_3)_2CHONa$ + CH_3Br $\longrightarrow$ $(CH_3)_2CHOCH_3$ + $NaBr$

9.73 Synthesis Using Alcohols: Sections 9.7-9.9

(a) $CH_3(CH_2)_3CH_2OH$ $\xrightarrow{PBr_3}$ $CH_3(CH_2)_3CH_2Br$ $\xrightarrow{NaOCH_3}$ $CH_3(CH_2)_3CH_2OCH_3$

(b) $CH_3(CH_2)_4CH_2Br$ $\xrightarrow{NaOH}$ $CH_3(CH_2)_4CH_2OH$ $\xrightarrow{PCC}$ $CH_3(CH_2)_4\overset{\overset{O}{||}}{C}H$

(c) $CH_3(CH_2)_3CH_2Cl$ $\xrightarrow{NaOH}$ $CH_3(CH_2)_3CH_2OH$ $\xrightarrow{Na}$

$CH_3(CH_2)_4CH_2OCH_2(CH_2)_3CH_3$ $\xleftarrow{CH_3(CH_2)_4CH_2Br}$ $CH_3(CH_2)_3CH_2ONa$

176

9.74 Synthesis Using Alcohols: Section 9.9

(a) $CH_3CH_2CH_2\overset{\overset{\displaystyle OH}{|}}{C}HCH_3$ + CrO_3/H^+ (b) $CH_3CH_2CH_2CH_2CH_2OH$ + PCC

(b) $CH_3CH_2CH_2CH_2CH_2OH$ + CrO_3/H^+

9.75 Synthesis Using Alcohols: Sections 9.7-9.8

(a) and H_2SO_4 (b) $CH_3(CH_2)_4CH_2OH$ and PBr_3 or HBr

(c) $CH_3(CH_2)_4CH_2OH \xrightarrow{PBr_3} CH_3(CH_2)_4CH_2Br$ $\longrightarrow$ $\begin{array}{c}CH_3OH\\ \swarrow Na\\ CH_3ONa\end{array}$

$CH_3(CH_2)_4CH_2OCH_3 \longleftarrow$

177

10

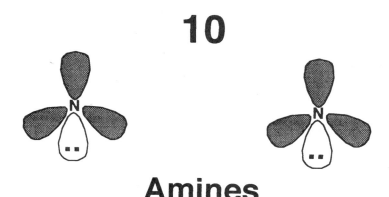

Amines

CHAPTER SUMMARY

Amines are derivatives of ammonia in which one or more hydrogens have been replaced by organic groups. Replacement of one hydrogen results in a **primary amine**. In a **secondary amine**, two hydrogens are replaced and in a **tertiary amine,** three hydrogens on ammonia are replaced . The nitrogen in alkyl amines is sp^3 hybridized, tetrahedral and has bond angles of about 109°.

Common nomenclature of simple amines involve following the name of the alkyl group with amine (such as propylamine). In **IUPAC nomenclature**, the name is based on the longest continuous chain of carbon atoms followed by the suffix -amine (such as 1-propanamine). Substituents on the carbon chain are located by a number; substituents on the nitrogen are located with N (such as N-methyl-1-propanamine). The simplest aromatic amine is **aniline**. In **unsaturated amines**, the double bond or triple bond is named with the usual suffix (en and yn respectively) and located by a number.

Melting points and boiling points of amines generally increase with molecular weight. Because of **hydrogen bonding**, amines have higher than expected boiling points and lower molecular weight amines are water soluble. Tertiary amines have no N-H bond and therefore cannot engage in hydrogen bonding.

Basicity is the characteristic property of amines; the presence of a **non-bonding** electron pair on nitrogen makes amines **Lewis bases.** Relative basicities are expressed using the **basicity constant, K_b,** which is defined as the concentrations of the ionized amine products in water divided by the concentration of the un-ionized amine. Larger K_b's mean greater acidity. **pK_b** is the negative logarithm of K_b; the smaller the pK_b the stronger the base.

Electron-releasing groups increase the availability of nitrogen's lone pair and, as a result, also increase the basicity of amines; **alkyl amines are more basic than ammonia.** **Electron-withdrawing groups** decrease the availability of the non-bonding electron pair and decrease basicity; **amides are much less basic than ammonia.** In aromatic amines, the non-bonding electron pair on nitrogen overlaps with the benzene pi-cloud by resonance decreasing the availability of the lone pair and stabilizing the compound. As a result, **aromatic amines are considerably less basic than aliphatic amines.** Electron-donating groups on an aromatic amine increase availability of the non-bonding pair whereas electron-withdrawing groups decrease availability; as a result **electron-donating groups on an aromatic ring increase basicity and electron-withdrawing groups decrease basicity.** The effect of these groups is greatest at the ortho and para positions.

Basicity can also be expressed with **acidity constants.** With amines **K_a** defines the equilibrium in the direction of the ammonium salt ionizing to the free amine and hydronium ion in water. Since this is the opposite of the definition of K_b, small K_a's and large pK_a's mean strong basicity.

Two common methods for preparing amines are alkylation of ammonia and reduction of other nitrogen compounds. **Alkylation** involves treating ammonia or an amine with an alkyl halide. The amine, as a Lewis base with a non-bonding electron pair, is a good nucleophile and displaces the halide ion from the alkyl halide; the reaction is **nucleophilic substitution** with a neutral nucleophile. Since alkylation tends to continue until four groups are bonded to the nitrogen, it has limited synthetic utility. Amines can be synthesized by **reduction of nitro compounds** with hydrogen and platinum catalyst or with a tin and hydrochloric acid solution. **Nitriles can be reduced** using hydrogen gas and nickel catalyst. Lithium aluminum hydride is used to **reduce amides.**

The characteristic reactions of amines depend on the availability of the non-bonding electron pair on nitrogen which makes the amines Lewis bases (nucleophiles). **Alkylation** involves displacement of halogen from an alkyl halide to form an **amine** and **acylation** involves displacement of chloride from a carboxylic acid chloride to form an **amide**.   **Sulfonamides** are formed from amines and **sulfonyl chlorides.**   The latter two reactions do not occur with a tertiary amine since there is no replaceable hydrogen.   All of these reactions represent some form of **nucleophilic substitution.**

Upon treatment with sodium nitrite and hydrochloric acid at $0°C$, primary aromatic amines can be converted to **aromatic diazonium salts.**   These salts are quite useful in organic synthesis as the diazonium group can be easily replaced by fluorine, chlorine, bromine, iodine, cyanide, hydroxy, and hydrogen. In these **diazonium replacement reactions,** nitrogen gas is evolved.   In **coupling reactions,** nitrogen is retained and actually bonds to an activated aromatic ring in an **electrophilic aromatic substitution reaction.**   This reaction is used to make **azo dyes.**

Heterocycles are cyclic compounds in which one or more of the ring atoms are not carbon. **Heterocyclic amines** have nitrogen as one of the ring atoms.   Although they are basic, their basicity can vary widely depending on structure and availability of nitrogen's non-bonding electron pair.   Aromatic heterocyclic amines in which the nitrogen's non-bonding electron pair is part of the aromatic pi-cloud are much less basic than those in which it is not.

Alkaloids are defined as plant-produced nitrogenous bases that have a physiological effect on humans.   They are often classified according to the heterocyclic amine present in the structure.

Connections 10.1 is about phenylalkylamines, a class of compounds that has adrenalin-like effects on humans and that is used as nasal decongestants, diet pills, and stimulants.

Connections 10.2 is about local anesthetics such as benzocaine and about the drug cocaine.

Connections 10.3 is about acetylcholine and neuromuscular blockade.

Connections 10.4 is about dyes and dyeing.   For a compound to be a dye it must have a chromophore group and extensive conjugation so it can exhibit color; in addition, it must have an auxochrome group which allows it to bind to a fabric.

SOLUTIONS TO PROBLEMS

10.1 Primary, Secondary, and Tertiary Amines

$$CH_3CH_2CH_2CH_2NH_2 \quad CH_3CH_2\overset{\displaystyle CH_3}{\underset{\displaystyle NH_2}{CHCH_3}} \quad CH_3\overset{\displaystyle CH_3}{\underset{}{CHCH_2NH_2}} \quad CH_3\overset{\displaystyle CH_3}{\underset{\displaystyle NH_2}{CCH_3}}$$

$$1° \qquad\qquad 1° \qquad\qquad 1° \qquad\qquad 1°$$

$$CH_3NHCH_2CH_2CH_3 \quad CH_3NH\overset{}{\underset{\displaystyle CH_3}{CHCH_3}} \quad CH_3CH_2NHCH_2CH_3 \quad CH_3\overset{}{\underset{\displaystyle CH_3}{NCH_2CH_3}}$$

$$2° \qquad\qquad 2° \qquad\qquad 2° \qquad\qquad 3°$$

10.2 Nomenclature of Amines

(a) 1-nonanamine;   (b) 2-hexanamine;   (c) N,N-dimethyl-3-pentanamine;

(d) N-ethyl-N-methyl-1-octanamine;   (e) 2-ethyl-N-methylcyclohexanamine;

(f) N-ethyl-N-methyl-3-nitroaniline

10.3 Nomenclature of Amines

Names of compounds in Problem 10.1 in order left to right.

First row: 1-butanamine;   2-butanamine;   2-methyl-1-propanamine;

2-methyl-2-propanamine

Second row: N-methyl-1-propanamine;   N-methyl-2-propanamine;

N-ethylethanamine;   N,N-dimethylethanamine.

10.4 Nomenclature of Amines

(a) 2-pentenamine;   (b) 5-methyl-3-hexynamine;

(c) 2,4-hexadien-1,6-diamine;   (d) N-methyl-3-penten-2-amine

10.5 Physical Properties

III<II<I<IV

No hydrogen bonding is possible in III. II has hydrogen bonding but since I has two N-H bonds it has a greater capability to hydrogen bond. IV has an O-H bond which is more polar than N-H bonds; thus it hydrogen bonds the best and has the highest boiling point.

10.6 Basicity of Amines

(a) $CH_3CH_2CH_2\overset{\displaystyle ..}{N}H_2$ +   HBr $\longrightarrow$ $CH_3CH_2CH_2NH_3^+$ Br$^-$

(b) $CH_3\overset{\cdot\cdot}{N}HCH_3$ + HNO_3 $\longrightarrow$ $CH_3\overset{+}{N}H_2CH_3\ NO_3^-$

(c) $(CH_3CH_2)_3N\colon$ + HCl $\longrightarrow$ $(CH_3CH_2)_3NH^+\ Cl^-$

10.7 Ammonium Fertilizers

$2\ NH_3$ + H_2SO_4 $\longrightarrow$ $(NH_4)_2SO_4$

$3\ NH_3$ + H_3PO_4 $\longrightarrow$ $(NH_4)_3PO_4$

10.8 Basicity Constants
Constants are shown in increasing basicity.
(a) $9.1 \times 10^{-10} < 5.6 \times 10^{-5} < 3.6 \times 10^{-4}$ (b) $9.1 < 4.3 < 3.2$

10.9 Basicity Constants
(a) diethylamine; (b) methylamine; (c) triethylamine; (d) p-methylaniline

10.10   Basicity of Amines
(a) **II < I < III** : alkyl groups are electron-donating and increase basicity.
(b) **I < II < III** : aromatic amines much less basic than alkyl amines; methyl on II increases basicity relative to I.
(c) **I < II < III** : electron-withdrawing nitro group decreases basicity especially when in ortho or para position; electron-releasing methyl group increases basicity.

10.11   Basicity of Amines
(a) The second compound is more basic because the electron-withdrawing C=O is further away; the first is an amide and is among the least basic nitrogen compounds.
(b) The first compound is more basic because it does not have an electron-withdrawing chlorine as does the first compound.

10.12   Expression of Basicity with Acidity Constants
(a) $8.3 \times 10^{-10} < 9.9 \times 10^{-10} < 2.3 \times 10^{-11}$ (b) $5.25 < 9.81 < 10.74$

10.13   Alkylation of Amines by Nucleophilic Substitution

a) $(CH_3)_3\overset{+}{N}CH_2$—⟨benzene⟩ $\overset{-}{Br}$ b) $CH_3CH_2\overset{+}{N}(CH_3)_3$ I^- c) CH_3CH_2 ⟩$\overset{N+}{}$⟨ CH_2CH_3 (pyrrolidine ring) Cl^-

d) $(CH_3)_4N^+ I^-$

10.14   Reduction of Nitro Compounds

(a) CH_3—⟨benzene⟩—NO_2 (b)   ⟨benzene with Br⟩—NO_2

10.15   Synthesis of Amines

(a) ⟨benzene⟩ $\xrightarrow[\text{FeCl}_3]{\text{Cl}_2}$ Cl—⟨benzene⟩ $\xrightarrow[\text{H}_2\text{SO}_4]{\text{HNO}_3}$ Cl—⟨benzene⟩—NO_2 $\xrightarrow[\substack{\text{HCl} \\ 2.\text{NaOH}}]{1.\text{Sn}}$ Cl—⟨benzene⟩—NH_2

(b) ⟨toluene⟩ $\xrightarrow[\text{H}_2\text{SO}_4]{\text{HNO}_3}$ ⟨toluene-NO$_2$⟩ $\xrightarrow[\text{FeBr}_3]{\text{Br}_2}$ ⟨toluene Br NO$_2$⟩ $\xrightarrow[\substack{\text{HCl} \\ 2.\text{NaOH}}]{1.\text{Sn}}$ ⟨toluene Br NH$_2$⟩

10.16   Synthesis of Amines

$CH_3(CH_2)_3CH_2Cl$ $\xrightarrow{\text{NaCN}}$ $CH_3(CH_2)_3CH_2CN$ $\xrightarrow[\text{Ni}]{2 \text{ H}_2}$ $CH_3(CH_2)_3CH_2CH_2NH_2$

10.17   Reduction of Amides

$$CH_3CH_2CH_2CH_2\overset{\overset{\displaystyle O}{\|}}{C}NHCH_2CH_3$$

OR

$$CH_3CH_2CH_2CH_2CH_2NH\overset{\overset{\displaystyle O}{\|}}{C}CH_3$$

$\xrightarrow[\text{2. H}_2\text{O}]{\text{1. LiAlH}_4}$ $CH_3CH_2CH_2CH_2CH_2NHCH_2CH_3$

10.18   Formation of Amides

(a) ⟨benzene⟩$\overset{\overset{\displaystyle O}{\|}}{C}$–N⟨pyrrolidine⟩   (b) $CH_3\overset{\overset{\displaystyle O}{\|}}{C}NHCH_2CH_3$ (c) $CH_3CH_2CH_2\overset{\overset{\displaystyle O}{\|}}{C}NH_2$ (d)CH_3—⟨benzene⟩$\overset{\overset{\displaystyle O}{\|}}{C}NCH_3$ (with phenyl)

10.19   Diazonium Salt Replacement Reactions

⟨aniline with NH$_2$ top, Cl bottom⟩ $\xrightarrow[\substack{\text{HCl} \\ 0^\circ\text{C}}]{\text{NaNO}_2}$ ⟨N_2^+ Cl^- top, Cl bottom⟩ $\longrightarrow$ products of replacement reactions

(a) — F / Cl benzene (b) — Cl / Cl (c) — Br / Cl (d) — I / Cl (e) — CN / Cl (f) — OH / Cl (g) — H / Cl

10.20   Synthesis Using Diazonium Salts

a) $\langle\text{ring}\rangle-NH_2 \xrightarrow[\substack{HCl \\ 0°C}]{NaNO_2} \langle\text{ring}\rangle-N_2^+Cl^- \xrightarrow{H_2O} \langle\text{ring}\rangle-OH$

b) $CH_3-\langle\text{ring}\rangle-NH_2 \xrightarrow[\substack{HCl \\ 0°C}]{NaNO_2} CH_3-\langle\text{ring}\rangle-N_2^+Cl^- \xrightarrow{HBF_4} CH_3-\langle\text{ring}\rangle-F$

c) $\langle\text{ring}\rangle-NO_2 \xrightarrow[HCl]{Sn} \langle\text{ring}\rangle-NH_2 \xrightarrow[\substack{HCl \\ 0°C}]{NaNO_2} \langle\text{ring}\rangle-N_2^+Cl^- \xrightarrow{CuCN} \langle\text{ring}\rangle-CN$

d) $\langle\text{ring}\rangle-NO_2 \xrightarrow[FeCl_3]{Cl_2} \langle\text{ring}\rangle-NO_2/Cl \xrightarrow[HCl]{Sn} \langle\text{ring}\rangle-NH_2/Cl \xrightarrow[\substack{HCl \\ 0°C}]{NaNO_2} \langle\text{ring}\rangle-N_2^+Cl^-/Cl \xrightarrow{CuCl} \langle\text{ring}\rangle-Cl/Cl$

10.21   Coupling Reactions of Diazonium Salts

$O_2N-\langle\text{ring}\rangle-NH_2 \xrightarrow[\substack{HCl \\ 0°C}]{NaNO_2} O_2N-\langle\text{ring}\rangle-N=N^+Cl^- \xrightarrow{\langle\text{ring}\rangle-NH_2}$

$O_2N-\langle\text{ring}\rangle-N=N-\langle\text{ring}\rangle-NH_2$

10.22   Aromatic Heterocyclic Amines

   Quinoline is aromatic. The ring system is flat, planar, and has a p-orbital on each carbon as a result of the "double bonds". There are a total of 6 pi electrons in the ring with the nitrogen. The non-bonding electron pair on nitrogen is not part of the aromatic sextet.

   Indole is also aromatic. Two double bonds and the non-bonding electron pair of nitrogen comprise the aromatic sextet. There is a p-orbital on each carbon of the ring system as a result of the "double bonds". To allow aromaticity, the lone pair of electrons on nitrogen also exists in a p-orbital and is part of the aromatic sextet.

10.23 IUPAC Nomenclature: Section 10.2

(a) 1-heptanamine;   (b) 2-butanamine;   (c) 1,8-octandiamine

10.24 IUPAC Nomenclature: Section 10.2

(a) cyclohexanamine;   (b) 4-methylcyclohexanamine;   (c) N-ethyi-4-methylcyclohexanamine;   (d) p-methylaniline;   (e) N-ethyl-p-methylaniline;

(f) N-ethyl-N-methylaniline;   (g) N,N-diethyl-2-bromo-1-propanamine;

(h) 7,7-dimethyl-N,N-dipropyl-2-octanamine

10.25 IUPAC Nomenclature: Section 10.2

(a) 4-heptyn-1-amine;   (b) 2-hexyn-1-amine;

(c) N-ethyl-2,4-hexadien-1-amine;   (d) N,N-dimethyl-2-butyn-1-amine;

(e) 3-cyclopenten-1-amine;   (f) 3-methyl-3-cyclopenten-1-amine;

(g) N,N-diethyl-3-methyl-3-cyclopenten-1-amine

10.26 IUPAC Nomenclature: Section 10.2

(a) [structure: cycloheptane with NH_2]   (b) $CH_3CH_2CH_2NHCH_2CH_3$ (c) $(CH_3CH_2CH_2CH_2)_3N$

(d) $CH_3CH_2\overset{\overset{\displaystyle CH_3}{|}}{N}CH(CH_3)_2$ (e) [phenyl]$-N(CH_3)_2$

(f) [structure: benzene ring with NH_2 and three Cl substituents]   (g) $CH_3(CH_2)_5CH_2NHCH_2CH_3$ (h) [structure: cyclopentane with N bearing H_3C, CH_2CH_3 and ring bearing $CH_2CH_2CH_3$]

10.27 Physical Properties: Sections 2.9, 9.2, and 10.3

(a) **I < II < III < IV :** increasing molecular weight in a homologous series.

(b) **III < I < II :** no hydrogen bonding in III; hydrogen bonding with OH is more effective than with NH because of increased bond polarity.

(c) **III < II < I:** these compounds are isomers; the number of NH bonds increases from zero to two in this order and thus hydrogen bonding increases.

(d) **III < II < I :** these compounds are isomers; the number of NH bonds increases from zero to two in this order and thus hydrogen bonding increases.

10.28 Physical Properties: Sections 2.9, 9.2, and 10.3

Methylamine has the lowest molecular weight and, though it has the most hydrogen bonding sites (two NH bonds), the low molecular weight gives it the lowest boiling point. Dimethylamine has one NH bond and thus hydrogen bonding and it is greater in molecular weight than methylamine. Although trimethylamine has the greatest molecular weight, it has no NH bonds and no hydrogen bonding; as a result it happens to fall in the middle. The boiling points of these compounds are close, they would be hard to predict, but they can be explained using hydrogen bonding and molecular weight.

10.29 Physical Properties: Section 2.9, 9.2, and 10.3

Pentane is non-polar and is incapable of hydrogen-bonding; this causes it to have the lowest boiling point. Butylamine has two NH bonds and diethylamine only one; the decreased ability to hydrogen bond gives diethylamine a lower boiling point. 1-butanol has an OH bond which is very polar and very effective in hydrogen bonding compared to amines; it has the highest boiling point as a result.

10.30 Basicity of Amines: Section 10.4A

(a) $CH_3CH_2NH_2$ + HCl $\longrightarrow$ $CH_3CH_2NH_3^+$ Cl$^-$

(b) $CH_3CH_2CH_2N(CH_3)_2$ + HBr $\longrightarrow$ $CH_3CH_2CH_2\overset{+}{N}H(CH_3)_2$ Br$^-$

(c) $(CH_3CH_2)_2NH$ + HNO_3 $\longrightarrow$ $(CH_3CH_2)_2NH_2^+ NO_3^-$

10.31 Basicity Constants: Section 10.4B

(a) $10^{-10} < 10^{-5} < 10^{-3}$; (b) $10 < 5 < 3$; (c) $11 < 6 < 3$; (d) $10^{-11} < 10^{-6} < 10^{-3}$

10.32 Acidity Constants: Sections 9.6A.1, 10.4C

(a) $10^{-12} < 10^{-8} < 10^{-3}$; (b) $10^{-3} < 10^{-8} < 10^{-12}$; (c) $12 < 8 < 3$;
(d) $13 < 9 < 4$; (e) $4 < 9 < 13$; (f) $10^{-13} < 10^{-9} < 10^{-4}$

(g) CH_3NH_2 + H_2O $\rightleftharpoons$ $CH_3NH_3^+$ + OH$^-$

(h) $CH_3NH_3^+$ + H_2O $\rightleftharpoons$ CH_3NH_2 + H_3O^+

10.33 Basicity of Amines: Section 10.4

(a) Propylamine is more basic than ammonia because the propyl group is an electron donating group and increases the electron availability around the nitrogen.   (b) Diethylamine is more basic than ethylamine because it has two electron-donating groups (the ethyls) whereas ethylamine has only one; the electron-donating groups increase electron availability.   (c) Aniline is an aromatic amine. Its non-bonding electron pairs are pulled into the benzene ring by resonance making them less available to acids. This makes it much less basic than cyclohexylamine which is an alkylamine.   (d) Both are a lot less basic than alkyl amines but the N-methylaniline has an electron-donating group on the nitrogen (methyl) which increases electron availability and thus basicity. (e) In N-phenylaniline, there are two benzene rings attached to the nitrogen. The non-bonding electron pair on nitrogen is drawn into both and this makes the compound less basic than aniline in which there is only one benzene ring. (f) Chlorine is an electron-withdrawing group. As such it makes the non-bonding electron pair on nitrogen less available. As a result, aniline is more basic.   (g)   Nitro groups are electron-withdrawing groups; they decrease electron availability and basicity.   There are two nitro groups on 2,4-dinitroaniline so it is less basic than p-nitroaniline where there is only one. (h) Chlorine withdraws electrons and decreases basicity; 2-chloropropanamine is less basic than propanamine for this reason. (i) Both of these compounds have a chlorine which is an electron-withdrawing group and which decreases basicity. In 3-chloropropanamine the chlorine is further away from the amine group than in 2-chloropropanamine and, because of this, it has a diminished effect. Thus 3-chloropropanamine is more basic.

10.34 Acidity and Basicity of Phenol and Aniline: Sections 9.6A.2, and 10.4B.3

Phenols are more acidic than alcohols because one of the non-bonding electron pairs on oxygen is drawn into the benzene ring by resonance. This stabilizes the phenoxide ion that is formed upon ionization and thus the acidity of phenol is enhanced by the phenomenon. This same withdrawal of electrons by the benzene ring stabilizes aniline and decreases the availability of the non-bonding electron pair on nitrogen. Both effects decrease the basicity of aniline relative to alkyl amines.

The withdrawal of electrons into the benzene ring makes both aniline and phenol more electron-rich. In electrophilic aromatic substitution, the ring is attacked by a positive electrophile; the more negative the ring, the more readily it reacts with an electrophile.

10.35 Alkylation of Amines: Section 10.5A

(a) $CH_3(CH_2)_4CH_2NH_2$ + 3 CH_3Br $\xrightarrow{Na_2CO_3}$ $CH_3(CH_2)_4CH_2N(CH_3)_3{}^+$ Br^-

(b) $CH_3CH_2CH_2NHCH_3$ + 2 CH_3CH_2Cl $\xrightarrow{Na_2CO_3}$ $CH_3CH_2CH_2\overset{+}{\underset{\underset{CH_3}{|}}{N}}(CH_2CH_3)_2$ Cl^-

(c) $\langle$O$\rangle$—$N(CH_2CH_3)_2$ + CH_3I $\longrightarrow$ $\langle$O$\rangle$—$\overset{+}{\underset{\underset{CH_3}{|}}{N}}(CH_2CH_3)_2$ I^-

(d) NH_3 + 4 CH_3Cl $\xrightarrow{Na_2CO_3}$ $(CH_3)_4N^+$ Cl^-

(e) $(CH_3)_3N$ + $CH_3(CH_2)_6CH_2Br$ $\longrightarrow$ $CH_3(CH_2)_6CH_2N(CH_3)_3{}^+$ Br^-

10.36 Alkylation of Amines: Section 10.5A

To make octanamine from bromooctane and ammonia, it is best to have ammonia in excess. This increases the probability that the bromooctane will only react with ammonia; the probability of it replacing a hydrogen on octanamine as it forms is decreased since there will be so much ammonia compared to octanamine. After the reaction, it is relatively easy to remove the excess ammonia because of its volatility.

10.37 S$_N$2 Alkylation Mechanism

transition state

10.38　Reduction of Nitro Compounds:　Section 10.5B.1

(a)　CH_3CH_2—⟨benzene ring⟩—NH_2

(b)　⟨benzene ring with Cl⟩—NH_2

10.39　Reduction of Nitro Compounds:　Sections 6.4 and 10.5B.1

(a)　Br—⟨benzene ring⟩—NO_2　$\xrightarrow[\text{2. NaOH}]{\text{1. Sn/HCl}}$　Br—⟨benzene ring⟩—NH_2

(b)　⟨benzene⟩　$\xrightarrow[\text{H}_2\text{SO}_4]{\text{HNO}_3}$　⟨benzene ring with NO_2⟩　$\xrightarrow[\text{FeBr}_3]{\text{Br}_2}$　⟨benzene ring with NO_2 and Br⟩　$\xrightarrow[\text{2. NaOH}]{\text{1. Sn/HCl}}$　⟨benzene ring with NH_2 and Br⟩

(c)　⟨benzene⟩　$\xrightarrow[\text{AlCl}_3]{\text{CH}_3\text{Cl}}$　⟨benzene ring with CH_3⟩　$\xrightarrow[\text{H}_2\text{SO}_4]{\text{HNO}_3}$　⟨benzene ring with CH_3 and NO_2⟩　$\xrightarrow[\text{2. NaOH}]{\text{1. Sn/HCl}}$　⟨benzene ring with CH_3 and NH_2⟩

10.40　Reduction of Nitriles: Sections 8.4 and 10.5B.2

$CH_3(CH_2)_4CH_2Br$　$\xrightarrow{\text{NaCN}}$　$CH_3(CH_2)_4CH_2CN$　$\xrightarrow[\text{Ni}]{2\,\text{H}_2}$　$CH_3(CH_2)_4CH_2CH_2NH_2$

10.41　Reduction of Amides:　Section 10.5B.3

(a)　$CH_3(CH_2)_4\overset{\text{O}}{\overset{\|}{C}}NH_2$　$\xrightarrow[\text{2. H}_2\text{O}]{\text{1. LiAlH}_4}$　$CH_3(CH_2)_4CH_2NH_2$

(b)　$CH_3CH_2CH_2\overset{\text{O}}{\overset{\|}{C}}NHCH_2CH_2CH_3$　$\xrightarrow[\text{2. H}_2\text{O}]{\text{1. LiAlH}_4}$　$CH_3CH_2CH_2CH_2NHCH_2CH_2CH_3$

(c)　$CH_3CH_2CH_2CH_2NH\overset{\text{O}}{\overset{\|}{C}}CH_2CH_3$　$\xrightarrow[\text{2. H}_2\text{O}]{\text{1. LiAlH}_4}$　$CH_3CH_2CH_2CH_2NHCH_2CH_2CH_3$

10.42　Reductions to form Amines:　Section 10.5B

(a)　$H_2N\overset{\text{O}}{\overset{\|}{C}}(CH_2)_4\overset{\text{O}}{\overset{\|}{C}}NH_2$　$\xrightarrow[\text{2. H}_2\text{O}]{\text{1. LiAlH}_4}$　$H_2NCH_2(CH_2)_4CH_2NH_2$

(b) $ClCH_2CH=CHCH_2Cl \xrightarrow{\text{2 NaCN}} NCCH_2CH=CHCH_2CN$

$H_2NCH_2(CH_2)_4CH_2NH_2 \xleftarrow{\text{5 H}_2 \; \text{Ni}}$

10.43 Formation of Amides: Section 10.6A

(a) $CH_3CH_2\overset{\overset{\displaystyle O}{\|}}{C}\underset{\underset{\displaystyle CH_3}{|}}{N}CH_2CH_3$ (b)   (c) $CH_3CH_2CH_2\overset{\overset{\displaystyle O}{\|}}{C}NHCH_3$

(d)

10.44 Reduction of Amides: Section 10.5B.3

(a) $CH_3CH_2CH_2\underset{\underset{\displaystyle CH_3}{|}}{N}CH_2CH_3$ (b) $\text{—}CH_2N$ (c) $CH_3CH_2CH_2CH_2NHCH_3$

(d) $Br\text{—}\text{—}CH_2NH\text{—}$

10.45 Reduction of Amides: Section 10.5B.3

$CH_3CH_2CH_2CH_2\overset{\overset{\displaystyle O}{\|}}{C}Cl \xrightarrow{\text{NH}_3} CH_3CH_2CH_2CH_2\overset{\overset{\displaystyle O}{\|}}{C}NH_2 \xrightarrow[\text{2. H}_2\text{O}]{\text{1. LiAlH}_4} CH_3CH_2CH_2CH_2CH_2\blacksquare$

10.46 Sulfonamides: Section 10.6B

(a) $SO_2NH(CH_2)_5CH_3$ (b) $SO_2N(CH_3)_2$ (c) SO_2N

10.47 Reactions of Amines: Sections 10.5A and 10.6

	I	II	III
Reagent	$CH_3CH_2CH_2NH_2$	$CH_3CH_2NHCH_3$	$\underset{}{\overset{CH_3}{|}}CH_3NCH_3$
a) HCl	$CH_3CH_2CH_2\overset{+}{N}H_3\overset{-}{Cl}$	$CH_3CH_2\underset{\underset{\displaystyle H+}{|}}{N}HCH_3 \; \overset{-}{Cl}$	$CH_3\underset{\underset{\displaystyle H}{|}}{\overset{\overset{\displaystyle CH_3}{|}}{N}}\overset{+}{C}H_3 \; \overset{-}{Cl}$

b) H_2SO_4 $(CH_3CH_2CH_2\overset{+}{N}H_3)_2SO_4^{2-}$ $(CH_3CH_2\overset{+}{N}HCH_3)_2SO_4^{2-}$ $\left(CH_3\overset{\underset{\displaystyle CH_3}{|}}{\overset{|}{N}}\overset{+}{H}\right)_2SO_4^{2-}$

 H CH_3

c) excess
    CH_3Br $CH_3CH_2CH_2\overset{\underset{\displaystyle CH_3}{|}}{\overset{\overset{\displaystyle CH_3}{|}}{N}}\overset{+}{C}H_3\ Br^-$ $CH_3CH_2\overset{\underset{\displaystyle CH_3}{|}}{\overset{\overset{\displaystyle CH_3}{|}}{N}}\overset{+}{C}H_3\ Br^-$ $CH_3\overset{\underset{\displaystyle CH_3}{|}}{\overset{\overset{\displaystyle CH_3}{|}}{N}}\overset{+}{C}H_3\ Br^-$

d) $CH_3\overset{\overset{\displaystyle O}{\|}}{C}Cl$ $CH_3CH_2CH_2NH\overset{\overset{\displaystyle O}{\|}}{C}CH_3$ $CH_3\overset{\overset{\displaystyle O}{\|}}{C}\overset{\underset{\displaystyle CH_3}{|}}{N}CH_2CH_3$ No Reaction

e) ⬡–SO_2Cl ⬡–$SO_2NHCH_2CH_2CH_3$ ⬡–$SO_2\overset{\underset{\displaystyle CH_3}{|}}{N}CH_2CH_3$ No Reaction

10.48 Reactions of Diazonium Salts: Section 10.7

⬡ NH_2 / CH_3 $\xrightarrow[\substack{HCl\\O°C}]{NaNO_2}$ ⬡ $N_2^+Cl^-$ / CH_3

a) ⬡ Cl/CH_3 b) ⬡ Br/CH_3 c) ⬡ I/CH_3 d) ⬡ CN/CH_3 e) ⬡ OH/CH_3 f) ⬡ F/CH_3 g) ⬡ H/CH_3

h) CH_3–⬡–N=N–⬡–OH   i) CH_3–⬡–N=N–⬡–$N(CH_3)_2$

10.49 Syntheses Using Diazonium Salts: Sections 6.4 and 10.7

a) ⬡ NH_2/CH_3 $\xrightarrow[\substack{HCl\\O°C}]{NaNO_2}$ ⬡ $N_2^+Cl^-$/CH_3 $\xrightarrow{HBF_4}$ ⬡ F/CH_3

b) ⬡ NH_2/Br $\xrightarrow[\substack{HCl\\O°C}]{NaNO_2}$ ⬡ $N_2^+Cl^-$/Br $\xrightarrow{KI}$ ⬡ I/Br

c)

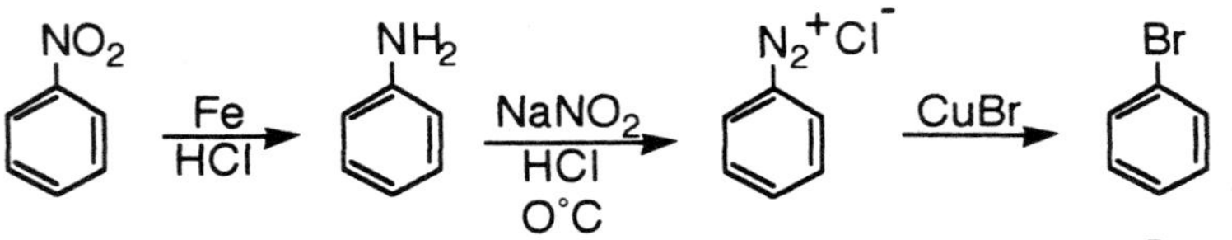

d)

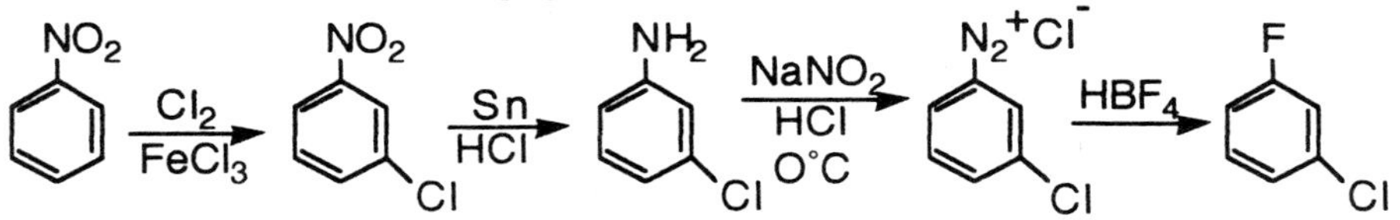

e)

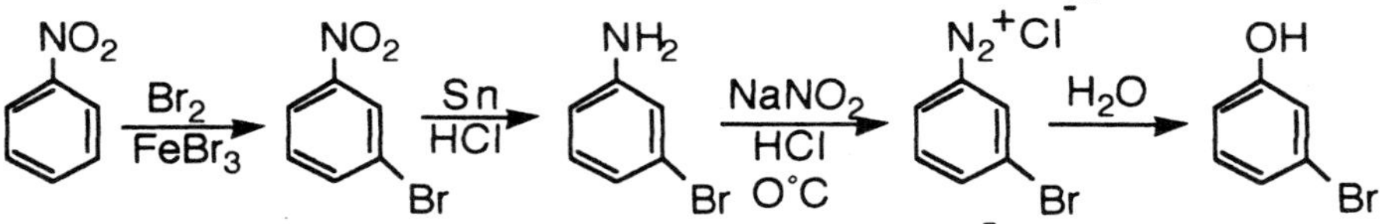

f)

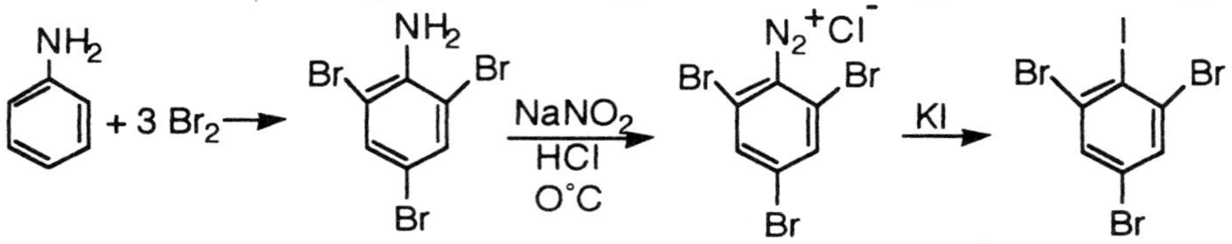

g)

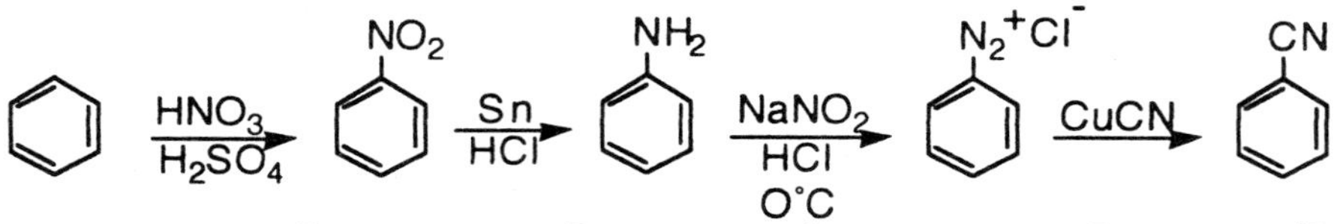

h)

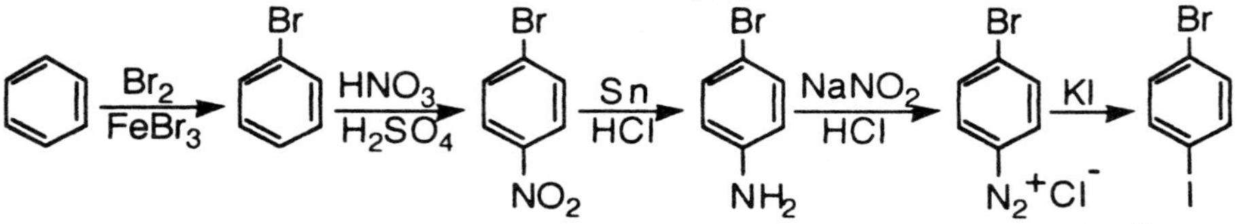

i)

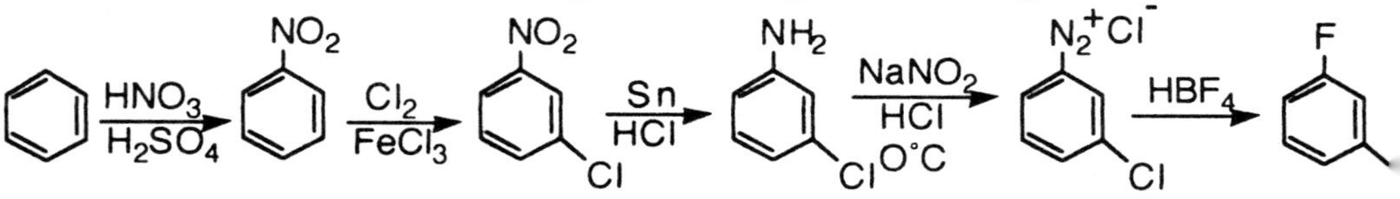

10.50   Diazonium Salts-Coupling Reactions: Section 10.7C

a)

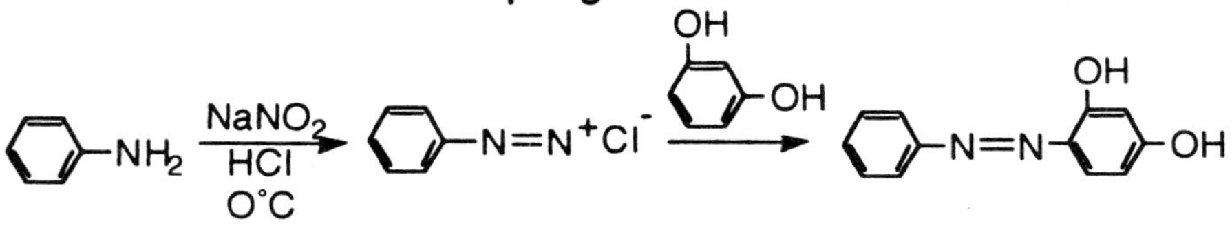

b)

10.51 Basicity of Aromatic Amines: Section 10.8A

Both compounds need six pi-electrons in the ring system and a p-orbital on each ring atom to be aromatic. In each case there are two double bonds which provide four pi-electrons and four p-orbitals. The final p-orbital and two pi-electrons are provided by a nitrogen in each case. Visualize the nitrogen not involved in a double bond as housing its non-bonding electron pair in a p-orbital that overlaps with the others in the ring. Because of this overlap, this non-bonding electron pair is not as available to acids and the basicity is drastically diminished. In imidazole, there is a second nitrogen. Its non-bonding electron pair is not part of the pi-system as the double bond provides the p-orbital at that location and there already exists the aromatic sextet of electrons. Consequently, this electron-pair is much more available than that of the other nitrogen or the pair on the nitrogen in pyrrole and imidazole is four million times more basic than pyrrole.

10.52 Aromaticity of Heterocyclic Compounds: Section 10.8A

Both compounds are cyclic, planar, have a p-orbital on each ring atom, and have six pi-electrons. Four of the electrons and four of the p-orbitals come from the double bonds. The last p-orbital and the final two pi-electrons are a result of one of the non-bonding electron pairs on oxygen and sulfur existing in a p-orbital that overlaps with the others and completes the aromatic system.

10.53 Dyes: Connections 10.4

Chromophore and auxochrome groups are listed early in the Connections essay. Look for these along with extensive conjugation in the structures of dyes presented.

11

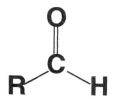 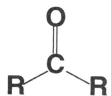

ALDEHYDES AND KETONES

CHAPTER SUMMARY

Aldehydes and **ketones** both have a **carbonyl group** (carbon-oxygen double bond); aldehydes have at least one carbon bonded to the carbonyl group, whereas in ketones the carbonyl is bonded to two carbons. In IUPAC nomenclature, the suffix for aldehydes is **-al** and for ketones, **-one**. The prefix for both is **oxo**. In polyfunctional compounds, the group highest in the following sequence is designated with a suffix and the others with prefixes: **aldehyde > ketone > alcohol > amine**. In naming, first determine the longest continuous carbon chain; insert the suffix an, en, or yn to designate all single bonds or one or more double bonds or triple bonds respectively; use the suffix ending for the functional group highest in the above sequence; name all other groups with prefixes; number the carbon chain to give the lowest number to the functional group.

Aldehydes and ketones are chemically distinguished by oxidation. Aldehydes are easily oxidized and ketones are not. Common tests include the **Tollens' silver mirror test** and **Benedicts** and **Fehlings tests.** In each test, aldehydes are oxidized to carboxylic acids and ketones are not oxidized.

The carbonyl group of aldehydes and ketones is reactive because it is polar, there is a pi bond, there are two non-bonding electron-pairs on oxygen, and it has a flat, open structure that makes it accessible to other reagents. Because of its polarity, the carbonyl group attracts nucleophiles to the partially positive carbon and electrophiles to the electron-rich oxygen. Because aldehydes have only one alkyl group compared to two for ketones (alkyl groups are large relative to hydrogen and hinder nucleophilic attack), they tend to be more reactive than ketones.

Addition is the characteristic reaction of aldehydes and ketones. When unsymmetrical reagents add, the positive part bonds to the partially negative carbonyl oxygen and the negative part bonds to the partially positive carbon. The reactions are not as simple as those of alkenes since the product of straight addition is often unstable and either exists in equilibrium with the original aldehyde or ketone or reacts further to form a more stable substance. Hydrogen and hydrogen cyanide usually form stable addition products. The addition products from water and hydrogen halides are in equilibrium with the original aldehyde or ketone; the equilibrium usually favors the starting materials. Most other adding reagents form an intermediate addition product that further reacts to form a stable substance.

Nucleophilic addition is the characteristic mechanism for addition reactions of aldehydes and ketones. It can be **base-initiated** in which a negative or neutral nucleophile attacks the carbonyl carbon generating a negative carbonyl oxygen that is subsequently neutralized. In the **acid-initiated mechanism**, hydrogen ion bonds to the carbonyl oxygen; a carbocation results which is neutralized by the nucleophile.

Hydrogen cyanide adds to aldehydes and ketones to form a simple addition product called a **cyanohydrin**. The mechanism is **base-initiated nucleophilic addition**.

Aldehydes and ketones undergo **catalytic hydrogenation** using hydrogen gas under pressure and a metal catalyst such as nickel. Primary alcohols result from the hydrogenation of aldehydes and secondary alcohols are prepared from ketones. Aldehydes and ketones can be **reduced** using **lithium aluminum hydride**. The reaction is **base-initiated** with hydride ion as the nucleophile. One mole of lithium aluminum hydride reduces four moles of aldehyde or ketone; the reaction mixture is then acidified to produce the neutral alcohol. **Sodium borohydride** can be used as well in these reductions.

Grignard reagents are prepared from the reaction of alkyl halides with magnesium in ether solvent. The alkyl group assumes a negative character and attacks carbonyl carbons in a **base-initiated nucleophilic addition.** Neutralization of the negative intermediate results in the preparation of an alcohol. Grignard reagents react with formaldehyde to form primary alcohols, with other aldehydes to form secondary alcohols, and with ketones to produce tertiary alcohols. Grignard reagents can react with other reagents also. With ethylene oxide, primary alcohols can be synthesized and with carbon dioxide, carboxylic acids are produced. Other **organometallic compounds** can be used in similar ways.

Aldehydes and ketones react with alcohols by **acid-initiated nucleophilic addition** to form **hemiacetals** which are usually unstable. Reaction with a second mole of alcohol produces an **acetal.**

Primary amines react with aldehydes and ketones to form **imines** by nucleophilic addition. Many of the products are crystalline derivatives which have been used to characterize the original carbonyl compounds.

Alpha hydrogens are hydrogens on carbons directly attached to a carbonyl group. They are **weakly acidic** and can be abstracted by base to form a **carbanion.** The carbanion is called an **enolate ion** and is resonance stabilized. Neutralization of the enolate ion results in an **enol**, a compound in which an alcohol group is directly bonded to a carbon involved in a carbon-carbon double bond. The enol is in equilibrium with the original aldehyde or ketone in an equilibrium referred to as **keto-enol tautomerism.** The equilibrium usually favors the keto form.

The **aldol condensation** involves the reaction of two molecules of an aldehyde or ketone that has alpha hydrogens. Abstraction of an alpha hydrogen by base produces a carbanion which attacks the carbonyl carbon of the other molecule by **base-initiated nucleophilic addition**; an alcohol group is formed. Often the alcohol dehydrates to form the final product, an unsaturated aldehyde or ketone. In a **crossed aldol condensation**, a carbonyl compound with alpha hydrogens reacts with one without alpha hydrogens.

Connections 11.1 is about formaldehyde and synthetic polymers.

SOLUTIONS TO PROBLEMS

11.1 Nomenclature of Aldehydes and Ketones

(a) hexanal; (b) 2-octanone; (c) cyclohexanone; (d) 4-bromo-2-pentanone;
(e) 4,4-dimethylpentanal

11.2 Nomenclature of Multifunctional Aldehydes and Ketones

(a) 1,3,5-cyclohexantrione; (b) 5-hydroxyhexanal; (c) 7-bromo-3-hydroxy-7-methyl-5-oxooctanal; (d) 6-amino-4-hydroxy-2-heptanone

11.3 Nomenclature of Unsaturated Aldehydes and Ketones

(a) 3-butynal; (b) 3-cyclopenten-1-one; (c) 7-hydroxy-2-methyl-4-oxo-5-octenal

11.4 Common Nomenclature of Aldehydes and Ketones

a) CH_3CHCH with CH_3 substituent, $=O$

b) CH_3CHCH with Cl substituent, $=O$

c) $CH_3CCH_2CH_2CH_3$, $=O$

d) CH_3C—(phenyl), $=O$

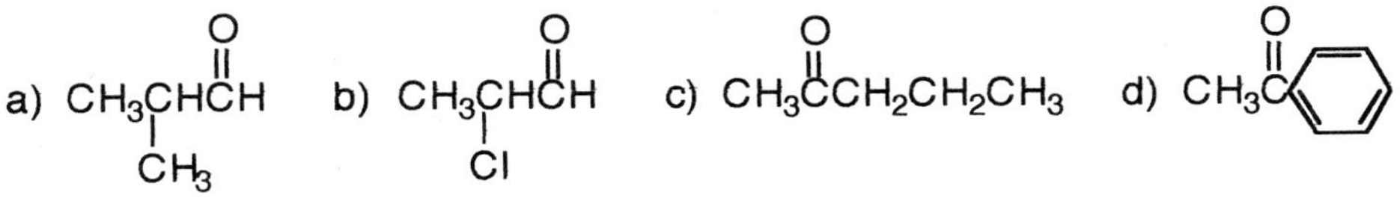

11.5 Oxidation of Aldehydes

a) CH_3CH_2CH ($=O$) $\xrightarrow[\text{Tollens}]{Ag(NH_3)_2OH}$ $CH_3CH_2CO_2H$ (neutralized) Ag mirror

CH_3CCH_3 ($=O$) $\xrightarrow{Ag(NH_3)_2OH}$ No reaction

b) CH_3CH_2CH ($=O$) $\xrightarrow[\text{Benedicts}]{Cu(OH)_2}$ $CH_3CH_2CO_2H$ (neutralized) Cu_2O red ppt

CH_3CCH_3 ($=O$) $\xrightarrow{\text{Benedicts}}$ No reaction

11.6 Addition of Water to Aldehydes and Ketones

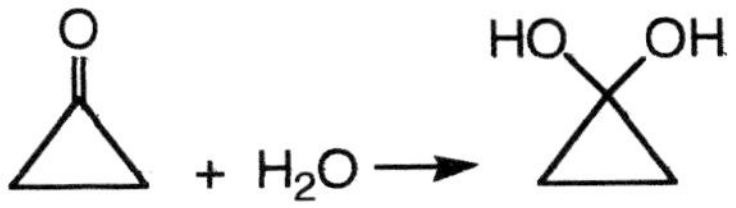

(cyclopropyl ketone, $=O$) $+ H_2O \longrightarrow$ (cyclopropane with HO and OH)

11.7 Nucleophilic Addition

Acid-initiated Nucleophilic Addition

Base-initiated Nucleophilic Addition

11.8 Addition of HCN to Aldehydes and Ketones

a) $CH_3\overset{O}{\overset{\|}{C}}CH_2CH_3 + HCN \longrightarrow CH_3\overset{OH}{\underset{CN}{\overset{|}{C}}}CH_2CH_3$

$+ HCN \longrightarrow$

b) $CH_3\overset{:\overset{..}{O}:}{\overset{\|}{C}H} \xrightarrow{:CN:} CH_3\overset{:\overset{..}{O}:^-}{\underset{CN}{\overset{|}{C}H}} \xrightarrow{HCN} CH_3\overset{:\overset{..}{O}H}{\underset{CN}{\overset{|}{C}H}} + CN^-$

11.9 Catalytic Hydrogenation

$CH_3CH_2\overset{O}{\overset{\|}{C}H} + H_2 \xrightarrow{Ni} CH_3CH_2CH_2OH$

$CH_3\overset{O}{\overset{\|}{C}}CH_3 + H_2 \xrightarrow{Ni} CH_3\overset{OH}{\overset{|}{C}H}CH_3$

3^0 alcohols can't be prepared in this way since a C=O can't have three alkyl groups attached to the carbon (there is no aldehyde or ketone precursor).

11.10 Lithium Aluminum Hydride Reductions

a) $CH_3CH_2\overset{O}{\overset{\|}{C}H} \xrightarrow{LiAlH_4} \xrightarrow[H^+]{H_2O} CH_3CH_2CH_2OH$

$CH_3\overset{O}{\overset{\|}{C}}CH_3 \xrightarrow{LiAlH_4} \xrightarrow[H^+]{H_2O} CH_3\overset{OH}{\overset{|}{C}H}CH_3$

b) $4\ CH_3\overset{\overset{\displaystyle \ddot{O}}{\|}}{C}H + LiAlH_4 \longrightarrow \left(CH_3\overset{\overset{\displaystyle :\!\ddot{O}\!:^-}{}}{\underset{\underset{\displaystyle H}{|}}{C}}H \right)_4 \overset{3++}{\underset{AlLi}{}} \xrightarrow{H_2O} CH_3\overset{\overset{\displaystyle :\!\ddot{O}H}{}}{\underset{\underset{\displaystyle H}{|}}{C}}H$

$$+\ LiOH\ +\ Al(OH)_3$$

11.11   Grignard Preparation of Alcohols

a) $CH_3Cl\ +\ Mg \xrightarrow{ether} CH_3MgCl$

b) $CH_3MgCl\ +\ H\overset{\overset{\displaystyle O}{\|}}{C}H \longrightarrow \xrightarrow[H^+]{H_2O} CH_3CH_2OH$

$CH_3MgCl\ +\ CH_3CH_2\overset{\overset{\displaystyle O}{\|}}{C}H \longrightarrow \xrightarrow[H^+]{H_2O} CH_3CH_2\overset{\overset{\displaystyle OH}{|}}{C}HCH_3$

$CH_3MgCl\ +\ CH_3\overset{\overset{\displaystyle O}{\|}}{C}CH_3 \longrightarrow \xrightarrow[H^+]{H_2O} CH_3\overset{\overset{\displaystyle OH}{|}}{\underset{\underset{\displaystyle CH_3}{|}}{C}}CH_3$

11.12   Grignard Reaction Mechanism

$H\overset{\overset{\displaystyle :\!\ddot{O}\!:}{\|}}{C}H + \overset{-\ \ +}{CH_3MgCl} \longrightarrow CH_3\overset{\overset{\displaystyle :\!\ddot{O}\!:^-\ \overset{+}{M}gCl}{}}{\underset{\underset{\displaystyle H}{|}}{C}}H \xrightarrow{H_2O} CH_3\overset{\overset{\displaystyle :\!\ddot{O}H}{}}{\underset{\underset{\displaystyle H}{|}}{C}}H + MgClOH$

11.13   Grignard Preparation of Alcohols

Method 1 $CH_3CH_2Cl \xrightarrow{Mg} CH_3CH_2MgCl$ $\overset{\overset{\displaystyle O}{\|}}{CH_3CCH_2CH_2CH_3}$

$CH_3\overset{\overset{\displaystyle OH}{|}}{\underset{\underset{\displaystyle CH_2CH_3}{|}}{C}}CH_2CH_2CH_3 \xleftarrow[H^+]{H_2O} CH_3\overset{\overset{\displaystyle OMgCl}{|}}{\underset{\underset{\displaystyle CH_2CH_3}{|}}{C}}CH_2CH_2CH_3$

Method 2 $CH_3I \xrightarrow{Mg} CH_3MgI$ $\overset{\overset{\displaystyle O}{\|}}{CH_3CH_2CCH_2CH_2CH_3}$

$CH_3CH_2\overset{\overset{\displaystyle OH}{|}}{\underset{\underset{\displaystyle CH_3}{|}}{C}}CH_2CH_2CH_3 \xleftarrow[H^+]{H_2O} CH_3CH_2\overset{\overset{\displaystyle OMgI}{|}}{\underset{\underset{\displaystyle CH_3}{|}}{C}}CH_2CH_2CH_3$

Method 3 $CH_3CH_2CH_2Br \xrightarrow{Mg} CH_3CH_2CH_2MgBr$

$$CH_3CH_2\overset{O}{\overset{\|}{C}}CH_3$$

$$\underset{\underset{CH_2CH_2CH_3}{|}}{CH_3CH_2\overset{OH}{\underset{|}{C}}CH_3} \xleftarrow[H^+]{H_2O} \underset{\underset{CH_2CH_2CH_3}{|}}{CH_3CH_2\overset{OMgBr}{\underset{|}{C}}CH_3}$$

11.14 Organometallic Compounds

Following is the preparation of the needed organometallic compounds. The Grignard shown could have been phenyllithium as well.

$$\text{Ph-Br} \xrightarrow[\text{Ether}]{\text{Mg}} \text{Ph-MgBr} \qquad \text{Ph-C}\equiv\text{CH} \xrightarrow{\text{NaNH}_2} \text{Ph-C}\equiv\text{CNa}$$

(a) $\text{Ph-MgBr} + \underset{CH_2-CH_2}{\overset{O}{\triangle}} \longrightarrow \text{Ph-CH}_2CH_2OMgBr \xrightarrow{H_2O/H^+} \text{Ph-CH}_2($

(b) $\text{Ph-MgBr} + O=C=O \longrightarrow \text{Ph-}\overset{O}{\overset{\|}{C}}OMgBr \xrightarrow{H_2O/H^+} \text{Ph-}\overset{O}{\overset{\|}{C}}OH$

(c) $\text{Ph-C}\equiv\text{CNa} + \overset{O}{\overset{\|}{HC}}\text{-Ph} \longrightarrow \text{Ph-C}\equiv\text{C-}\underset{H}{\overset{ONa}{\underset{|}{\overset{|}{C}}}}\text{-Ph} \xrightarrow[H^+]{H_2O} \text{Ph-C}\equiv\text{C-}\underset{H}{\overset{OH}{\underset{|}{\overset{|}{C}}}}$

11.15 Hemiacetals and Acetals

(a) $\underset{OCH_2CH_3}{\overset{OH}{\underset{|}{\overset{|}{Ph-CH}}}}$

(b) $\underset{OCH_2CH_3}{\overset{OCH_2CH_3}{\underset{|}{\overset{|}{Ph-CH}}}}$

11.16 Hemiacetal and Acetal Formation Reaction Mechanism

$$\text{C}_6\text{H}_5\text{-CH=O} \overset{\text{H}^+}{\rightleftharpoons} \text{C}_6\text{H}_5\text{-CH-OH}^+ \overset{\text{CH}_3\text{CH}_2\text{OH}}{\longleftarrow} \text{C}_6\text{H}_5\text{-CH(OH)(OCH}_2\text{CH}_3)\text{H}^+$$

$$\overset{\text{H}^+}{\text{C}_6\text{H}_5\text{-CH(OCH}_2\text{CH}_3)_2} \overset{\text{CH}_3\text{CH}_2\text{OH}}{\rightleftharpoons} \text{C}_6\text{H}_5\text{-CH}^+\text{-OCH}_2\text{CH}_3 \overset{-\text{H}_2\text{O}}{\rightleftharpoons} \overset{\text{H}^+}{\text{C}_6\text{H}_5\text{-CH(OH)(OCH}_2\text{CH}_3)}$$

$$\overset{-\text{H}^+}{\longrightarrow} \quad \text{C}_6\text{H}_5\text{-CH(OCH}_2\text{CH}_3)_2 \quad \textbf{acetal} \qquad \text{C}_6\text{H}_5\text{-CH(OH)(OCH}_2\text{CH}_3) \quad \textbf{hemiacetal}$$

11.17 Acetal Formation

$$\text{cyclohexanone} = \text{O} \ + \ \text{HOCH}_2\text{CH}_2\text{OH}$$

11.18 Hydrolysis of Acetals

$$\underset{\text{O-CH}_2}{\overset{\text{O-CH}_2}{\bigcirc}} \overset{\text{H}^+}{\rightleftharpoons} \underset{\text{O-CH}_2}{\overset{\text{O-CH}_2}{\bigcirc}}\text{H}^+ \rightleftharpoons \bigcirc^+ \underset{\text{O-CH}_2}{\overset{\text{HO-CH}_2}{}} \overset{\text{H}_2\text{O}}{\longleftarrow}$$

$$\bigcirc=\text{O} \underset{-\text{H}^+}{\rightleftharpoons} \bigcirc^+\text{-OH} \rightleftharpoons \underset{\text{O-CH}_2}{\overset{\text{HO} \quad \text{H-O-CH}_2}{\bigcirc}}\text{H}^+ \rightleftharpoons \underset{\text{O-CH}_2}{\overset{\text{HO} \quad \text{H-O-CH}_2}{\bigcirc}}$$

$$+ \quad \text{HOCH}_2\text{CH}_2\text{OH}$$

11.19 Reaction of Aldehydes and Ketones with Primary Amines

a) $\text{C}_6\text{H}_5\text{-CH=O} + \text{H}_2\text{NNH-}\text{C}_6\text{H}_3(\text{NO}_2)_2 \longrightarrow \text{C}_6\text{H}_5\text{-CH=NNH-}\text{C}_6\text{H}_3(\text{NO}_2)_2 + \text{H}_2\text{O}$

b) 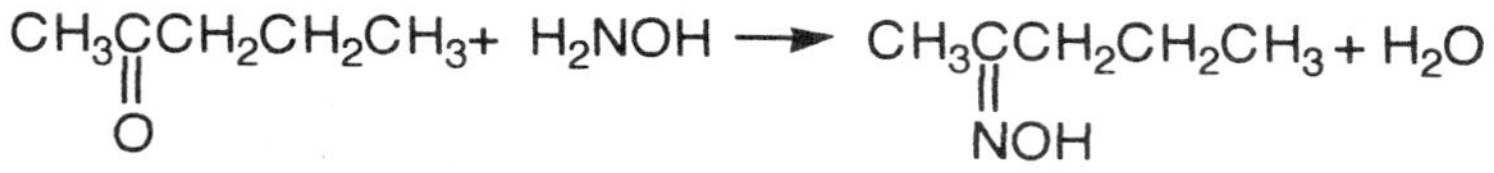 $CH_3CCH_2CH_2CH_3 + H_2NOH \longrightarrow CH_3CCH_2CH_2CH_3 + H_2O$
      ‖   ‖
      O   NOH

11.20   Keto and Enol Forms

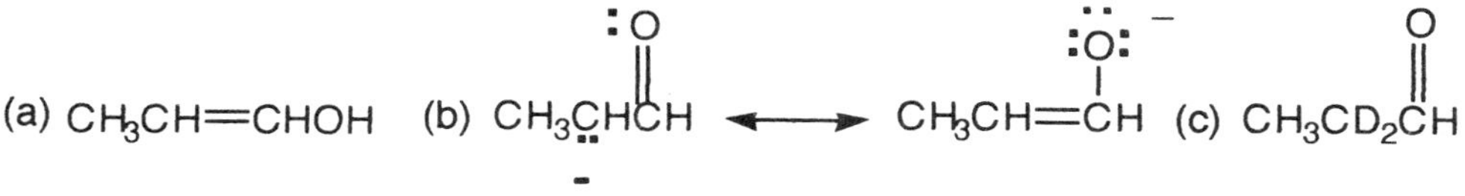

(a) $CH_3CH=CHOH$ (b) $CH_3CHCH \longleftrightarrow CH_3CH=CH$ (c) CH_3CD_2CH

11.21   Aldol Condensation

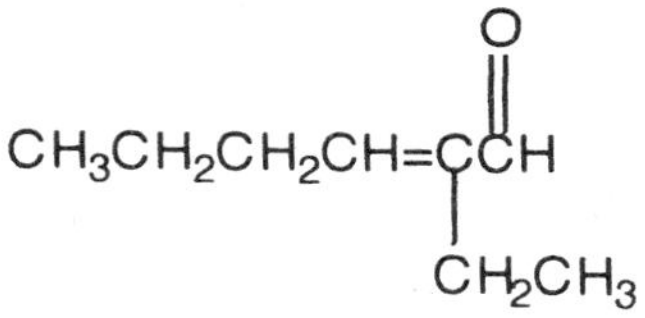

$$CH_3CH_2CH_2CH=CCH$$
$$\overset{|}{CH_2CH_3}$$

11.22   Aldol Condensation Mechanism

Only the aldol mechanism is shown; not the subsequent dehydration.

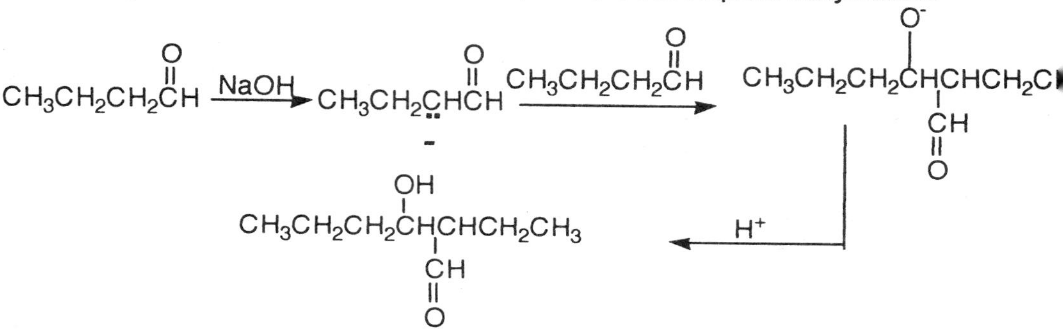

$$CH_3CH_2CH_2CH \xrightarrow{NaOH} CH_3CH_2CHCH \longrightarrow CH_3CH_2CH_2CHCHCH_2CH_3$$

$$CH_3CH_2CH_2CHCHCH_2CH_3 \xleftarrow{\ H^+\ }$$

(aldol intermediates with OH and CH=O substituents shown)

11.23   Crossed Aldol Condensation

Aldol Product

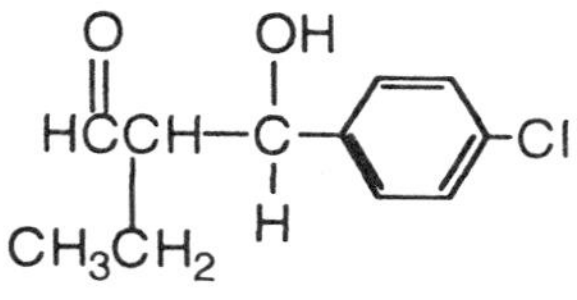

$$HCCH-C-\bigcirc-Cl$$
$$\overset{|}{CH_3CH_2}\ \overset{|}{H}$$

Dehydration Product

$$HCC=C-\bigcirc-Cl$$
$$\overset{|}{CH_3CH_2}\ \overset{|}{H}$$

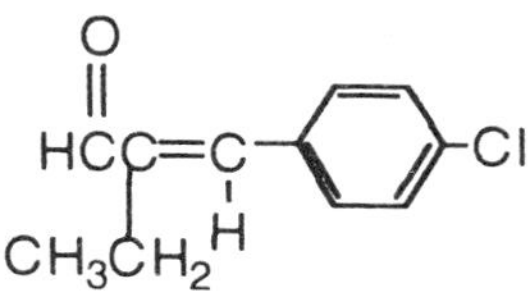

11.24　Mechanism of Aldol Condensation

Only the aldol condensation mechanism is shown, not the dehydration.

$$CH_3CH_2CH_2\overset{\overset{\displaystyle O}{\|}}{C}H \xrightarrow{\;OH^-\;} CH_3CH_2\overset{\overset{\displaystyle O}{\|}}{C}H\overset{\displaystyle ..}{C}H \longrightarrow$$

$$H\overset{\overset{\displaystyle O}{\|}}{C}CH-\overset{\overset{\displaystyle O^-}{|}}{\underset{\underset{\displaystyle H}{|}}{C}}-\underset{\displaystyle CH_3CH_2}{} -\!\!\!\bigcirc\!\!\!-Cl \xrightarrow{\;H^+\;} H\overset{\overset{\displaystyle O}{\|}}{C}CH-\overset{\overset{\displaystyle OH}{|}}{\underset{\underset{\displaystyle H}{|}}{C}}-\!\!\!\bigcirc\!\!\!-Cl$$

$$\underset{\displaystyle CH_3CH_2}{}\qquad\qquad\qquad\underset{\displaystyle CH_3CH_2}{}$$

11.25　IUPAC Nomenclature of Aldehydes:　Section 11.2A

(a) decanal;　(b) butanal;　(c) 4-methylpentanal;　(d) 5-ethyl-3-methylheptanal;　(e) p-methylbenzaldehyde;　(f) 1,6-hexandial

11.26　IUPAC Nomenclature of Ketones:　Section 11.2A

(a) 2-pentanone;　(b) 3-heptanone;　(c) 2-methyl-4-heptanone;

(d) 4-methylcyclohexanone;　(e) 2,4,6-heptantrione;　(f) 4,5-dibromo-1,3-cyclopentandione

11.27　IUPAC Nomenclature of Aldehydes and Ketones:

　　　Section 11.2A

(a) propanal;　(b) 3-pentanone;　(c) 3-oxobutanal;　(d) 3,5-dihydroxyhexanal;　(e) 4-amino-2-pentanone;　(f) 5-hydroxy-3-oxohexanal;

(g) 4-hydroxy-2,6-octandione;　(h) 3-amino-5-methylhexanal;

(i) 4-hydroxycyclohexanone

11.28　IUPAC Nomenclature of Aldehydes and Ketones:

　　　Section 11.2A

(a) 3-butenal;　(b) 1-N,N-dimethylamino-3-butyn-2-one;　(c) 2,4-hexadienal;

(d) 6-hydroxy-3-oxo-4,7-octadienal;　(e) 3-hepten-2,5-dione;

(f) 4-oxo-2-hexen-5-ynal

11.29　IUPAC Nomenclature:　Section 11.2A

a) $CH_3CH_2\overset{\overset{\displaystyle O}{\|}}{C}CH_2CH_2CH_2CH_3$　　　b) $CH_3CH_2CH_2CH_2CH_2CH_2CH_2\overset{\overset{\displaystyle O}{\|}}{C}H$

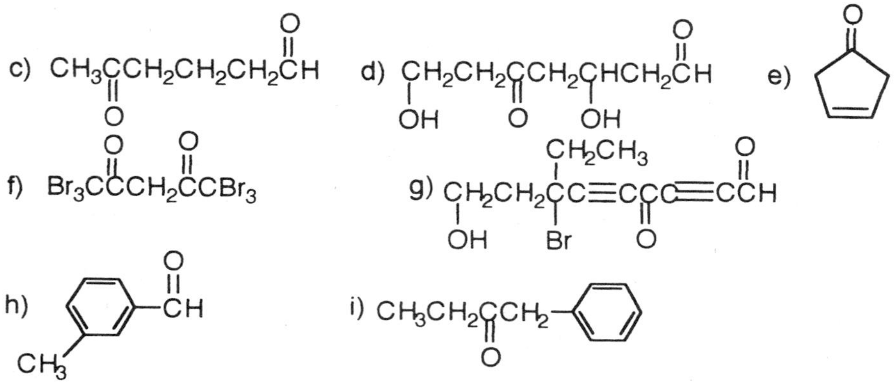

11.30 Common Nomenclature

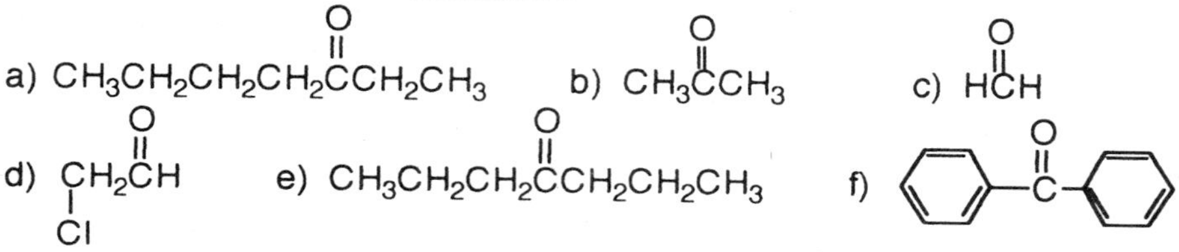

11.31 Preparations of Aldehydes and Ketones: Section 11.3

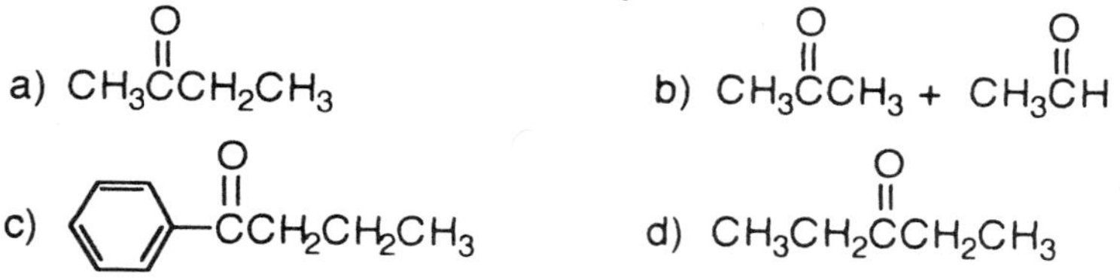

11.32 Preparations of Aldehydes and Ketones: Section 11.3

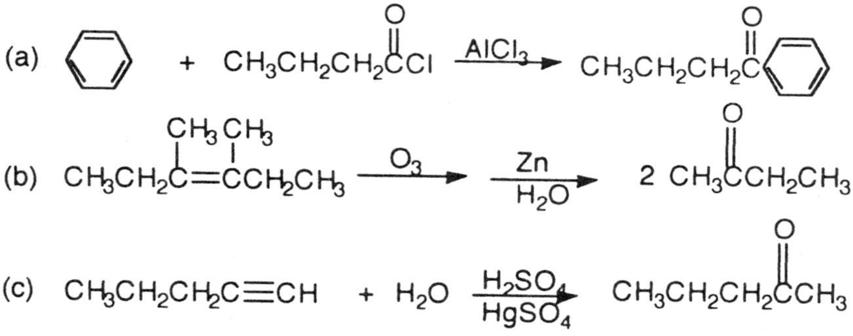

(d)

$$\text{OH} \quad \xrightarrow{\text{Na}_2\text{Cr}_2\text{O}_7} \quad$$

(with cyclopentane ring bearing OH and CH$_3$ → cyclopentanone bearing CH$_3$)

(e) $CH_3(CH_2)_8CH_2OH \xrightarrow{\text{PCC}} CH_3(CH_2)_8\overset{\text{O}}{\overset{\|}{C}}H$

11.33 Reactions of Aldehydes and Ketones: Section 11.5-11.6

	I. Products from $\phi\!-\!CHO$	II. Products from $\phi\!-\!COCH_3$
Reagent		
a) Tollens' Reagent $Ag(NH_3)_2OH$	$\phi\!-\!CONH_4 + Ag$	No reaction
b) Benedict's Reagent Cu^{2+}, NaOH	$\phi\!-\!CONa$	No reaction
c) HCN	$\phi\!-\!\underset{H}{\overset{OH}{C}}\!-\!CN$	$\phi\!-\!\underset{CH_3}{\overset{OH}{C}}\!-\!CN$
d) H_2/Ni	$\phi\!-\!CH_2OH$	$\phi\!-\!\underset{OH}{CH}CH_3$
e) $LiAlH_4$, then H_2O	$\phi\!-\!CH_2OH$	$\phi\!-\!\underset{OH}{CH}CH_3$
f) CH_3MgCl, then H_2O	$\phi\!-\!\underset{OH}{CH}CH_3$	$\phi\!-\!\underset{OH}{\overset{CH_3}{C}}\!-\!CH_3$
g) $\phi\!-\!MgBr$, then H_2O	$\phi\!-\!\underset{OH}{CH}\!-\!\phi$	$\phi\!-\!\underset{OH}{\overset{\phi}{C}}\!-\!CH_3$
h) $\phi\!-\!NHNH_2$	$\phi\!-\!CH\!=\!NNH\!-\!\phi$	$\phi\!-\!\underset{CH_3}{C}\!=\!NNH\!-\!\phi$
i) H_2NOH	$\phi\!-\!CH\!=\!NOH$	$\phi\!-\!\underset{CH_3}{C}\!=\!NOH$

j) CH$_3$OH/H$^+$

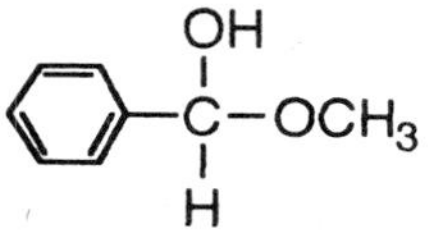

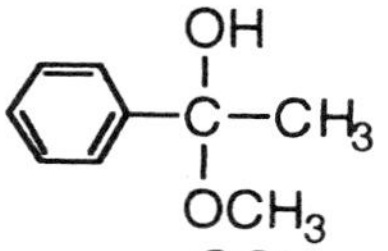

k) 2 CH$_3$OH/H$^+$

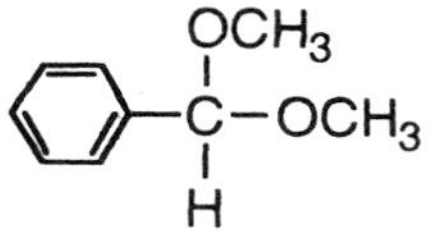

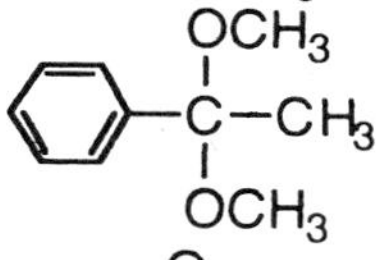

l) D$_2$O, NaOD   No Reaction

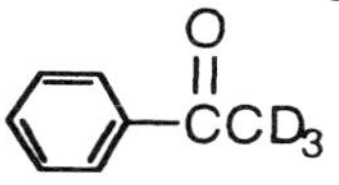

11.34 Grignard Synthesis of Alcohols: Section 11.5F.3

a) CH$_3$CH$_2$CH$_2$Br $\xrightarrow{\text{Mg}}$ CH$_3$CH$_2$CH$_2$MgBr $\xrightarrow{\text{HCH}}$

CH$_3$CH$_2$CH$_2$CH$_2$OH $\xleftarrow[\text{H}^+]{\text{H}_2\text{O}}$ CH$_3$CH$_2$CH$_2$CH$_2$OMgBr

b) CH$_3$Br $\xrightarrow{\text{Mg}}$ CH$_3$MgBr $\xrightarrow{\text{CH}_3\text{CH}}$ CH$_3$CHCH$_3$ $\xrightarrow[\text{H}^+]{\text{H}_2\text{O}}$ CH$_3$CHCH$_3$
OMgBr   OH

c) **Method 1**

CH$_3$CH$_2$CH$_2$Cl + Mg $\longrightarrow$ CH$_3$CH$_2$CH$_2$MgCl

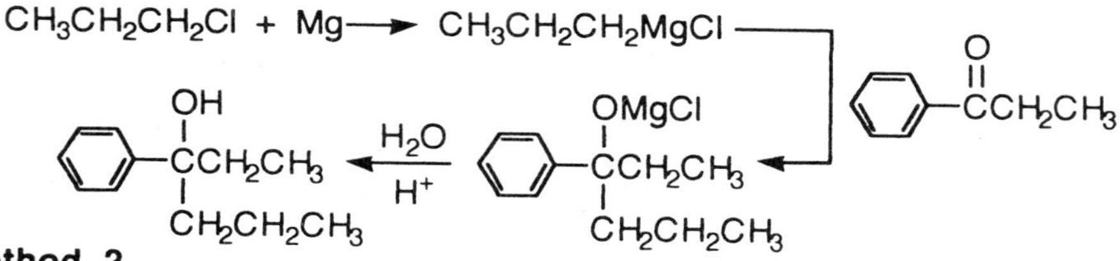

Method 2

CH$_3$CH$_2$I + Mg $\longrightarrow$ CH$_3$CH$_2$MgI

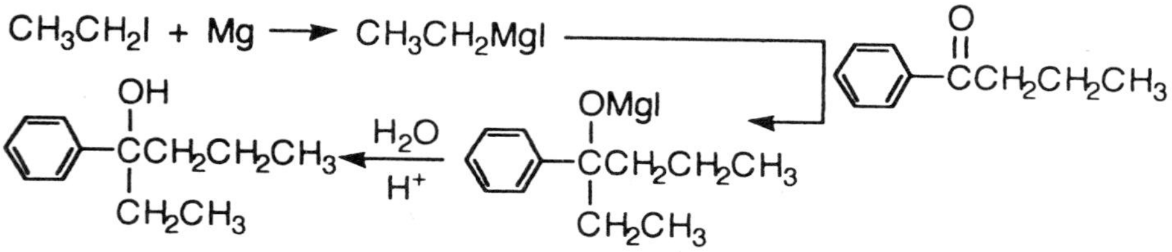

Method 3

$$\text{C}_6\text{H}_5\text{-Br} + \text{Mg} \longrightarrow \text{C}_6\text{H}_5\text{-MgBr}$$

$$\text{CH}_3\text{CH}_2\overset{\overset{\displaystyle O}{\|}}{\text{C}}\text{CH}_2\text{CH}_2\text{CH}_3$$

$$\underset{\underset{\displaystyle \text{C}_6\text{H}_5}{|}}{\overset{\overset{\displaystyle \text{OH}}{|}}{\text{CH}_3\text{CH}_2\text{C}}}\text{CH}_2\text{CH}_2\text{CH}_3 \xleftarrow[\text{H}^+]{\text{H}_2\text{O}} \underset{\underset{\displaystyle \text{C}_6\text{H}_5}{|}}{\overset{\overset{\displaystyle \text{OMgBr}}{|}}{\text{CH}_3\text{CH}_2\text{C}}}\text{CH}_2\text{CH}_2\text{CH}_3$$

11.35 Organometallic Chemistry: Section 11.5F

a) $\text{CH}_3\text{CH}_2\overset{\overset{\displaystyle \text{OH}}{|}}{\text{CH}}\text{-C}_6\text{H}_5$

b) $\text{C}_6\text{H}_5\text{CH}_2\overset{\overset{\displaystyle \text{OH}}{|}}{\underset{\underset{\displaystyle \text{CH}_3}{|}}{\text{C}}}\text{-CH}_3$

c) $\text{CH}_3\text{C}\equiv\text{C}\overset{\overset{\displaystyle \text{OH}}{|}}{\underset{\underset{\displaystyle \text{CH}_3}{|}}{\text{C}}}\text{CH}_3$

d) $\text{CH}_3\overset{\overset{\displaystyle \text{OH}}{|}}{\underset{\underset{\displaystyle \text{C}_6\text{H}_5}{|}}{\text{C}}}\text{CH}_2\text{CH}_3$

e) $\text{CH}_3\overset{\overset{\displaystyle \text{CH}_3}{|}}{\text{CH}}\text{CH}_2\text{CH}_2\text{OH}$

f) $\text{C}_6\text{H}_5\overset{\overset{\displaystyle O}{\|}}{\text{C}}\text{OH}$

11.36 Aldol Condensation: Section 11.6B

(a)
$$\text{CH}_3\text{CH}_2\text{CH}_2\text{CH}_2\overset{\overset{\displaystyle O}{\|}}{\underset{\underset{\displaystyle H}{|}}{\text{C}}}\text{—}\underset{\underset{\displaystyle \text{CH}_2\text{CH}_2\text{CH}_3}{|}}{\overset{\overset{\displaystyle O}{\|}}{\text{CHCH}}} \xrightarrow{\text{OH}^-} \text{CH}_3\text{CH}_2\text{CH}_2\text{CH}_2\text{—}\underset{\underset{\displaystyle H}{|}}{\overset{\overset{\displaystyle \boxed{\text{OH}}}{|}}{\text{C}}}\text{—}\underset{\underset{\displaystyle \text{CH}_2\text{CH}_2\text{CH}_3}{|}}{\overset{\overset{\displaystyle \boxed{\text{H}}}{|}}{\text{C}}}\text{—}\overset{\overset{\displaystyle O}{\|}}{\text{CH}}$$

$$\text{CH}_3\text{CH}_2\text{CH}_2\text{CH}_2\overset{}{\underset{\underset{\displaystyle H}{|}}{\text{C}}}\text{=}\underset{\underset{\displaystyle \text{CH}_2\text{CH}_2\text{CH}_3}{|}}{\text{C}}\text{—}\overset{\overset{\displaystyle O}{\|}}{\text{CH}} \xleftarrow{\text{H}^+ \quad (\text{-H}_2\text{O})}$$

(b)
$$(\text{CH}_3)_2\text{CHCH}_2\overset{\overset{\displaystyle O}{\|}}{\underset{\underset{\displaystyle H}{|}}{\text{C}}}\text{—}\underset{\underset{\displaystyle \text{CH}(\text{CH}_3)_2}{|}}{\overset{\overset{\displaystyle O}{\|}}{\text{CHCH}}} \xrightarrow{\text{OH}^-} (\text{CH}_3)_2\text{CHCH}_2\text{—}\underset{\underset{\displaystyle H}{|}}{\overset{\overset{\displaystyle \boxed{\text{OH}}}{|}}{\text{C}}}\text{—}\underset{\underset{\displaystyle \text{CH}(\text{CH}_3)_2}{|}}{\overset{\overset{\displaystyle \boxed{\text{H}}}{|}}{\text{C}}}\text{—}\overset{\overset{\displaystyle O}{\|}}{\text{CH}}$$

$$(\text{CH}_3)_2\text{CHCH}_2\overset{}{\underset{\underset{\displaystyle H}{|}}{\text{C}}}\text{=}\underset{\underset{\displaystyle \text{CH}(\text{CH}_3)_2}{|}}{\text{C}}\text{—}\overset{\overset{\displaystyle O}{\|}}{\text{CH}} \xleftarrow{\text{H}^+ \quad (\text{-H}_2\text{O})}$$

(c)

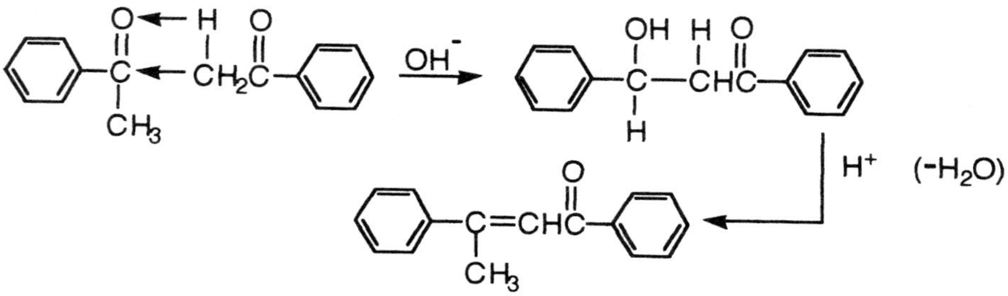

11.37 Crossed Aldol Condensation: Section 11.6B.3

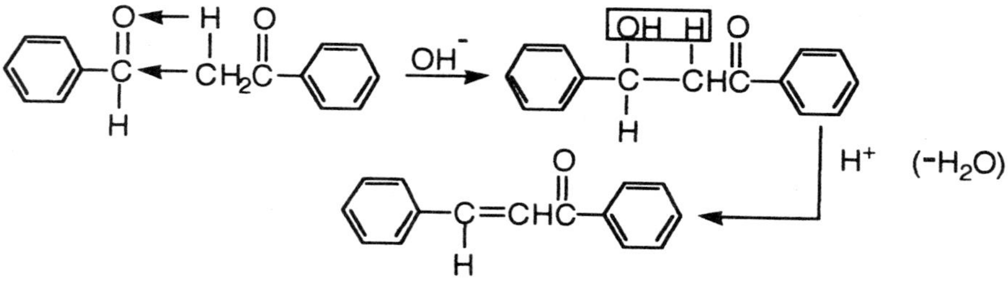

11.38 Aldol Condensation: Section 11.6B

The starting aldehydes or ketones are shown. These substances are exposed to base, NaOH, to effect reaction. The carbon-carbon double bond in the product shown in the text is the point of connection between the condensing molecules.

a) 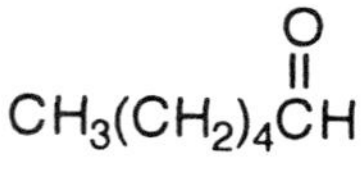 $CH_3(CH_2)_4\overset{\overset{\displaystyle O}{\|}}{C}H$

b)

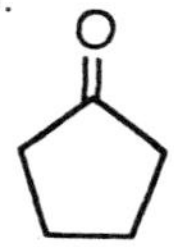

c) 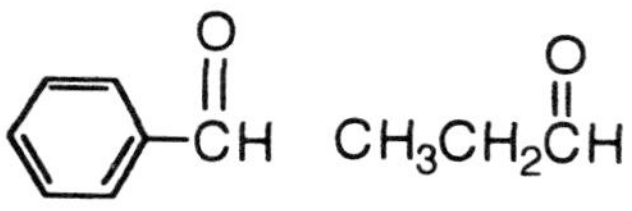 $\overset{\overset{\displaystyle O}{\|}}{C}H$ $CH_3CH_2\overset{\overset{\displaystyle O}{\|}}{C}H$

11.39 Enolate Ions: Section 11.6A

(a) 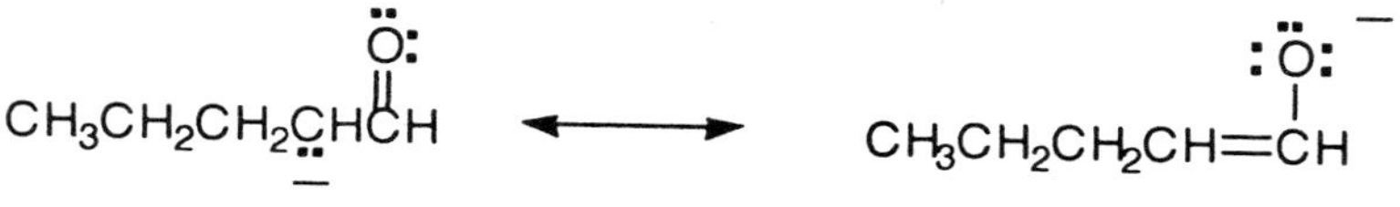 $CH_3CH_2CH_2\overset{..}{\underset{..}{C}}H\overset{\overset{\displaystyle \overset{..}{\overset{..}{O}}:}{}}{C}H$ ⟷ $CH_3CH_2CH_2CH{=}\overset{\overset{\displaystyle :\overset{..}{\overset{..}{O}}:}{|}}{C}H$ $^-$

(b) $(CH_3)_2CH\underset{..}{C}H\overset{\overset{\displaystyle \overset{..}{\overset{..}{O}}:}{\|}}{C}H$ ⟷ $(CH_3)_2CHCH{=}\overset{\overset{\displaystyle :\overset{..}{\overset{..}{O}}:}{|}}{C}H$ $^-$

(c) $\overset{\overset{\displaystyle :\overset{..}{\overset{..}{O}}}{\|}}{C}\underset{..}{C}H_2$ ⟷ $\overset{\overset{\displaystyle :\overset{..}{\overset{..}{O}}:}{|}}{C}{=}CH_2$ $^-$

208

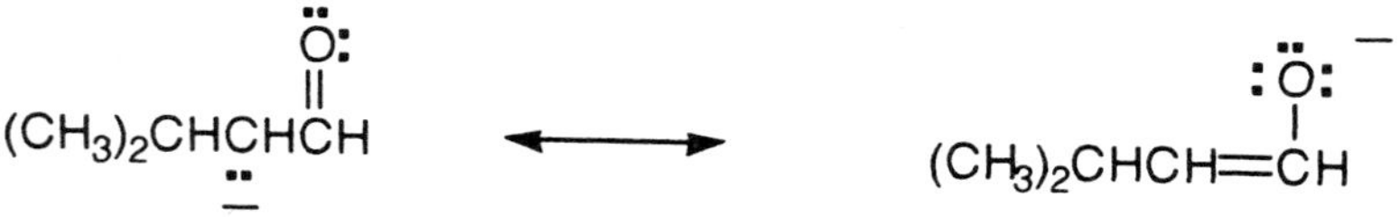

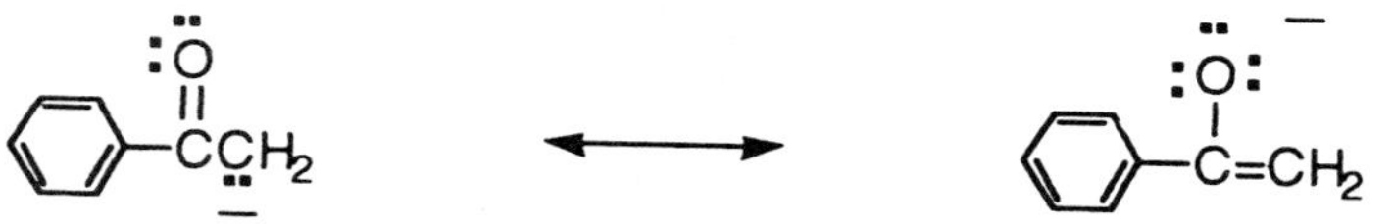

11.40 Keto-Enol Tautomerism: Section 11.6A

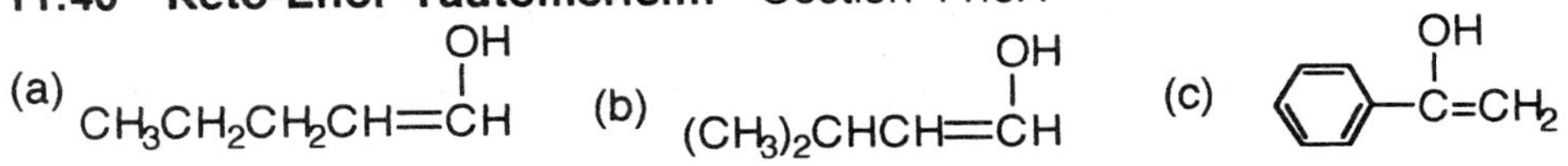

(a)
$$CH_3CH_2CH_2CH{=}\overset{\displaystyle OH}{\overset{|}{C}}H$$

(b)
$$(CH_3)_2CHCH{=}\overset{\displaystyle OH}{\overset{|}{C}}H$$

(c)

11.41 Acetal Formation: Section 11.5G

a)
$$CH_3CH_2\overset{\displaystyle O}{\overset{||}{C}}H + 2\ CH_3CH_2OH \xrightarrow{\ H^+\ } CH_3CH_2\overset{\displaystyle OCH_2CH_3}{\underset{\displaystyle H}{\overset{|}{\underset{|}{C}}}}{-}OCH_2CH_3 + H_2O$$

b)
$$CH_3\overset{\displaystyle O}{\overset{||}{C}}CH_3 + 2\ CH_3OH \xrightarrow{\ H^+\ } CH_3\overset{\displaystyle OCH_3}{\underset{\displaystyle OCH_3}{\overset{|}{\underset{|}{C}}}}CH_3 + H_2O$$

c)

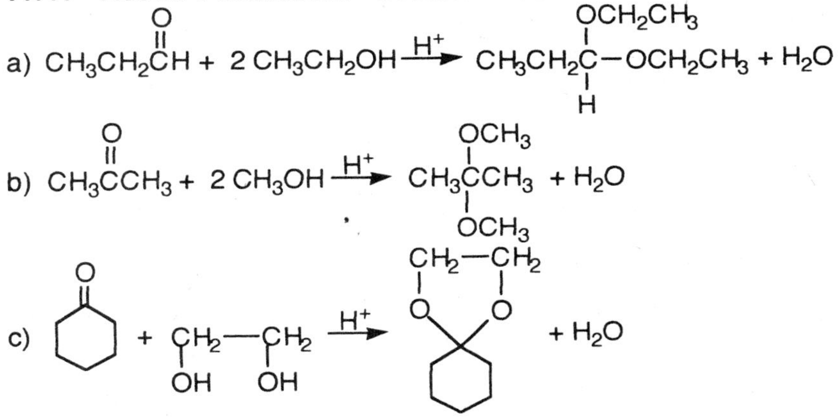

11.42 Acetal Formation: Section 11.5G

   One mole of alcohol comes internally from the alcohol group on the hydroxy aldehyde. This causes the cyclic structure. The second mole comes from the methanol.

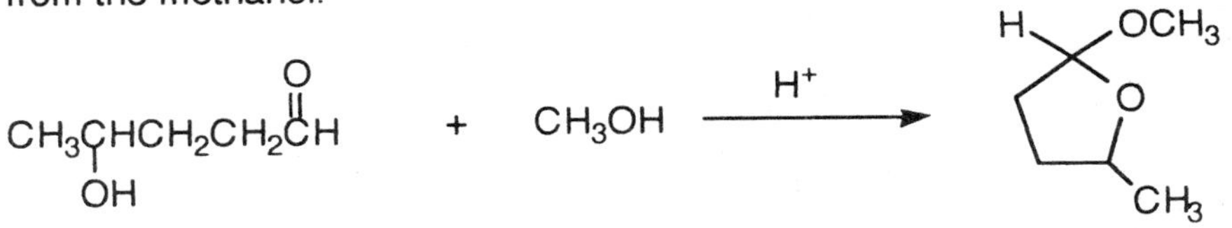

11.43 Synthesis of Familiar Compounds

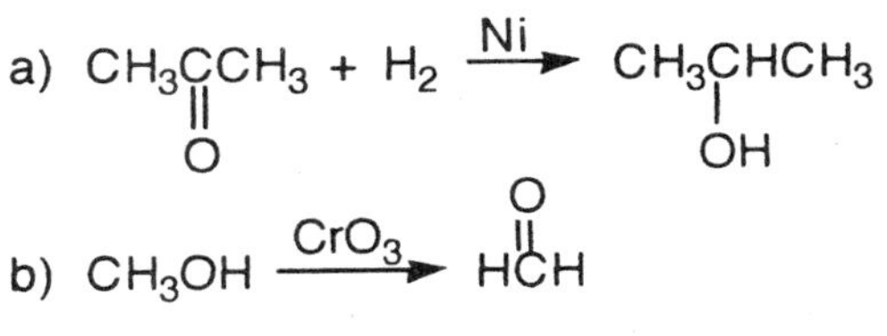

a)
$$CH_3\overset{\displaystyle }{\underset{\displaystyle O}{\overset{||}{C}}}CH_3 + H_2 \xrightarrow{\ Ni\ } CH_3\overset{\displaystyle }{\underset{\displaystyle OH}{\overset{|}{C}}}HCH_3$$

b)
$$CH_3OH \xrightarrow{\ CrO_3\ } H\overset{\displaystyle O}{\overset{||}{C}}H$$

c) $CH_3CH_2CH_2\overset{\overset{\displaystyle O}{\|}}{C}H \xrightarrow{OH^-} CH_3CH_2CH_2\overset{\overset{\displaystyle OH}{|}}{C}H\overset{\underset{\displaystyle CH_3CH_2}{|}}{C}H\overset{\overset{\displaystyle O}{\|}}{C}H \xrightarrow[Ni]{H_2} CH_3CH_2CH_2\overset{\overset{\displaystyle OH}{|}}{C}H\overset{\underset{\displaystyle CH_3CH_2}{|}}{C}HCH_2OH$

d) $CH_3\overset{\overset{\displaystyle O}{\|}}{C}H \xrightarrow[H^+]{K_2Cr_2O_7} CH_3\overset{\overset{\displaystyle O}{\|}}{C}OH$

e) ⬡—MgBr + CH_2—CH_2 (epoxide with O) → ⬡—CH_2CH_2OMgBr →

⬡—$CH_2CH_2OH \xleftarrow[H^+]{H_2O}$ ←

11.44 Preparation of Alcohols: Sections 11.5D-F
Grignard Preparations

$CH_3Br \xrightarrow{Mg} CH_3MgBr \xrightarrow{CH_3CH_2CH_2\overset{\overset{\displaystyle O}{\|}}{C}H} \xrightarrow[H^+]{H_2O} CH_3CH_2CH_2\overset{\overset{\displaystyle OH}{|}}{C}HCH_3$

$CH_3CH_2CH_2Br \xrightarrow{Mg} CH_3CH_2CH_2MgBr \xrightarrow{CH_3\overset{\overset{\displaystyle O}{\|}}{C}H} \xrightarrow[H^+]{H_2O} CH_3CH_2CH_2\overset{\overset{\displaystyle OH}{|}}{C}HCH_3$

Reductions

$CH_3CH_2CH_2\overset{\overset{\displaystyle O}{\|}}{C}CH_3 + H_2 \xrightarrow{Ni} CH_3CH_2CH_2\overset{\overset{\displaystyle OH}{|}}{C}HCH_3$

$4\ CH_3CH_2CH_2\overset{\overset{\displaystyle O}{\|}}{C}CH_3 + LiAlH_4 \longrightarrow \xrightarrow[H^+]{H_2O} CH_3CH_2CH_2\overset{\overset{\displaystyle OH}{|}}{C}HCH_3$

11.45 Keto-Enol Tautomerism: Section 11.6A

a) $CH_3CH_2\underset{\underset{\displaystyle O}{\|}}{C}CH_2CH_3$ $CH_3CH=CHCH_2CH_3$ b) (cyclopentanone) (cyclopentenol with OH)

c) $CH_3\overset{\overset{\displaystyle O}{\|}}{C}H$

$CH_2=\overset{\underset{\displaystyle OH}{|}}{C}H$

11.46 Tautomerism: Section 11.6A

$CH_3CH=NCH_3$

11.47　Reaction Mechanisms:　Sections 11.5-11.6

a) $CH_3CH_2\overset{\overset{\displaystyle\cdot\cdot}{O}}{C}H + CH_3\overset{-}{:}\ \overset{+}{MgCl} \longrightarrow CH_3CH_2\overset{\overset{\displaystyle\cdot\cdot}{:O}:\ \overset{+}{MgCl}}{\underset{H}{C}}CH_3 \xrightarrow[H^+]{H_2O} CH_3CH_2\overset{OH}{\underset{H}{C}}CH_3$

b) $CH_3CH_2\overset{\overset{\displaystyle\cdot\cdot}{O}}{C}H \quad \overset{+}{Na}\ \overset{-}{:}CN: \longrightarrow CH_3CH_2\overset{\overset{\displaystyle\cdot\cdot}{:O}:\ Na^+}{\underset{CN:}{C}}H \xrightarrow[H_2O]{H^+} CH_3CH_2\overset{OH}{\underset{CN}{C}}H$

c) $4\ CH_3CH_2\overset{\overset{\displaystyle\cdot\cdot}{O}}{C}H + LiAlH_4 \longrightarrow (CH_3CH_2\overset{\overset{\displaystyle:\overset{-}{O}:}{}}{\underset{H}{C}})_4Al^{3+}Li^+ \xrightarrow[H^+]{H_2O} 4\ CH_3CH_2CH_2OH$

d) $CH_3CH_2\overset{\overset{\displaystyle\cdot\cdot}{O}}{C}H \xrightarrow{H^+} CH_3CH_2\overset{OH}{\underset{H}{C}}+ \xrightarrow{\overset{\displaystyle\cdot\cdot}{H_2N}OH} CH_3CH_2\overset{:\overset{\cdot\cdot}{O}H}{\underset{H}{C}}\div\overset{H}{\underset{H}{\overset{+}{N}}}OH \longrightarrow$

$CH_3CH_2\overset{\overset{\displaystyle\overset{+}{H}}{\overset{|}{:OH}}}{\underset{H}{C}}-\overset{H}{\underset{H}{N}}OH \xrightarrow{-H_2\overset{\cdot\cdot}{O}:} CH_3CH_2\overset{+}{\underset{H}{C}}-\overset{H}{\underset{H}{N}}OH \xrightarrow{-H^+} CH_3CH_2CH=NOH$

e) $CH_3CH_2\overset{\overset{\displaystyle O}{||}}{C}H \xrightarrow{OH^-} CH_3\overset{\overset{\displaystyle O}{||}}{\underset{\cdot\cdot}{C}}H\overset{\overset{\displaystyle O}{||}}{C}H \xrightarrow{CH_3CH_2\overset{\overset{\displaystyle\cdot\cdot}{O}}{C}H} CH_3CH_2\overset{:\overset{-}{O}:}{C}:\overset{\overset{\displaystyle O}{||}}{\underset{CH_3}{C}}H\overset{}{\underset{}{C}}H$

$\xrightarrow{CH_3CH_2\overset{\overset{\displaystyle O}{||}}{C}H} CH_3CH_2\overset{OH}{C}H-\overset{\overset{\displaystyle O}{||}}{\underset{CH_3}{C}}H-\overset{\overset{\displaystyle O}{||}}{C}H + CH_3\overset{\cdot\cdot}{C}H\overset{\overset{\displaystyle O}{||}}{C}H$

$\left(\xrightarrow{\text{dehydration}} CH_3CH_2CH=\overset{\overset{\displaystyle O}{||}}{\underset{CH_3}{C}}CH\right)$

f) $CH_3CH_2\overset{\overset{\displaystyle :\overset{..}{O}:}{\|}}{C}H \underset{}{\overset{H^+}{\rightleftharpoons}} CH_3CH_2\overset{\overset{\displaystyle :\overset{..}{O}H}{|}}{\underset{H}{C}}^+ \underset{}{\overset{CH_3\overset{..}{O}H}{\rightleftharpoons}} CH_3CH_2\overset{\overset{\displaystyle :\overset{..}{O}H}{|}}{\underset{H}{C}}-\overset{..}{O}CH_3 \rightleftharpoons CH_3CH_2\overset{\overset{\displaystyle +\overset{H}{O}H}{|}}{\underset{H}{C}}-\overset{..}{O}CH_3$

$\overset{-H_2\overset{..}{O}:}{\rightleftharpoons} CH_3CH_2\overset{+}{\underset{H}{C}}-\overset{..}{O}CH_3 \overset{CH_3\overset{..}{O}H}{\rightleftharpoons} CH_3CH_2\overset{\overset{\displaystyle +:\overset{..}{O}CH_3}{|}}{\underset{H}{C}}-\overset{..}{O}CH_3 \overset{-H^+}{\rightleftharpoons} CH_3CH_2\overset{\overset{\displaystyle :\overset{..}{O}CH_3}{|}}{\underset{H}{C}}-\overset{..}{O}CH_3$

11.48 Reaction Mechanisms Involving Organometallic Reactions:

Section 11.5F.4

a) $CH_3-\langle\rangle-\overset{\overset{\displaystyle O}{\|}}{C}H \xrightarrow{CH_3CH_2\overset{-}{M}gCl} CH_3-\langle\rangle-\overset{\overset{\displaystyle O^-Mg^+Cl}{|}}{\underset{CH_2CH_3}{C}}H \xrightarrow{H_2O/H^+} CH_3-\langle\rangle-\overset{\overset{\displaystyle OH}{|}}{\underset{CH_2CH_3}{C}}H$

b)

c) $CH_3\overset{\overset{\displaystyle O}{\|}}{C}H \xrightarrow{CH_3CH_2C\equiv CNa} CH_3\overset{\overset{\displaystyle O^-Na^+}{|}}{\underset{C\equiv CCH_2CH_3}{C}}H \xrightarrow{H_2O/H^+} CH_3\overset{\overset{\displaystyle OH}{|}}{C}HC\equiv CCH_2CH_3$

11.49 Acidity of Alpha Hydrogens: Section 11.6A

$CH_3CH_2\overset{\overset{\displaystyle O}{\|}}{C}CH_2\overset{\overset{\displaystyle O}{\|}}{C}CH_2CH_3$ **most acidic c>b>a   least acidic**
   a   b   c

Hydrogen c is next to two electron withdrawing groups, carbonyl groups, and is the most acidic; b is adjacent to one and a is not adjacent to any.

11.50 Aldol-Type Condensations: Section 11.6B

a) $\langle\rangle-\overset{\overset{\displaystyle O}{\|}}{C}H + CH_3NO_2 \xrightarrow{\text{base}} \langle\rangle-\overset{\overset{\displaystyle OH}{|}}{C}HCH_2NO_2 \xrightarrow{-H_2O} \langle\rangle-CH=CHNO_2$

b) $\langle\rangle-\overset{\overset{\displaystyle O}{\|}}{C}H + CH_3CN \xrightarrow{\text{base}} \langle\rangle-\overset{\overset{\displaystyle OH}{|}}{C}HCH_2CN \xrightarrow{-H_2O} \langle\rangle-CH=CHCN$

c) $\langle\rangle-\overset{\overset{\displaystyle O}{\|}}{C}H + \overset{CO_2CH_3}{\underset{CO_2CH_3}{C}H_2} \xrightarrow{\text{base}} \langle\rangle-\overset{\overset{\displaystyle OH}{|}}{C}H-\overset{\overset{\displaystyle CO_2CH_3}{|}}{\underset{CO_2CH_3}{C}}H \xrightarrow{-H_2O} \langle\rangle-CH=C\overset{CO_2CH_3}{\underset{CO_2CH_3}{}}$

11.51 Reaction Mechanisms of Aldol-Type Condensations:
Section 11.6

a) $CH_3CCH_3 \xrightarrow{OH^-} CH_3CCH_2^- \xrightarrow{CH_3CCH_3} CH_3CCH_2CCH_3 \xrightarrow{H_2O}$

$CH_3C=CHCCH_3 \xleftarrow{\text{dehydration}} CH_3CCH_2CCH_3$

b) $CH_3CH_2CH \xrightarrow{OH^-} CH_3CH_2CH \xrightarrow{} \langle C_6H_5 \rangle-C-CHCH- \xrightarrow{H_2O}$

$\langle C_6H_5 \rangle-CH=CCH \xleftarrow{\text{dehydration}} \langle C_6H_5 \rangle-CHCHCH$

c) $CH_3NO_2 \xrightarrow{OH^-} CH_2NO_2 \xrightarrow{} \langle C_6H_5 \rangle-CHCH_2NO_2 \xrightarrow{H_2O}$

$\langle C_6H_5 \rangle-CH=CHNO_2 \xleftarrow{\text{dehydration}} \langle C_6H_5 \rangle-CHCH_2NO_2$

11.52 Resonance in Carbanions: Section 11.6A

a) $CH_3CH-CH \longleftrightarrow CH_3CH=CH$

b) $\langle C_6H_5 \rangle-C-CH_2 \longleftrightarrow \langle C_6H_5 \rangle-C=CH_2$

c) $HCCHCH \longleftrightarrow HC=CHCH \longleftrightarrow HC-CH=CH$

11.53 Organic Qualitative Analysis

a)   Propanal is an aldehyde and will give a positive silver mirror test when treated with Tollens' reagent or a brick-red precipitate when treated with Benedict's reagent. Propanone is a ketone and reacts with neither of these reagents.

b)   Propanone is a ketone and will give a positive 2,4 DNP test. A colored precipitate will form when 2,4-dinitrophenylhydrazine is mixed with propanone. 2-Propanol is an alcohol and will not react with 2,4-DNP.

c)   Butanal and butanone being an aldehyde and ketone respectively give a positive 2,4-DNP test. Butanol is an alcohol and will not react with the 2,4-DNP reagent since the test is specific for carbonyl compounds. Butanal can be distinguished from butanone with Tollens' or Benedict's tests both of which are specific for aldehydes.

11.54 Organic Qualitative Analysis

Ozonolysis produces aldehydes and ketones from alkenes. Both give positive 2,4 DNP tests and aldehydes also give positive Tollens tests. The structure of the unknown can be determined from the ozonolysis products.

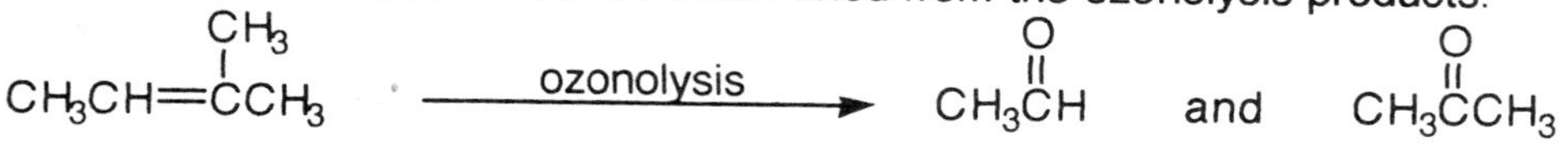

11.55 Carbohydrate Chemistry: Section 11.5G

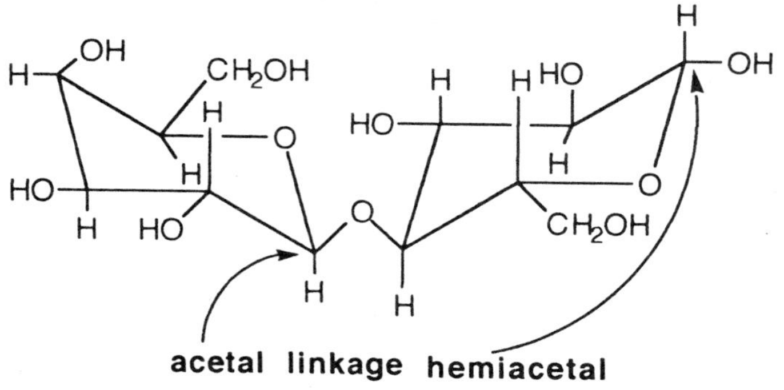

acetal linkage hemiacetal

Lactose

12

R—C(=O)—O—H   (O=)C—R, H—O

CARBOXYLIC ACIDS

CHAPTER SUMMARY

Carboxylic acids are structurally characterized by the **carboxyl group**, a carbon-oxygen double bond with a directly attached OH group. This is a very reactive functional group because there are three polar bonds, the carbon-oxygen double and single bonds and the oxygen-hydrogen bond; the double bond has electrons in a pi-bond; and there are two unshared electron pairs on each oxygen. Carboxylic acids have an unpleasant odor and taste; they are found widely in nature.

Carboxylic acids are almost always named using a suffix. The suffix **oic acid** is attached to the name for the longest continuous carbon chain. If the acid group is attached to a ring the suffix **carboxylic acid** is used. A general procedure for naming organic compounds is given in this chapter since this is the last major functional group covered. **The procedure for naming organic compounds is: (1)** name the longest continuous carbon chain; **(2)** identify carbon-carbon double and triple bonds with suffixes; **(3)** name the highest priority functional group with a suffix (acid > aldehyde > ketone > alcohol > amine) and the others with prefixes; **(4)** number the carbon chain giving preference to the functional group named by a suffix, then multiple bonds (carbon-carbon double bonds take priority over triple bonds when making a choice is necessary), then groups named with prefixes; **(5)** name all other groups with prefixes and number them. Carboxylic acids are also named with common names that often describe a familiar source or property of the compound.

The boiling points of carboxylic acids are high relative to other classes of compounds due to **hydrogen bonding;** carboxylic acid molecules can hydrogen bond in two places and as a result often exist as dimers.

Acidity is the characteristic property of carboxylic acids; they react with strong bases like sodium hydroxide and weaker bases such as sodium bicarbonate. The ability to be neutralized by sodium bicarbonate distinguishes carboxylic acids from phenols. The acidity of carboxylic acids is the result of **resonance stabilization** in the **carboxylate anion** formed upon ionization or neutralization of the acid.

The acidity of carboxylic acids is described by the **acidity constant, K_a,** and its negative logarithm, **pK_a.** Large K_a's and small pK_a's denote high acidities. Acid strength is influenced by substituents on the carboxylic acid molecule. **Electron-withdrawing groups** disperse the negative character of the carboxylate ion and increase acidity whereas **electron-releasing groups** intensify the negative charge and decrease acidity. The **strength, number, and proximity** of electron-withdrawing groups can have dramatic effects on relative acidities.

Salts of carboxylic acids are named by changing the **oic acid** suffix of the acid to **ate** and preceding it by the name of the **cation.**

Carboxylic acids can be prepared by **oxidation of alkylbenzenes, oxidation of primary alcohols, hydrolysis of nitriles, and reaction of Grignard reagents with carbon dioxide.**

Carboxylic acids can be converted into a variety of derivatives including **acid chlorides, acid anhydrides, esters, and amides.**

Connections 12.1 is about food preservatives.

SOLUTIONS TO PROBLEMS

12.1 IUPAC Nomenclature
(a) heptanoic acid;   (b) 4,4,4-tribromobutanoic acid;   (c) 1,5-pentandioic acid;   (d) cyclohexanecarboxylic acid   (e) m-methylbenzoic acid;
(f) 3-chlorocyclobutanecarboxylic acid

12.2 IUPAC Nomenclature

(a) 2-hydroxypropanoic acid;   (b) 1-cyclohexenecarboxylic acid;

(c) 3-hexyn-1,6-dioic acid

12.3 IUPAC Nomenclature

(a) 4-amino-2-pentenoic acid;   (b) 4-hydroxy-2,5-cyclohexadien-1-one;

(c) 3,5-dioxohexanoic acid

12.4 Water Solubility of Carboxylic Acids

most soluble ethanoic acid > pentanoic acid > decanoic acid **least soluble**

As the molecular weight of the acids increase the proportion of non-polar
hydrocarbon to polar carboxyl group increases. As a result, the solubility in the
polar solvent, water, decreases.

12.5 Hydrogen Bonding

12.6 Neutralization of Carboxylic Acids

(a) ⟨⟩-CO_2H +   NaOH   ⟶   ⟨⟩-CO_2Na + H_2O

(b) $2CH_3CH_2CO_2H$ + $Ca(OH)_2$ ⟶ $(CH_3CH_2CO_2)_2Ca$ + $2H_2O$

(c) $CH_3CH=CH-CH=CHCO_2H$ +   KOH ⟶ $CH_3CH=CH-CH=CHCO_2K$ + H_2O

(d) $HO_2CCHCH_2CH_2CO_2H$ +   NaOH   ⟶ $HO_2CCHCH_2CH_2CO_2Na$ + H_2O
    | |
    NH_2 NH_2

12.7 Relative Acidities of Carboxylic Acids and Phenols

Both phenols and carboxylic acids are neutralized by the strong base
sodium hydroxide. However, phenols are a lot less acidic than carboxylic acids
and will not react with the weak base sodium bicarbonate. If one adds sodium
bicarbonate to the two aqueous solutions described, bubbles of carbon dioxide

will be visible as the bicarbonate and propanoic acid neutralize one another. Since the phenol does not react no CO_2 evolution will be observed.

12.8 Relative Acidities

 butane < butanol < phenol < butanoic acid < nitric acid

12.9 Relative Acidities

(a) II < III < V < IV < I   (b) III < IV < II < I

12.10   Relative Acidities

(a) III < II < IV < I   (b) IV < I < III < II   (c) I < II < III

12.11   Nomenclature of Carboxylic Acid Salts

(a) Sodium ethanoate;   (b) calcium propanoate;   (c) potassium 2-butenoate
(d) ammonium p-bromobenzoate

12.12   Preparations of Carboxylic Acids

(a)

(b)

12.13 Preparations of Carboxylic Acids

The important thing to remember here is that the methyl group is an ortho-para director whereas the carboxylic acid group directs meta. For the para isomer one would oxidize the methyl after introducing the nitro group; just the opposite would be done to obtain the meta isomer.

12.14 Preparations of Carboxylic Acids

Directive effects are important. The methyl directs the nitro para. The methyl and nitro both direct to the desired position for the bromine. Oxidation gives the requested compound.

12.15 Preparations of Carboxylic Acids

12.16 Preparations of Carboxylic Acids

$$CH_3(CH_2)_3CH_2Br \xrightarrow{NaCN} CH_3(CH_2)_3CH_2CN \xrightarrow[H^+]{H_2O} CH_3(CH_2)_3CH_2CO_2H$$

12.17 Preparations of Carboxylic Acids

12.18 Preparations of Carboxylic Acids

$$CH_3(CH_2)_3CH_2Br \xrightarrow[ether]{Mg} CH_3(CH_2)_3CH_2MgBr$$

$$\xrightarrow{CO_2} CH_3(CH_2)_3CH_2CO_2MgBr \xrightarrow[H^+]{H_2O} CH_3(CH_2)_3CH_2CO_2H$$

12.19 Preparations of Carboxylic Acids

$$BrCH_2CH_2CH_2CH_2Br \xrightarrow{NaCN} NCCH_2CH_2CH_2CH_2CN$$

$$\xrightarrow[H^+]{H_2O} HO_2CCH_2CH_2CH_2CH_2CO_2H$$

12.20 Nomenclature of Carboxylic Acids: Section 12.2A

(a) nonanoic acid;   (b) pentanoic acid;   (c) 4-methylpentanoic acid;
(d) 3-ethyl-5-methylhexanoic acid;   (e) 1,4-butandioic acid

(f) 2,2,2-trichloroethanoic acid

12.21   Nomenclature of Carboxylic Acids: Section 12.2A
(a) cycloheptanecarboxylic acid;   (b) cyclobutane -1,3-dicarboxylic acid;
(c) 3-ethylcyclopentanecarboxylic acid

12.22   Nomenclature of Carboxylic Acids: Section 12.2A
(a) 2,4-dichlorobenzoic acid;   (b) p-nitrobenzoic acid;
(c) m-methylbenzoic acid

12.23   Nomenclature of Polyfunctional Carboxylic Acids:
Section 12.2B

(a) 2-hexenoic acid;   (b) 5-hydroxyhexanoic acid;
(c) 4-oxocyclohexanecarboxylic acid;   (d) 1-cyclobutenecarboxylic acid;
(e) 2-buten-1,4-dioic acid;   (f) 3,5-dioxohexanoic acid;
(g) 2,4-hexadienoic acid;   (h) 4-oxo-2-butynoic acid;
(i) 4-amino-2-butenoic acid

12.24   Nomenclature of Organic Compounds: Section 12.2C
(a) 2-butenal;   (b) 6-amino-2,4-hexandien-1-ol;   (c) 3-hydroxy-2,4-
pentandione;   (d) 3-hexyn-2,5-dione;   (e) 4-hydroxy-2-cyclohexen-1-one;
(f) N,N-dimethyl-2-cyclopentenamine; (g) 4-hydroxy-2-heptenoic acid;
(h) 8-bromo-7-hydroxy-2-octen-5-yn-4-one

12.25   Nomenclature of Carboxylic Acid Salts: Section 12.4D
(a) sodium butanoate;   (b) calcium ethanoate;   (c) potassium 4,4,4-
tribromobutanoate;   (d) ammonium 2,4-dibromobenzoate;   (e) sodium
cyclopentanecarboxylate;   (f) sodium 4-oxo-2-pentenoate

12.26   IUPAC Nomenclature: Section 12.2

(a) $CH_3CHCH_2CO_2H$ (with CH_3 substituent)   (b) $CH_3CHC\equiv CCH_2CO_2H$ (with Br substituent)   (c) $CH_3CCH_2CH_2CO_2H$ (with O)

(d) [cyclooctatetraene]—CO_2H (e) $CH_3CHCCH_2CCH_3$ (with OH, O, O)   (f) $CH_3CH_2CH_2CH_2CH_2CO_2Na$

12.27   Preparations of Carboxylic Acids: Section 12.5

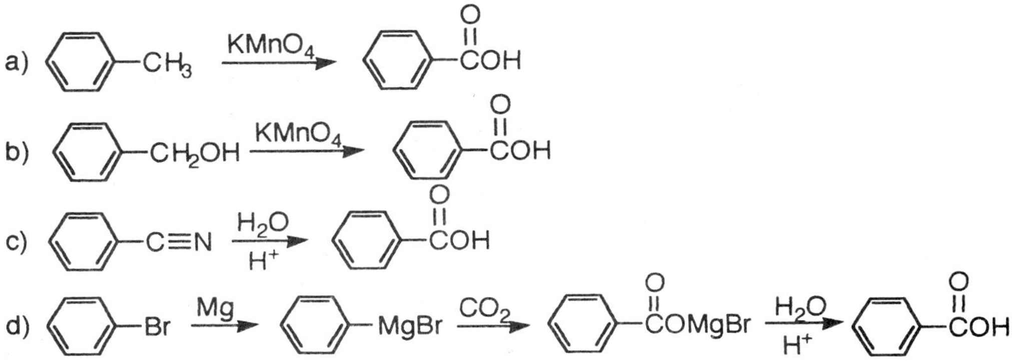

a)

b)

c)

d)

12.28 Preparations of Carboxylic Acids: Section 12.5

(a) $CH_2CH_2CH_2CH_2OH \xrightarrow{HBr} CH_3CH_2CH_2CH_2Br$ NaCN

$CH_3CH_2CH_2CH_2CO_2H \xleftarrow[H^+]{H_2O} CH_3CH_2CH_2CH_2CN$

(b)

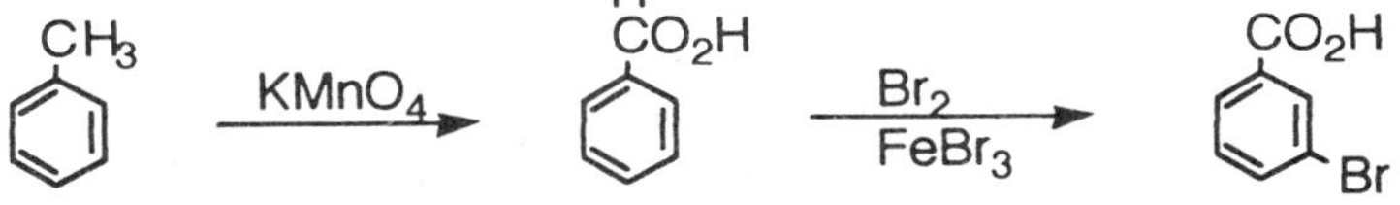

(c) $CH_3(CH_2)_4\underset{Cl}{CHCH_3} \xrightarrow[ether]{Mg} CH_3(CH_2)_4\underset{MgCl}{CHCH_3}$ CO_2

$CH_3(CH_2)_4\underset{CO_2H}{CHCH_3} \xleftarrow[H^+]{H_2O} CH_3(CH_2)_4\underset{CO_2MgCl}{CHCH_3}$

(d) $CH_3(CH_2)_3\overset{O}{\overset{\|}{C}}H \xrightarrow[H^+]{NaCN} CH_3(CH_2)_3\overset{OH}{\underset{}{C}}HCN \xrightarrow[H^+]{H_2O} CH_3(CH_2)_3\overset{OH}{\underset{}{C}}HCO_2H$

(e) $CH_3CH_2CH_2CH_2CH_2CH_2CH_2OH \xrightarrow[H^+]{CrO_3} CH_3CH_2CH_2CH_2CH_2CH_2CO_2H$

12.29 Physical Properties: Section 12.3

a) $H\overset{O}{\overset{\|}{C}}OCH_3 < HOCH_2\overset{O}{\overset{\|}{C}}H < CH_3\overset{O}{\overset{\|}{C}}OH$

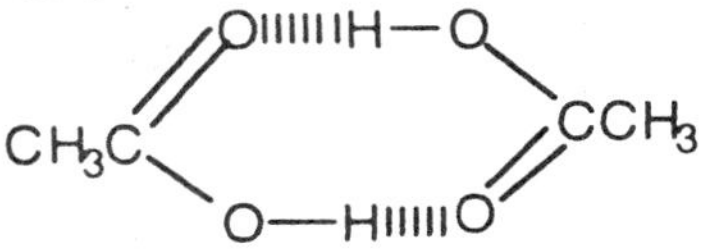

      The lowest boiling compound has no O-H bonds and cannot hydrogen bond. In the compound with the highest boiling point, the O-H bond is polarized by the C=O making the hydrogen bonding even stronger. Also in ethanoic acid, two molecules can orient so that hydrogen bonding can occur in two positions.

b) **Lowest Boiling**   **Highest Boiling**

$$CH_3OCCH_2COCH_3 < CH_3OCCH_2CH_2COH < HOC(CH_2)_3COH$$

The lowest boiling compound cannot hydrogen bond; the next has one site for hydrogen bonding (the acid group) and the highest boiling has two sites (both acid groups).

12.30 Physical Properties: Section 12.3

The dramatic difference in boiling points between chloroethane and ethanoic acid is a result of the ability of ethanoic acid to hydrogen bond. Bromoethane has a higher boiling point than chloroethane because bromoethane has a higher molecular weight. However, this increase in molecular weight doesn't come close to offsetting the hydrogen bonding ability of ethanoic acid in influencing boiling points.

12.31 Acidity: Section 12.4B-C

                    Least Acidic   —>   **Most Acidic**

     a) $4 < 2 < 3 < 1$ b) $3 < 2 < 1$ c) $2 < 4 < 1 < 3$

     d) $4 < 1 < 3 < 2$ e) $3 < 2 < 1$ f) $3 < 1 < 4 < 2$

12.32 Neutralization Reactions of Carboxylic Acids: Section 12.4A

a) $CH_3(CH_2)_5CO_2^-Na^+$

b) 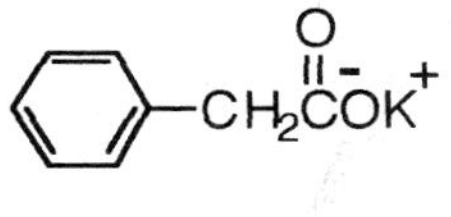 C_6H_5—$CH_2\overset{\overset{O}{\|}}{C}O^-K^+$

c) $(^-O_2C(CH_2)_3CO_2^-)_2Ca^{2+}$

d) $CH_3CO_2^-NH_4^+$

12.33 Reactions of Carboxylic Acids: Section 12.6

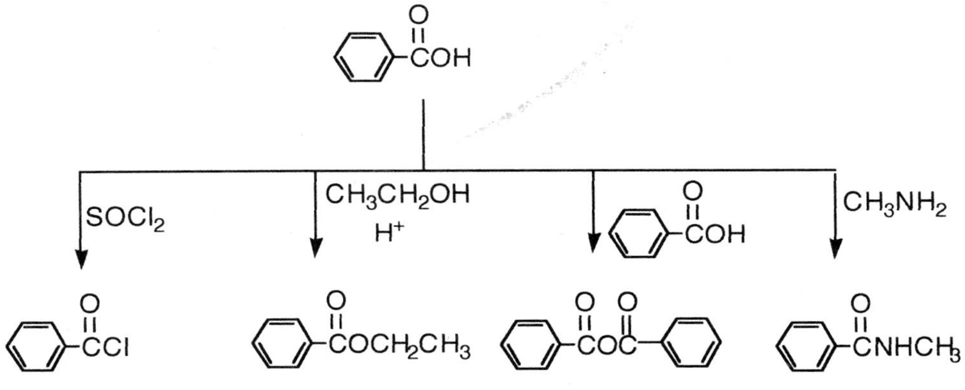

$SOCl_2$ CH_3CH_2OH / H^+ benzoic acid CH_3NH_2

13

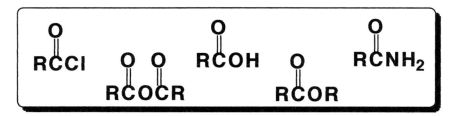

DERIVATIVES OF CARBOXYLIC ACIDS

CHAPTER SUMMARY

Carboxylic acids and their derivatives can be expressed as variations of a single formula in which an electronegative atom - oxygen, nitrogen, or halogen - is attached to a carbon-oxygen double bond.   The fundamental acid derivatives and their electronegative groups are: **acid chlorides**, Cl; **acid anhydrides**, O_2CR; **carboxylic acids**, OH; **esters**, OR; and **amides**, NH_2, NHR, or NR_2. Acid chlorides and anhydrides are very reactive and are used in synthetic organic chemistry; the others are found abundantly in nature.

The reactivity of carboxylic acids and their derivatives is a result of **polar bonds** and **non-bonding electron pairs**. The carboxyl carbon can attract both **nucleophiles** and **electrophiles**.

Carboxylic acids are named by adding the suffix **oic acid** the the base name. **Acid chlorides** are name by changing the **ic acid** of the parent carboxylic acid to **yl chloride**.   The suffix **acid** is change to **anhydride** to name **acid anhydrides**.   **Esters** are named like acid salts by changing **ic acid** to **ate** and preceding the name by the name of the **alkyl group**.   To name **amides**, the **oic acid** is changed to **amide**.

A characteristic reaction of carboxylic acid derivatives is **nucleophilic acyl substitution.** In this reaction a negative or neutral nucleophile replaces a leaving group to form a substitution product. The **leaving groups** and **nucleophiles** are the groups that define the various acid derivatives; as a result, the reaction usually involves the conversion of one acid derivative into another. The **order of reactivity** of acid derivatives is: **acid chloride > anhydride > acid or ester > amide.** In general, reaction of any of these derivatives with water produces acids, with alcohols esters result, and with amines, amides are formed.

Nucleophilic acyl substitution can be **initiated by a negative or neutral nucleophile** attacking the partially positive carbonyl carbon. In this usually two-step mechanism, a **tetrahedral intermediate** is formed; loss of the leaving group produces the new acid derivative. Alternatively, the reaction can be **acid catalyzed.** In this mechanism, a hydrogen ion bonds to the carbonyl oxygen to form a carbocation. The nucleophile bonds to the carbocation to form the **tetrahedral intermediate**; eventually the leaving group departs to form the new acid derivative.

Carboxylic acid chlorides are synthesized by treating a carboxylic acid with thionyl chloride. Acid chlorides react with sodium salts of carboxylic acids to form anhydrides, with alcohols to form esters, with water to form acids, and with amines to form amides. These reactions generally proceed by a nucleophile initiated mechanism.

Acid anhydrides are synthesized from acid chlorides and carboxylic acid salts. In some cases, heating a dicarboxylic acid can cause the elimination of water to form an anhydride. Anhydrides react with alcohols to form esters, with water to form carboxylic acids, and with amines to form amides. The reaction mechanism is usually nucleophile initiated.

Carboxylic acids react with thionyl chloride to form acid chlorides; heating some acids can result in acid anhydrides. Reaction with alcohols gives esters, and with amines, amides are formed. Many nucleophilic acyl substitution reactions of carboxylic acids are acid-initiated.

Esters can be converted to acids with water; reaction with an alcohol produces a new ester by a process called **transesterification.** Esters react with amines to form amides. Esters react by both acid and nucleophile initiated mechanisms.

Amides are the least reactive of the carboxylic acid derivatives; they can be prepared from any of the other acid derivatives. **Hydrolysis**, either acid or base catalyzed, to form acids is the only nucleophilic acyl substitution reaction.

Polyamides such as Nylon are formed from dicarboxylic acids and diamines. **Polyesters** such as Dacron are formed from the reaction of dicarboxylic acids or diesters with dialcohols.

Carboxylic acid derivatives can also undergo **nucleophilic addition** reactions. By a combination of nucleophilic acyl substitution and nucleophilic addition, all of the acid derivatives except amides can be reduced to primary alcohols using **lithium aluminum hydride.** Reduction of amides produces amines. Esters and acid chlorides react with **Grignard reagents** to form tertiary alcohols.

Certain acid derivatives are capable of reactions involving intermediate carbanions. The **malonic ester synthesis** is used to synthesize substituted acetic acids and the **Claisen condensation** produces keto esters.

Connections 13.1 is about Aspirin and pain relievers.

Connections 13.2 is about barbiturates.

SOLUTIONS TO PROBLEMS

13.1 Nomenclature of Acid Chlorides
(a) pentanoyl chloride;   (b) 2-propenoyl chloride;   (c) p-nitrobenzoyl chloride

13.2 Nomenclature of Acid Anhydrides
(a) ethanoic anhydride;   (b) pentanoic anhydride;
(c) ethanoic pentanoic anhydride

13.3 Nomenclature of Esters
(a) methyl ethanoate; (b) ethyl 2-propenoate;   (c) isopropyl m-chlorobenzoate

13.4 Nomenclature of Amides
(a) ethanamide;   (b) 2-propenamide;   (c) p-bromo-N-methylbenzamide;
(d) N-methylethanamide;   (e) N,N-dimethyl-2-propenamide;
(f) p-bromo-N-ethyl-N-methylbenzamide

13.5 Reactions of Acid Chlorides

$$CH_3\overset{O}{\overset{\|}{C}}Cl + \text{Reagents a-f} \longrightarrow \underline{\hspace{3cm}} + HCl \quad (\text{NaCl in c})$$

a) $CH_3\overset{O}{\overset{\|}{C}}OH$
    b) $CH_3\overset{O}{\overset{\|}{C}}OCH_2CH_3$
    c) $CH_3\overset{O}{\overset{\|}{C}}O\overset{O}{\overset{\|}{C}}CH_3$

d) $CH_3\overset{O}{\overset{\|}{C}}NH_2$
    e) $CH_3\overset{O}{\overset{\|}{C}}NHCH_3$
    f) $CH_3\overset{O}{\overset{\|}{C}}N(CH_2CH_3)_2$

13.6 Nucleophilic Acyl Substitution Mechanism

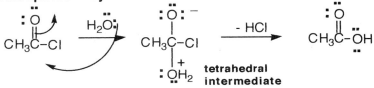

tetrahedral intermediate

13.7 Reactions of Acid Anhydrides

$$CH_3\overset{O}{\overset{\|}{C}}O\overset{O}{\overset{\|}{C}}CH_3 + \text{Reagents a-d} \longrightarrow \underline{\hspace{3cm}} + CH_3\overset{O}{\overset{\|}{C}}OH$$

a) $CH_3\overset{O}{\overset{\|}{C}}OH$
    b) $CH_3\overset{O}{\overset{\|}{C}}OCH_2CH_3$
    c) $CH_3\overset{O}{\overset{\|}{C}}NH_2$
    d) $CH_3\overset{O}{\overset{\|}{C}}NHCH_3$

13.8 Nucleophilic Acyl Substitution Mechanism

$$CH_3\overset{O}{\overset{\|}{C}}O\overset{O}{\overset{\|}{C}}CH_3 \xrightarrow[\substack{\text{neutral} \\ \text{nucleophile}}]{:NH_3} \underset{\substack{+NH_3 \\ \text{tetrahedral} \\ \text{intermediate}}}{CH_3\overset{\overset{-O}{\|}}{C}O\overset{O}{\overset{\|}{C}}CH_3} \xrightarrow{- CH_3\overset{O}{\overset{\|}{C}}OH} CH_3\overset{O}{\overset{\|}{C}}NH_2$$

13.9 Reactions of Carboxylic Acids

$$CH_3CH_2\overset{O}{\overset{\|}{C}}OH + \text{reagents a-d} \longrightarrow \underline{\hspace{2cm}} + H_2O \; (SO_2 + HCl \text{ in part a})$$

(a) $CH_3CH_2\overset{O}{\overset{\|}{C}}Cl$
    (b) $CH_3CH_2\overset{O}{\overset{\|}{C}}OCH_3$
    (c) $CH_3CH_2\overset{O}{\overset{\|}{C}}NH_2$
    (d) $CH_3CH_2\overset{O}{\overset{\|}{C}}NHCH_3$

13.10 Preparations of Amines from Carboxylic Acids

$$\text{C}_6\text{H}_5\text{CO}_2\text{H} \quad \text{and} \quad (CH_3)_2NH$$

13.11   Acid Catalyzed Esterification Mechanism
The Reaction

$$
\text{C}_6\text{H}_5-\overset{\displaystyle O}{\overset{\|}{C}}-OH \;+\; CH_3OH \;\xrightarrow{\;H^+\;}\; \text{C}_6\text{H}_5-\overset{\displaystyle O}{\overset{\|}{C}}-OCH_3 \;+\; H_2O
$$

The Mechanism

Ph–C(=O)–OH $\xrightleftharpoons{\;H^+\;}$ Ph–C$^+$(OH)–OH $\xrightleftharpoons{\;CH_3OH\;}$ Ph–C(OH)(OH)–OCH$_3$, H$^+$ **tetrahedral intermediate**

Ph–C(OH)(OH)(OCH$_3$) H$^+$ $\xrightarrow{\;-H_2O\;}$ Ph–C$^+$(OH)–OCH$_3$ $\xrightleftharpoons{\;-H^+\;}$ Ph–C(=O)–OCH$_3$

13.12   Reactions of Esters

$$
CH_3\overset{\displaystyle O}{\overset{\|}{C}}OCH_2CH_3 + \text{Reagents a-d} \longrightarrow \underline{\hspace{3cm}} + CH_3CH_2OH
$$

a) $CH_3\overset{\displaystyle O}{\overset{\|}{C}}OH$ b) $CH_3\overset{\displaystyle O}{\overset{\|}{C}}OCH_3$ c) $CH_3\overset{\displaystyle O}{\overset{\|}{C}}NH_2$ d) $CH_3\overset{\displaystyle O}{\overset{\|}{C}}N(CH_2CH_3)_2$

13.13   Acid and Base Catalyzed Ester Hydrolysis
The Reaction

$$
\text{Ph}-\overset{\displaystyle O}{\overset{\|}{C}}-OCH_3 \;+\; H_2O \;\longrightarrow\; \text{Ph}-\overset{\displaystyle O}{\overset{\|}{C}}-OH \;+\; CH_3OH
$$

(a) Acid Catalyzed Hydrolysis
Note that this mechanism is the opposite of the esterification shown in problem 13.11.
Both processes are equilibriums.

Ph–C(=O)–OCH$_3$ $\xrightleftharpoons{\;-H^+\;}$ Ph–C$^+$(OH)–OCH$_3$ $\xrightleftharpoons{\;-H_2O\;}$ Ph–C(OH)(OH)(OCH$_3$) H$^+$ **tetrahedral intermediate**

Ph–C(OH)(OH)(OCH$_3$) H$^+$ $\xrightleftharpoons{\;CH_3OH\;}$ Ph–C$^+$(OH)–OH $\xrightleftharpoons{\;H^+\;}$ Ph–C(=O)–OH

(b) Base Catalyzed Hydrolysis: Saponification

13.14 Synthesis of Esters

(a) CH_3CH_2COH + CH_3CH_2OH

(b) $Br-\!\!\bigcirc\!\!-COH$ + $CH_3CH_2CH_2OH$

13.15 Preparations of Esters

(a) CH_3CH_2CCl + CH_3CH_2OH $\longrightarrow$ $CH_3CH_2COCH_2CH_3$ + HCl

(b) $Br-\!\!\bigcirc\!\!-CCl$ + $CH_3CH_2CH_2OH$ $\longrightarrow$ $Br-\!\!\bigcirc\!\!-COCH_2CH_2CH_3$ + HCl

13.16 Hydrolysis of Amides

CH_3CNH_2 + H_2O

$\xrightarrow{H^+}$ CH_3COH + NH_4+

$\xrightarrow{OH^-}$ CH_3CO^- + NH_3

13.17 Polyamides

$H_2N-\!\!\bigcirc\!\!-NH_2$ $ClC-\!\!\bigcirc\!\!-CCl$

13.18 Polyesters

13.19 Reduction of Acid Chlorides with Lithium Aluminum Hydride

$$\langle\!\!\bigcirc\!\!\rangle\!\!-CH_2\overset{O}{\overset{||}{C}}Cl \xrightarrow{LiAlH_4} \xrightarrow{H_2O/H^+} \langle\!\!\bigcirc\!\!\rangle\!\!-CH_2CH_2OH \ + \ HCl$$

13.20 LiAlH₄ Reduction of Amides to Amines

(a) $CH_3CH_2NH_2$ (b) $CH_3CH_2CH_2CH_2NHCH_3$ (c) $\langle\!\!\bigcirc\!\!\rangle\!\!-CH_2N(CH_2CH_3)_2$

13.21 Reaction of Grignard Reagents with Esters

13.22 Reaction of Esters with Grignard Reagents

13.23 Malonic Ester Synthesis

13.24 Claisen Condensation

$$\underset{\underset{CH_3}{|}}{CH_3CH_2\overset{\overset{O}{||}}{C}\,CH\,\overset{\overset{O}{||}}{C}\,OCH_2CH_3}$$

13.25 Drawing Esters and Acids: Section 13.1

CH₃CH₂CH₂CO₂H $\underset{}{\overset{\overset{CH_3}{|}}{CH_3CHCO_2H}}$ CH₃CH₂CO₂CH₃

CH₃CO₂CH₂CH₃   HCO₂CH₂CH₂CH₃ $\underset{}{\overset{\overset{CH_3}{|}}{HCO_2CHCH_3}}$

13.26 Drawing Amides: Section 13.1

$CH_3CH_2\overset{\overset{O}{||}}{C}NH_2$ $CH_3\overset{\overset{O}{||}}{C}NHCH_3$ $H\overset{\overset{O}{||}}{C}NHCH_2CH_3$ $H\overset{\overset{O}{||}}{C}N(CH_3)_2$

13.27 Drawing Anhydrides: Section 13.1

$CH_3\overset{\overset{O}{||}}{C}O\overset{\overset{O}{||}}{C}CH_3$ $CH_3CH_2\overset{\overset{O}{||}}{C}O\overset{\overset{O}{||}}{C}CH_2CH_3$ $CH_3\overset{\overset{O}{||}}{C}O\overset{\overset{O}{||}}{C}CH_2CH_3$

13.28 Drawing Acid Chlorides: Section 13.1

$CH_3CH_2CH_2\overset{\overset{O}{||}}{C}Cl$ $\underset{\underset{CH_3}{|}}{CH_3CH\overset{\overset{O}{||}}{C}Cl}$

13.29 Nomenclature of Carboxylic Acids: Section 12.2

(a) butanoic acid;   (b) 7-methyloctanoic acid;   (c) 2-pentenoic acid;
(d) 4-oxopentanoic acid;   (e) 2-aminoethanoic acid;   (f) 6-hydroxy-2,4-heptadienoic acid;   (g) m-bromobenzoic acid;   (h) cyclopentanecarboxylic acid;   (i) 1,6-hexandioic acid

13.30 Nomenclature of Acid Chlorides: Section 13.1B.1

(a) butanoyl chloride;   (b) 2-butenoyl chloride;   (c) 3-oxopentanoyl chloride;
(d) p-chlorobenzoyl chloride

13.31 Nomenclature of Acid Anhydrides: Section 13.1B.2
(a) butanoic anhydride;   (b) propanoic anhydride;
(c) butanoic propanoic anhydride

13.32 Nomenclature of Esters: Section 13.1B.3
(a) methyl pentanoate;   (b) ethyl butanoate;   (c) propyl propanoate;
(d) butyl ethanoate;   (e) pentyl methanoate; (f) isopropy propanoate;
(g) methyl p-nitrobenzoate;   (h) butyl 2,4-hexadienoate

13.33 Nomenclature of Amides: Section 13.1B.4
(a) pentanamide;   (b) N-methylbutanamide;   (c) N-ethylpropanamide;
(d) N,N-dimethylpropanamide; (e) N-ethyl-N-methylethanamide;
(f) N,N-diethyl-m-methylbenzamide;   (g) 4-hyroxy-N-propyl-2-pentenamide

13.34 Nomenclature of Carboxylic Acid Derivatives: Section 13.1

(a) $CH_3CH_2CH_2CO_2H$ (b) $H_2N-\langle\bigcirc\rangle-CO_2H$ (c) $H_2N-\langle\bigcirc\rangle-\overset{\overset{O}{\parallel}}{C}OCH_2CH_3$

(d) $CH_3CH_2CH_2\overset{\overset{O}{\parallel}}{C}OCH_2CH_2CH_2CH_2CH_3$ (e) $CH_3CH=CH-CH=CHCO_2K$

(f) $\langle\bigcirc\rangle-\overset{\overset{O}{\parallel}}{C}NH_2$ with OH   (g) $H\overset{\overset{O}{\parallel}}{C}N(CH_3)_2$ (h) $\langle\bigcirc\rangle-\overset{\overset{O}{\parallel}}{C}N(CH_2CH_3)_2$ with CH_3

(i) $CH_3(CH_2)_2\overset{\overset{O}{\parallel}}{C}O\overset{\overset{O}{\parallel}}{C}(CH_2)_4CH_3$ (j) $CH_3(CH_2)_2\overset{\overset{O}{\parallel}}{C}O\overset{\overset{O}{\parallel}}{C}(CH_2)_2CH_3$ (k) $\langle\bigcirc\rangle-\overset{\overset{O}{\parallel}}{C}Cl$ with Cl

13.35 Reactions of Acid Derivatives: Section 13.3-13.7

(a) $CH_3CH_2\overset{\overset{O}{\parallel}}{C}Cl$ + H_2O $\longrightarrow$ $CH_3CH_2\overset{\overset{O}{\parallel}}{C}OH$ + HCl

(b) $CH_3CH_2\overset{\overset{O}{\parallel}}{C}O\overset{\overset{O}{\parallel}}{C}CH_2CH_3$ + H_2O $\longrightarrow$ $CH_3CH_2\overset{\overset{O}{\parallel}}{C}OH$ + $CH_3CH_2\overset{\overset{O}{\parallel}}{C}OH$

(c) $CH_3CH_2\overset{\overset{O}{\parallel}}{C}OCH_3$ + H_2O $\longrightarrow$ $CH_3CH_2\overset{\overset{O}{\parallel}}{C}OH$ + CH_3OH

(d) $CH_3CH_2\overset{\overset{O}{\parallel}}{C}NH_2$ + H_2O $\longrightarrow$ $CH_3CH_2\overset{\overset{O}{\parallel}}{C}OH$ + NH_3

13.36 Reactions of Acid Chlorides: Section 13.3

The by-product in all of these reactions is HCl except in (a) where it is CH_3CO_2Na.

(a) $C_6H_5{-}\overset{O}{\overset{\|}{C}}{-}O{-}\overset{O}{\overset{\|}{C}}{-}CH_2CH_3$

(b) $C_6H_5{-}\overset{O}{\overset{\|}{C}}{-}OH$

(c) $C_6H_5{-}\overset{O}{\overset{\|}{C}}{-}OCH_3$

(d) $C_6H_5{-}\overset{O}{\overset{\|}{C}}{-}NH_2$

(e) $C_6H_5{-}\overset{O}{\overset{\|}{C}}{-}NHCH_3$

(f) $C_6H_5{-}\overset{O}{\overset{\|}{C}}{-}N(CH_3)CH_2CH_3$

13.37 Reactions of Acid Anhydrides: Section 13.4

The by-product of each of these reactions is $CH_3CH_2CO_2H$.

(a) $CH_3CH_2\overset{O}{\overset{\|}{C}}OH$

(b) $CH_3CH_2\overset{O}{\overset{\|}{C}}OCH(CH_3)$ [OCHCH$_3$ with CH$_3$ below]

(c) $CH_3CH_2\overset{O}{\overset{\|}{C}}NH_2$

(d) $CH_3CH_2\overset{O}{\overset{\|}{C}}{-}N$ (pyrrolidine ring)

13.38 Reactions of Carboxylic Acids

The by-product of these reactions is water except in (d) where it is $SO_2 + HCl$.

(a) $C_6H_5{-}\overset{O}{\overset{\|}{C}}{-}OCH_2CH_2CH_3$

(b) $C_6H_5{-}\overset{O}{\overset{\|}{C}}{-}NH_2$

(c) $C_6H_5{-}\overset{O}{\overset{\|}{C}}{-}N(CH_3)_2$

(d) $C_6H_5{-}\overset{O}{\overset{\|}{C}}{-}Cl$

13.39 Reactions of Esters: Section 13.6

The by-product of all of these reactions is methanol, CH_3OH.

(a) $CH_3CH_2\overset{O}{\overset{\|}{C}}OH$

(b) $CH_3CH_2\overset{O}{\overset{\|}{C}}NH_2$

(c) $CH_3CH_2\overset{O}{\overset{\|}{C}}NHCH_3$

(d) $CH_3CH_2\overset{O}{\overset{\|}{C}}O(CH_2)_4CH_3$

13.40 Reactions of Amides: Section 13.7

Products are shown   (a) $C_6H_5{-}\overset{O}{\overset{\|}{C}}{-}OH$ + NH_3 (b) $CH_3CH_2\overset{O}{\overset{\|}{C}}OH$ + CH_3NHCH_3

13.41 Preparations of Amides: Section 13.7

To make the amide in 13.40a treat any of the following compounds with NH_3.

(a) $C_6H_5\overset{O}{\overset{\|}{C}}Cl$

(b) $C_6H_5\overset{O}{\overset{\|}{C}}O\overset{O}{\overset{\|}{C}}C_6H_5$

(c) $C_6H_5\overset{O}{\overset{\|}{C}}OH$

(d) $C_6H_5\overset{O}{\overset{\|}{C}}OCH_3$

To make the amide in 13.40b treat any of the following compounds with (CH$_3$)$_2$NH.

(a) CH$_3$CH$_2$CCl   (b) CH$_3$CH$_2$COCCH$_2$CH$_3$ (c) CH$_3$CH$_2$COH   (d) CH$_3$CH$_2$COCH$_3$

13.42 Preparations of Esters: Section 13.6

To prepare the ester in problem 13.39, treat any of the following compounds with methanol, CH$_3$OH, or with methanol and acid catalyst in (c).

(a) CH$_3$CH$_2$CCl   (b) CH$_3$CH$_2$COCCH$_2$CH$_3$ (c) CH$_3$CH$_2$COH

13.43 Preparations of Esters: Section 13.6C

(a) CH$_3$CH$_2$CH$_2$CH$_2$CO$_2$H + CH$_3$OH   (b) CH$_3$CH$_2$CH$_2$CO$_2$H + CH$_3$CH$_2$OH

(c) CH$_3$CH$_2$CO$_2$H + CH$_3$CH$_2$CH$_2$OH   (d) CH$_3$CO$_2$H + CH$_3$CH$_2$CH$_2$CH$_2$OH

(e) HCO$_2$H + CH$_3$CH$_2$CH$_2$CH$_2$CH$_2$OH   (f) CH$_3$CH$_2$CO$_2$H + (CH$_3$)$_2$CHOH

(g) O$_2$N⟨ ⟩-CO$_2$H + CH$_3$OH   (h) CH$_3$CH=CH-CH=CHCO$_2$H + CH$_3$CH$_2$CH$_2$CH

13.44 Lactones: Section 13.6

13.45 Reactions of Diacids: Section 13.4

13.46   Nucleophilic Acyl Substitution Mechanisms:

Sections 13.2-13.7

(a) Neutral Nucleophile Initiated

(b) Neutral Nucleophile Initiated

(c) Neutral Nucleophile Initiated

(d) Negative Nucleophile Initiated: Saponification

(e) Acid Catalyzed Esterification

(f) Negative Nucleophile Initiated

13.47   Reactions with LiAlH$_4$: Section 13.9A

(a) (b)   (c) $CH_3(CH_2)_{12}CH_2OH$

(d) $CH_3CH_2CH_2CH_2CH_2NH_2$ (e)

13.48   Acid Derivatives and Grignard Reagents: Section 13.9B

(a)   (b)   (c)

13.49   Nucleophilic Addition Mechanisms: Section 13.9

(a) This mechanism shows the reaction of hydride as a negative nucleophile but does not show the coordination of intermediates with aluminum.

(b) As with the previous mechanism, this one involves nucleophilic acyl substitution and nucleophilic addition.

13.50 Grignard Synthesis of Alcohols: Section 13.9B

13.51 Hydrolysis of Urea: Section 13.7

Urea is an amide (actually a diamide) and is subject to hydrolysis by water, the "solvent" in urine. Water is the nucleophile in the substitution and the result can be visualized as carbonic acid and ammonia. The carbonic acid is unstable and decomposes to carbon dioxide.

13.52 Decomposition of Aspirin: Section 13.6

Aspirin is an ester. In the presence of moisture, such as a humid climate, under prolonged conditions, it can hydroxyze by nucleophilic acyl substitution to an acid and alcohol. The acid formed is acetic acid, vinegar acid.

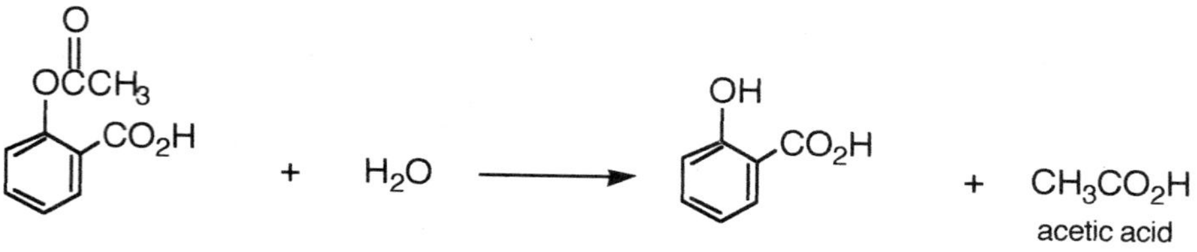

$$+ \quad H_2O \quad \longrightarrow \qquad\qquad + \quad CH_3CO_2H$$

acetic acid

13.53 Hydrolysis of Salol: Section 13.6

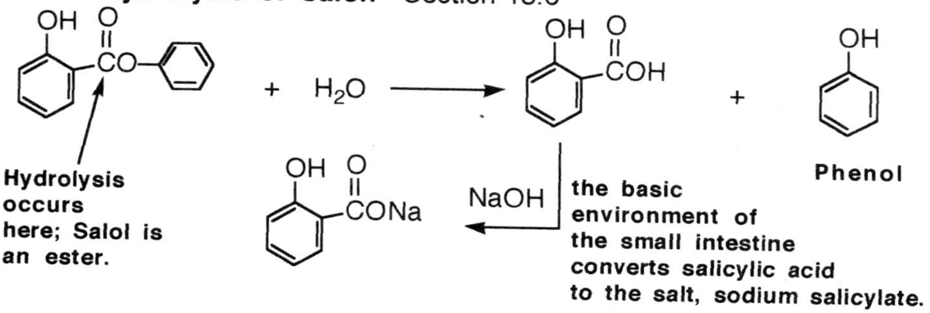

$$+ \quad H_2O \quad \longrightarrow$$

Phenol

Hydrolysis occurs here; Salol is an ester.

NaOH

the basic environment of the small intestine converts salicylic acid to the salt, sodium salicylate.

13.54 Acidity of Carboxylic Acids: Section 12.4A

$$CH_3(CH_2)_{16}CO_2H \quad + \quad NaOH \quad \longrightarrow \quad CH_3(CH_2)_{16}CO_2Na \quad + \quad H_2O$$

13.55 Condensation Polymers: Section 13.8

a) nylon 6-10

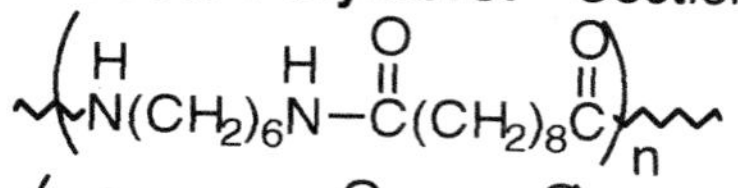

b) nylon 4-6

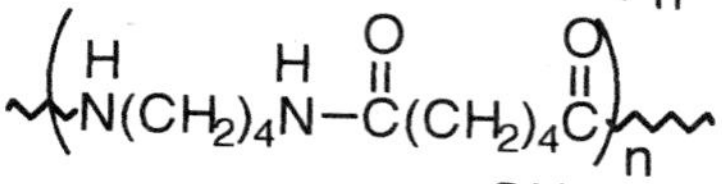

c) polycarbonates

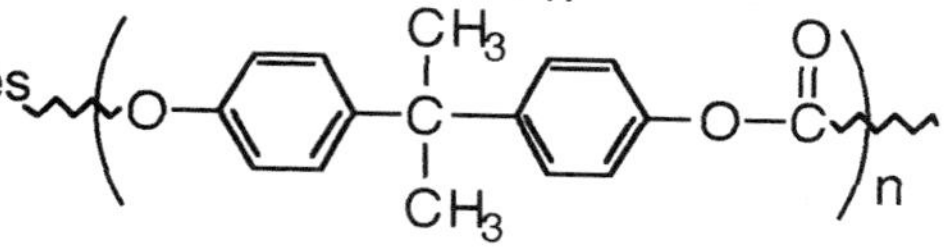

13.56 Polyurethanes: Section 13.8

These polymers appear to be structurally both polyamides and polyesters.

13.62 Preparation of Medicinal Compounds: Chapters 12 and 13

a) 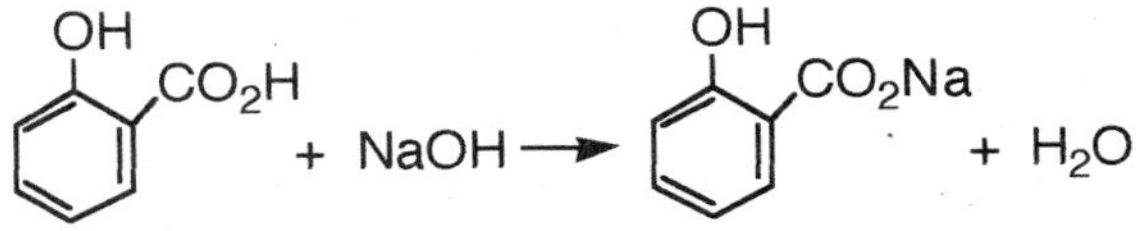 2-hydroxybenzoic acid + NaOH → sodium salt + H_2O

b) CH_3CH_2O—⟨ ⟩—NH_2 + $CH_3\overset{O}{\overset{\|}{C}}O\overset{O}{\overset{\|}{C}}CH_3$ →

CH_3CO_2H + CH_3CH_2O—⟨ ⟩—$NH\overset{O}{\overset{\|}{C}}CH_3$

c) HO—⟨ ⟩—NH_2 + $CH_3\overset{O}{\overset{\|}{C}}O\overset{O}{\overset{\|}{C}}CH_3$ → HO—⟨ ⟩—$NH\overset{O}{\overset{\|}{C}}CH_3$ + CH_3CO_2H

d) H_2N—⟨ ⟩—$\overset{O}{\overset{\|}{C}}OH$ + CH_3CH_2OH $\xrightarrow{H^+}$ H_2N—⟨ ⟩—$\overset{O}{\overset{\|}{C}}OCH_2CH_3$ + H_2O

e) + CH_3OH $\xrightarrow{H^+}$ + H_2O

f) + NH_3 $\xrightarrow{heat}$ + H_2O

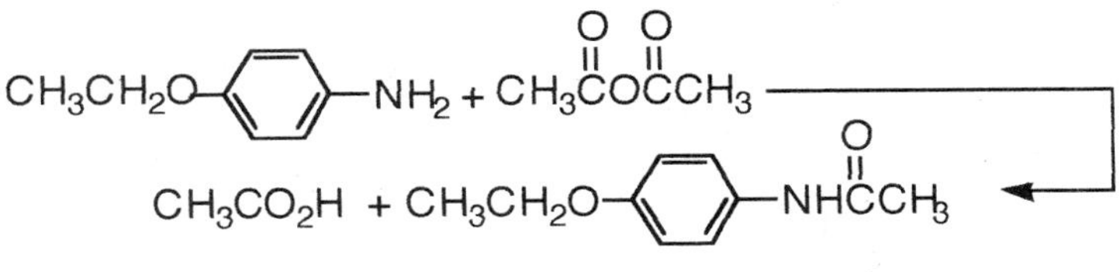

g) + $CH_3\overset{O}{\overset{\|}{C}}O\overset{O}{\overset{\|}{C}}CH_3$ → + CH_3CO_2H

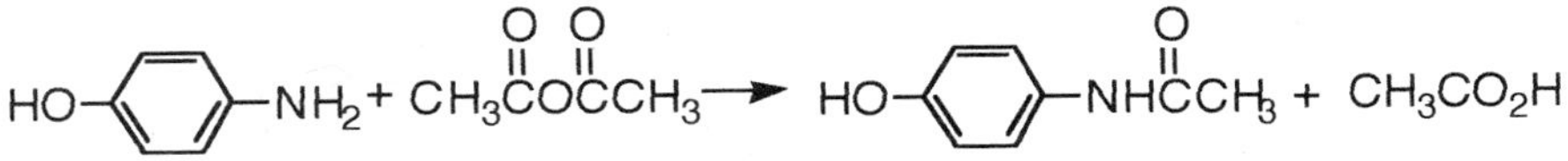

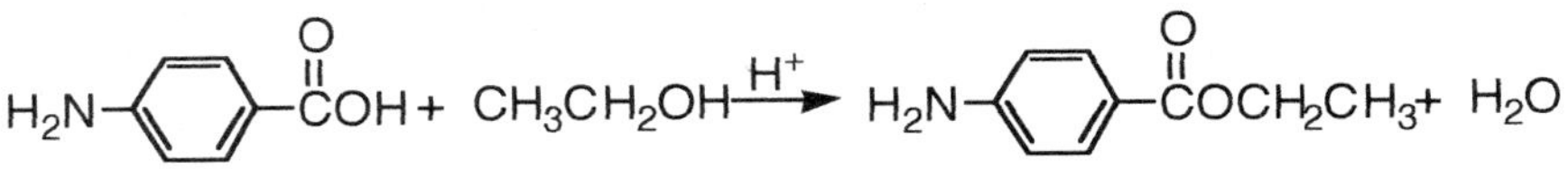

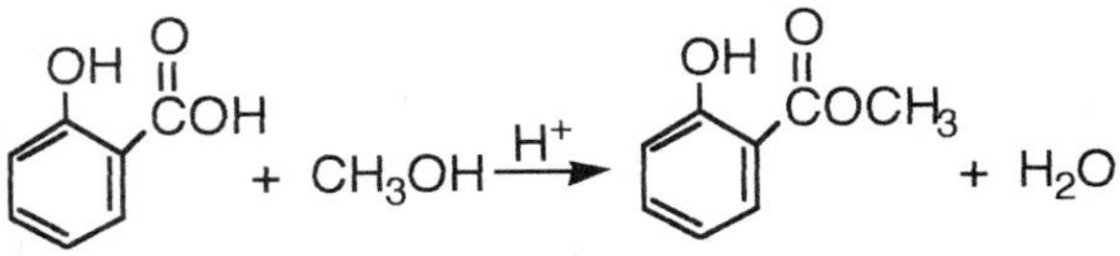

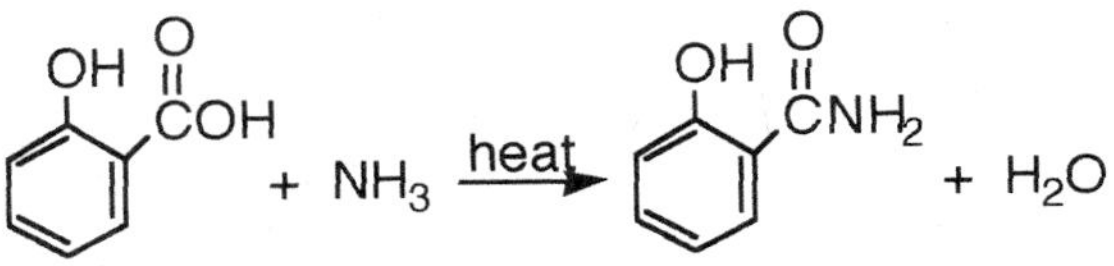

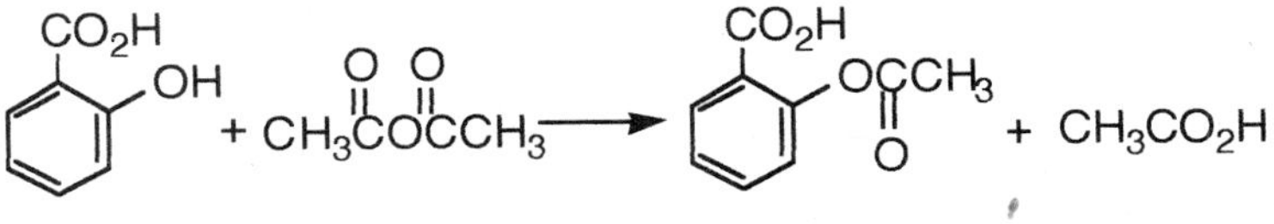

h)

241

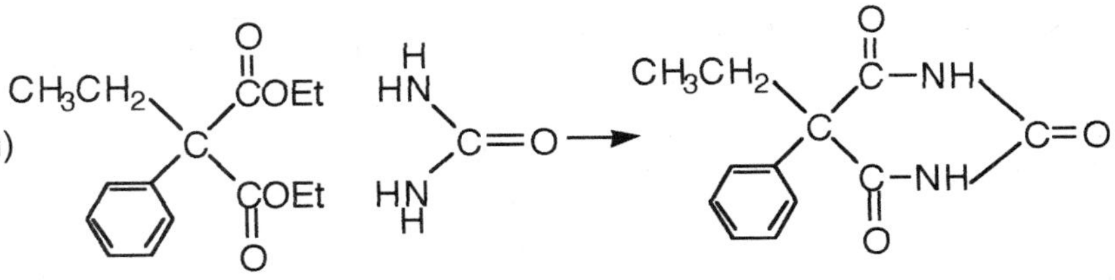

13.63   **Reaction Mechanisms:**   Section 13.6B

In the starting materials, the methanol has the labeled oxygen. Since the ester oxygen has the labeled oxygen in the ester product, the oxygen of the ester must have come from the alcohol.

14

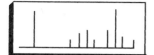
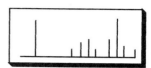

Spectroscopy

CHAPTER SUMMARY

Spectroscopy is an instrumental method for determining the structure of organic compounds by measuring and interpreting their interaction with electromagnetic radiation. Radiation can cause a measurable transformation or pertubation in molecules such as molecular rotation, bond vibration, promotion of electrons to higher energy levels, or even permanent disruption of the molecule.

Energy is described in wavelengths or frequency. The **wavelength** is the distance between two maxima in an energy wave. **Frequency** is the number of waves per unit distance or cycles per second. The energy of a electromagnetic radiation is directly proportional to frequency (the greater the frequency, the greater the energy), and inversely proportional to wavelength (the shorter the wavelength, the greater the energy).

Spectroscopy is possible because molecules absorb exactly the wavelength of energy necessary for a particular permutation and the absorption of these wavelengths is often characteristic of a particular structural feature. It is not possible either to accumulate radiation of lower energies to attain the total needed for a molecular transition or to extract it from higher energy radiation; it must be the exact wavelength or frequency corresponding to the energy of the transition. A **spectrophotometer** is an instrument that measures the absorption of energy by a chemical compound.

In **infrared spectroscopy**, the interaction of compounds with infrared radiation in the 2-15 micrometer wavelength range or frequencies in the 5000 cm^{-1} to 670 cm^{-1} range is measured. This relatively weak radiation causes vibration of bonds in the molecule including stretching, scissoring, bending, rocking, twisting, or wagging. Infrared spectroscopy is useful in identifying functional groups in molecules; this is especially evident in the 1400-3500 cm^{-1} region where the characterizing bonds in alkenes, alkynes, aldehydes, ketones, alcohols, and acids stretch. The remainder of the spectrum, in conjunction with the functional group region, gives a "fingerprint" that is often unique for a compound.

Ultraviolet-visible spectroscopy utilizes the 200-750 nanometer region of the electromagnetic spectrum. Radiation of these wavelengths causes the promotion to higher energy levels of loosely held electrons such as non-bonding electrons or electrons involved in pi-bonds. For absorption in this particular region there must be conjugation of double bonds.

In **nuclear magnetic resonance spectroscopy** energy in the radiofrequency range causes the nuclei some atoms such as ^{1}H and ^{13}C to flip from alignment of their magnetic moments with an external field to non-alignment.

There are three important aspects to ^{1}**H or proton nuclear magnetic resonance.** The number of different signals that appear is often equal to the number of different hydrogens in a molecule. The location of a signal is characteristic of hydrogens in specific chemical environments and is described by **chemical shift**; chemical shift is measured in delta units and in proton nmr most signals come between 0 and 15. Chemical shifts are compared to **tetramethylsilane (TMS)** which has a shift defined as zero. The area under an nmr peak can be integrated. Comparison of the **integration** of the signals on an nmr spectrum gives the ratio of hydrogen types in a molecule; if the molecular formula is known, the actual number of each type of hydrogen can be determined. **Splitting** is caused by the influence of the magnetic fields generated by hydrogens on adjacent carbons on the field felt by a proton. The number of peaks into which a signal is split is one more that the total number of hydrogens on directly adjacent carbons.

Carbon-13 NMR requires sophisticated instrumentation since ^{13}C is only 1.1% of naturally occurring carbon. ^{13}C NMR is useful in the following ways. (1) The number of peaks in a spectrum is the number of non-equivalent carbons in the molecule. (2) The chemical shift provides information about the structural environment of each carbon. The range in ^{13}C NMR is more than 200 delta units. (3) The number of peaks into which a signal is split is one more than the number of hydrogens bonded to that carbon.

Using **mass spectrometry** it is possible to determine the molecular weight and molecular formula of a compound. The structure of the compound is determined by breaking the molecule into smaller identifiable fragments with an electron beam, separating the fragments by mass in a magnetic field, and piecing the identified fragments back together, like a puzzle. The most intense peak in a mass spectrum is called the **base peak.** The peak equal to the molecular weight of the compound is called the **molecular ion.** The molecular formula of a compound is determined using the ratios of natural occurring isotopes of an element. For example, carbon-13 is 1.1% of natural carbon. For every carbon in the molecular ion (M), the M+1 peak is 1.1% of M. For chlorine, the M+2 peak is 33% of M for each chlorine; for bromine it is 50% for each bromine; and for sulfur the M+2 peak is 4.5% of M for each sulfur. If the molecular ion has an odd mass number, there are an odd number of nitrogens in the molecule. When a molecule fragments upon exposure to a beam of electrons, the most common **fragment ions** are those that are most stable; they generally follow carbocation stability principles. By understanding the most likely fragmentation points, the structures of fragment ions can be deduced from their masses and pieced together to determine the structure of the original molecule.

Connections 15.1 is about MRI, Magnetic Resonance Imaging.

SOLUTIONS TO PROBLEMS

14.1 Infrared Spectroscopy: Section 14.2

a)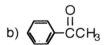

cyclohexene has a C=C stretch around 1600-1670 cm^{-1} and cyclohexane does not.

b)

both compounds have characteristic benzene peaks but this compound has a C=O stretch around 1660-1780 cm^{-1}

c) $CH_3C{\equiv}CH$ as an alkyne, this compound has a triple bond stretch at 2100-2260 cm^{-1} and a C-H (triple bond carbon) stretch around 3300 cm^{-1}

d) $CH_3CH_2\overset{\overset{\textstyle O}{\|}}{C}OH$ both compounds have the carbonyl, C=O stretch but this one also has the O-H stretch for the carboxylic acid as a broad band 2500-3300 cm^{-1}

e) $CH_3CH_2CH_2NH_2$ this is a primary amine and has an N-H stretch appearing as a doublet around 3300-3500 cm^{-1}

$CH_3NHCH_2CH_3$ as a secondary amine, the N-H stretch appears as a singlet around 3300-3500 cm^{-1}

$(CH_3)_3N$ this is a tertiary amine, there is no N-H stretch

f) $CH_3CH_2C{\equiv}N$ the carbon-nitrogen triple bond stretch is at 2210-2260 cm^{-1}

g)   both compounds are similar in that they have carbonyl groups and benzene rings; this one has the C-H bond of the aldehyde group that stretches at 2700-2820 cm^{-1}

h) CH_3CH_2OH there is the characteristic O-H stretch around 3400-3650 cm^{-1} in the alcohol but absent in the ether.

i) CH_3NO_2 the NO$_2$ group gives two strong bands at 1500-1570 cm^{-1} and 1300-1370 cm^{-1}

j)   both have carbonyl and benzene signals but the acid has a broad O-H stretch between 2500 and 3300 cm^{-1}

k) $CH_3CH_2NHCH_3$ N-H stretch as single peak at 3300-3500 cm^{-1}

14.2 ^{1}H Nuclear Magnetic Resonance: Section 14.4

Approximate NMR data follows.

a) CH_4 $\partial = 0.9$ singlet   b) CH_3OCH_3 $\partial = 3.3\text{-}5$ singlet   c) $CH_3\overset{\overset{\displaystyle O}{\|}}{C}CH_3$
                                                                 $\partial = 2\text{-}2.5$ singlet

d) ⬡ $\partial = 7\text{-}8$ singlet   e) CH_3Br $\partial = 2.7\text{-}3.8$ f) $CHBr_3$ $\partial = 4.5\text{-}6$

g) CH_3OH $\partial = 3\text{-}3.5$ singlet, 3H   and $\partial =$ variable singlet, 1H

h) ⬡–CH_3 a: $\partial = 7\text{-}8$ singlet, 5H   i) CH_3–⬡–CH_3 a: $\partial = 2.3\text{-}2.9$ singlet, 6H
   a b   b: $\partial = 2.3\text{-}2.9$ singlet, 3H   a   b   b: $\partial = 7\text{-}8$ singlet, 4H

j) ⬡–$CH_2O\overset{\overset{\displaystyle O}{\|}}{C}CH_3$ a: 7-8 singlet, 5H   k) Cl_2CHCH_2Cl a: $\partial = 5.5\text{-}6.6$ triplet
  a   b   c   b: 5.5 singlet, 2H   a   b   b: $\partial = 2.3\text{-}2.6$ doublet
                           c: 2.3 singlet, 3H

l) CH_3CHBr_2 a: $\partial = 1.3$ doublet, 3H   m) $(CH_3)_2CHOCH(CH_3)_2$ a: $\partial = 1.3$ doublet, 12H
   a   b   b: $\partial = 5.2$ quartet, 1H   a   b   b   a   b: $\partial = 3.3\text{-}5$ heptet,   2H

n) $BrCH_2CH_2CH_2Br$ a: $\partial = 2.7\text{-}3.8$ triplet, 4H
    a   b   a   b: $\partial = 1\text{-}1.6$ pentet, 2H

o) ⬡–$CH_2\overset{\overset{\displaystyle O}{\|}}{C}CH_2Cl$ a: $\partial = 7\text{-}8$ singlet, 5H
         a   b   c   b: $\partial = 3.4$ singlet, 2H
                         c: $\partial = 4.5$ singlet, 2H

p)
            c   d
          CH_2CH_3
              | a: $\partial = 7\text{-}8$ singlet, 5H;   b: $\partial = 3.8$ singlet, 2H
⬡–$CH_2NCH_2CH_3$ c: $\partial = 2.2\text{-}3$ quartet, 4H;   d: $\partial = 0.9\text{-}1.6$ triplet, 6H
 a   b   c   d

q) $CH_2{=}C(OCH_2CH_3)_2$ a: $\partial = 4.5\text{-}6$ singlet, 2H   r) ⬡–$\underset{\underset{\displaystyle Br}{|}}{C}HCH_3$ a: $\partial = 7\text{-}8$ singlet, 5H
  a   b   c   b: $\partial = 3.3\text{-}5$ quartet, 4H   a   b c   b: $\partial = 4.5$ quartet, 1H
                        c: $\partial = 0.9\text{-}1.6$ triplet, 6H   c: $\partial = 1.3$ doublet, 3H

14.3 ¹H Nuclear Magnetic Resonance: Section 14.4

a) $CH_3\overset{\overset{\displaystyle O}{\|}}{C}CH_3$

There is only one signal in the NMR and this compound has only one type of hydrogen. The other compound has three types of hydrogens and would have three signals.

b) ⟨benzene ring⟩$-\overset{\overset{\displaystyle O}{\|}}{C}OCH_3$

Each compound has only two types of hydrogens which appear as singlets. The difference is in the chemical shift of the methyl group. Here it is connected to oxygen and comes at $\partial = 3.9$. In the other compound it is connected to a carbonyl and would appear at $\partial = 2\text{-}2.5$.

c) $\underset{a}{CH_3}O\underset{b}{\overset{\overset{\displaystyle c}{\underset{\displaystyle |}{CH_3}}}{C}}\underset{c}{HCH_3}$

This compound has three types of hydrogens as shown and three signals: a at 3.1, b at 3.5, and c at 1.1. The other compound has only two types of hydrogens and would have only two signals one of which would be a triplet and the other a quartet.

d) $\underset{a}{CH_3}\underset{b}{CH_2}-\overset{c}{⟨benzene⟩}-\underset{b}{CH_2}\underset{a}{CH_3}$

This is the only compound that gives three signals; the others give two. In addition, this compound shows splitting whereas the others give only singlets. a is at 1.3, b at 2.7, and c at 7.2.

e) $\underset{a}{CH_3}O-⟨benzene⟩-\underset{c}{CH_2}\underset{d}{CH_3}$
$\underset{b}{}$

a is at $\partial = 3.7$, b at 7.0, c at 2.6, and d at 1.2. The third compound gives only three peaks with no splitting and can be eliminated. The first two each give four peaks with identical splitting patterns. The difference is in the chemical shifts. The methyl singlet (a) comes at 3.7 in this compound as it is connected to an oxygen. In the other compound it is connected to the benzene ring and would come at 2.3-2.9. Likewise, the CH_2 quartet comes at 2.6 here since it is connected to the benzene ring. In the other compound, it is bonded to oxygen and would come at 3.3-5.

14.4 ^{1}H Nuclear Magnetic Resonance: Section 14.4

a) CH$_3$COCH$_3$
   (O double bond above first C)
   a   b

a: ∂ = 2.0, singlet, 3H
b: ∂ = 3.7, singlet, 3H

b) CH$_3$COCCH$_3$ b
   (O double bond, CH$_3$ groups labeled b)
   a   CH$_3$
           b

a: ∂ = 2.1, singlet, 3H
b: ∂ = 1.4, singlet, 9H

c) CH$_3$CH$_2$OH
   a   b   c

a: ∂ = 1.1, triplet, 3H
b: ∂ = 4.4, quartet, 2H
c: ∂ = 3.6, singlet, 1H

d) CH$_3$CCH$_2$CH$_3$
   (O double bond above second C)
   a   b   c

a: ∂ = 2.1, singlet, 3H
b: ∂ = 2.4, quartet, 2H
c: ∂ = 1.1, triplet, 3H

e) CH$_3$CHCH$_3$
   a   b   a
   Br

a: ∂ = 1.7, doublet, 6H
b: ∂ = 3.4, heptet, 1H

f) CH$_3$COH
   (O double bond above C)
   a   b

a: ∂ = 2.0, singlet, 3H
b: ∂ = 11.4, singlet, 1H

g) ClCH$_2$CHCl
   (Cl above second C)
   a   b

a: ∂ = 3.9, doublet, 2H
b: ∂ = 5.8, triplet, 1H

h) (benzene ring)—CH$_3$
   a   b

a: ∂ = 7.2, singlet, 5H
b: ∂ = 2.3, singlet, 3H

i) (benzene ring)—CH—(benzene ring)
   a   b   a
            Cl

a: ∂ = 7.3, singlet, 10H
b: ∂ = 6.1, singlet, 1H

j) (benzene ring)—CHCCH$_3$
   a   b O c
   (second benzene ring below)
            a

a: ∂ = 7.0, singlet, 10H
b: ∂ = 5.0, singlet, 1H
c: ∂ = 2.1, singlet, 3H

k) CH$_3$CHCOH
   a   b   O
   Cl   c

a: ∂ = 1.8, doublet, 3H
b: ∂ = 4.5, quartet, 1H
c: ∂ = 11.2, singlet, 1H

l) CH$_3$CH$_2$OCCHCl$_2$
   a   b   (O double bond) c

a: ∂ = 1.4, triplet, 3H
b: ∂ = 4.3, quartet, 2H
c: ∂ = 6.9, singlet, 1H

m) CH$_3$CH$_2$OCCH$_2$COCH$_2$CH$_3$
   a   b   (O) c   (O) b   a

a: ∂ = 1.3, triplet, 6H
b: ∂ = 4.2, quartet, 4H
c: ∂ = 3.4, singlet, 2H

n) $\langle \rangle$—CH_2CH_3
   a   b   c

a: $\partial = 7.2$, singlet, 5H
b: $\partial = 2.7$, quartet, 2H
c: $\partial = 1.3$, triplet, 3H

14.5 Carbon-13 NMR Without Splitting: Section 14.5

(a) C_3H_7Br

$CH_3CH_2CH_2Br$ This compound has three different types of carbons and is the one with three signals at 36, 26, and 13.

CH_3CHCH_3 The two methyl carbons are equivalent here so there are only two
   $\underset{Br}{}$ different types of carbons and two signals, 45 and 28.

(b) C_4H_9Cl

Following are the four compounds with the formula C_4H_9Cl and the number of C-13 NMR signals each would give (equal to the number of different types of hydrogens).

a   b   c   d   a   b   c   d   a b $\overset{a}{C}H_3$ c   a b $\overset{a}{C}H_3$ a
$CH_3CH_2CH_2CH_2Cl$ $CH_3CH_2CHCH_3$ CH_3CHCH_2Cl CH_3CCH_3
                              $\underset{Cl}{}$ $\underset{Cl}{}$

   4 signals   4 signals   3 signals   2 signals

(c) Dibromobenzenes

Each isomer has a different number of different types of carbons and thus can easily be identified by C-13 NMR.

ortho: 134, 128, 125
3 types of C's, 3 signals

meta: 134, 131, 130, 123
4 types of C's, 4 signals

para: 133, 121
2 types of C's, 2 signals

(d) trimethylbenzenes

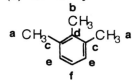

6 types of carbons
6 NMR signals

9 types of carbons
9 NMR signals

3 types of carbons
3 NMR signals

(e) tetramethylbenzenes

5 types of carbons
5 signals
135,134,127,21,16

7 types of carbons
7 signals
136,134,132,128,21,20,15

3 types of carbons
3 signals
134,131,19

The methyl groups carbons are the signals in the range of 15-21; the chemical shifts around 127-136 are aromatic carbons.

(f) hexamethylbenzene

CH₃—[ring]—CH₃ ... (hexamethylbenzene structure)

In hexamethylbenzene, the six methyl carbons are equivalent and the six benzene carbons are equivalent. As a result, there are only two carbon-13 NMR signals in the spectrum.

(g) C₄H₁₀O Isomers

a b c d
CH₃CH₂CH₂CH₂OH

4 types of C
4 signals

a b c d
CH₃CH₂CHCH₃
　　　OH

4 types of C
4 signals

a
CH₃ c
CH₃CHCH₂OH
　b

3 types of C
3 signals

a
CH₃
CH₃CCH₃ a
　OH b

2 types of C
2 signals

a b c d
CH₃CH₂CH₂OCH₃

4 types of C
4 signals

a
CH₃ c
CH₃CHOCH₃
　b

3 types of C
3 signals

a b b a
CH₃CH₂OCH₂CH₃

2 types of C
2 signals

251

(h) C_5H_{12} isomers

$$\overset{a}{C}H_3\overset{b}{C}H_2\overset{c}{C}H_2\overset{b}{C}H_2\overset{a}{C}H_3$$

3 types of carbons
3 signals

$$\overset{a}{C}H_3\overset{\overset{a}{C}H_3}{\underset{|}{\overset{b}{C}H}}\overset{c}{C}H_2\overset{d}{C}H_3$$

4 types of carbons
4 signals

$$a\;CH_3\overset{\overset{a}{C}H_3}{\underset{\underset{a}{C}H_3}{\overset{|}{\underset{|}{C}}}}CH_3\;a$$

2 types of carbons
2 signals

(i) $C_5H_{10}O$ ketones

$$\overset{a}{C}H_3\overset{b}{C}H_2\overset{c}{C}H_2\overset{d}{\underset{\underset{O}{\|}}{C}}\overset{e}{C}H_3$$

5 types of carbons
5 signals

$$\overset{a}{C}H_3\overset{b}{C}H_2\overset{c}{\underset{\underset{O}{\|}}{C}}\overset{b}{C}H_2\overset{a}{C}H_3$$

3 types of carbons
3 signals: 212,35,8

$$\overset{a}{C}H_3\overset{b}{\underset{|}{\overset{\overset{a}{C}H_3}{C}}}H\overset{c}{\underset{\underset{O}{\|}}{C}}\overset{d}{C}H_3$$

4 types of carbons
4 signals

(j) C_8H_{18} isomer

$$\overset{a}{C}H_3 \overset{a}{C}H_3$$
$$\overset{a}{C}H_3—\overset{b}{C}—\overset{b}{C}—\overset{a}{C}H_3$$
$$\overset{}{C}H_3 \overset{}{C}H_3$$
$$\underset{a}{} \underset{a}{}$$

This is the most symmetrical of the 18 isomers. There are only two kinds of carbons and thus only two signals in the carbon-13 NMR spectrum. (There is only one kind of hydrogen and only one peak in the proton NMR spectrum.)

14.6 ^{13}C NMR with Splitting: Section 14.5

(a) C_2H_6O isomers

$\underset{a}{C}H_3\underset{b}{C}H_2OH$ There are two kinds of carbons and two signals. a is a quartet since there are three attached hydrogens and b is a triplet since there are two hydrogens,

$\underset{a}{C}H_3O\underset{a}{C}H_3$ There is only one kind of carbon and thus only one signal. Since the carbons have three attached hydrogens the one signal appears as a quartet.

(b) C_3H_8O isomers

$$CH_3CH_2CH_2OH$$
a   b   c

3 signals
a: quartet
b: triplet
c: triplet

OH
$$CH_3\overset{|}{C}HCH_3$$
a   b   a

2 signals
a: quartet
b: doublet

$$CH_3CH_2OCH_3$$
a   b   c

3 signals
a: quartet
b: triplet
c: quartet

(c) C_4H_8O

O
||
$$CH_3CH_2CH_2CH$$
a   b   c   d

4 signals
a: quartet   b: triplet
c: triplet   d: doublet

O
||
$$CH_3CH_2CCH_3$$
a   b   c   d

4 signals
a: quartet   b: triplet
c: singlet   d: quartet

    b   O
    | ||
$$CH_3\overset{|}{C}HCH$$
a   a CH_3 c

3 signals
a: quartet   b: doublet
c: doublet

(d) C_4H_{10}

a   b   b   a
$$CH_3CH_2CH_2CH_3$$

2 signals
a: quartet   b: triplet

       a
     CH_3
a | a
$$CH_3\overset{|}{C}HCH_3$$
      b

2 signals
a: quartet   b: doublet

(e) C_5H_{12} isomers

a   b   c   b   a
$$CH_3CH_2CH_2CH_2CH_3$$

3 signals
a: quartet   b: triplet
c: triplet

       a
     CH_3
a |
$$CH_3\overset{|}{C}HCH_2CH_3$$
     b   c   d

4 signals
a: quartet   b: doublet
c: triplet   d: quartet

      a CH_3
        |b
a $CH_3\overset{|}{\underset{|}{C}}CH_3$ a
      CH_3 a

2 signals
a: quartet
b: singlet

14.7 ^{13}C NMR: Section 14.5

         c
    O CH_3
    || b | c
   $HCOCCH_3$
   a
        |
       CH_3
       c

a: $\partial = 161$; doublet

b: $\partial = 81$; singlet

c: $\partial = 28$; quartet

14.8 ^{13}C NMR: Section 14.5

```
                                 133
                      115
              165                          190
                 \     \       /         /
        55  ──→  CH₃O─        ─CH
                 /     /       \
              115          133   131
```

14.9 Mass Spectrometry: Section 14.6

a) C_8H_{18} $\dfrac{\text{No. of}}{\text{carbons}} = \dfrac{M+1}{0.011 \times M} = \dfrac{8.8}{0.011 \times 100} = 8$

b) C_2H_5Cl $\dfrac{\text{No. of}}{\text{carbons}} = \dfrac{M+1}{0.011 \times M} = \dfrac{2.2}{0.011 \times 100} = 2$

The M + 2 peak is 33% of M indicating one chlorine.

M = 2 C's (24) + 1 Cl (35) + 5 H (5) = 64

c) C_3H_5BrO $\dfrac{\text{No. of}}{\text{carbons}} = \dfrac{M+1}{0.011 \times M} = \dfrac{1.3}{0.011 \times 40} = 3$

The M + 2 peak is almost equal to the M indicating the presence of one bromine (or three chlorines).

M = 3 C's (36) + 1 Br (79) + 1 O (16) + 5 H (5) = 136

d) CH_4S $\dfrac{\text{No. of}}{\text{carbons}} = \dfrac{M+1}{0.011 \times M} = \dfrac{1.1}{0.011 \times 100} = 1$

The M + 2 peak is 4.5% of M indicating one sulfur.

M = 1 C (12) + 1 S (32) + 4 H (4) = 48

e) $C_2H_2Cl_2$ $\dfrac{\text{No. of}}{\text{carbons}} = \dfrac{M+1}{0.011 \times M} = \dfrac{1.8}{0.011 \times 80} = 2$

$\dfrac{M+2}{M} = \dfrac{54}{80} = 0.675$

The M + 2 peak is 0.675 of M peak indicating the presence of two chlorines.

M = 2 C's (24) + 2 Cl's (70) + 2 H (2) = 96

254

f) $C_6H_4Br_2$ No. of carbons $= \dfrac{M+1}{0.011 \times M} = \dfrac{3.3}{0.011 \times 50} = 6$

$$\frac{M+2}{M} = \frac{99}{50} = 1.98$$

The M + 2 peak is twice M indicating the presence of two bromines.

$$M = 6 \text{ C's } (72) + 2 \text{ Br's } (158) + 4 \text{ H } (4) = 234$$

g) C_3H_9N No. of carbons $= \dfrac{M+1}{0.011 \times M} = \dfrac{2.5}{0.011 \times 75} = 3$

The M peak is odd indicating the presence of an odd number of nitrogens.

$$M = 3 \text{ C's } (36) + 1 \text{ N } (14) + 9 \text{ H } (9) = 59$$

h) C_7H_5OCl No. of carbons $= \dfrac{M+1}{0.011 \times M} = \dfrac{2.3}{0.011 \times 30} = 7$

$$\frac{M+2}{M} = \frac{10}{30} = 0.33$$

The M + 2 is one third of M indicating one chlorine.

M = 7 C's (84) + 1 Cl (35) + 1 O (16) + 5 H (5) = 140

i) $C_2H_4S_2$ $\dfrac{\text{No. of}}{\text{carbons}} = \dfrac{M + 1}{0.011 \times M} = \dfrac{1.5}{0.011 \times 70} = 2$

$\dfrac{M + 2}{M} = \dfrac{6.3}{70} \times 100\% = 9\%$

M + 2 is 9% of M indicating two sulfurs (4.5% of M for each).

M = 2 C's (24) + 2 S's (64) + 4 H (4) = 92

14.10   Mass Spectrometry: Section 14.6

(a) The shortest alkyl group is 29: CH_3CH_2+

The longest acyl group is 99:

$$CH_3CH_2CH_2CH_2CH_2\overset{\overset{\textstyle O}{\|}}{C}+$$

The ketone is:

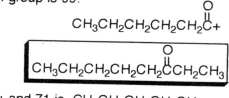

$$CH_3CH_2CH_2CH_2CH_2\overset{\overset{\textstyle O}{\|}}{C}CH_2CH_3$$

57 is $CH_3CH_2\overset{\overset{\textstyle O}{\|}}{C}+$ and 71 is $CH_3CH_2CH_2CH_2CH_2+$

b) $R_1CH_2CH=CHCH_2R_2$ Alkenes fragment by losing R_1 and R_2.

98 - 83 = 15 indicating one R is a CH_3

98 - 69 = 29 indicating other R is an CH_2CH_3

The alkene is $CH_3CH_2CH=CHCH_2CH_2CH_3$ —>

$+CH_2CH=CHCH_2CH_2CH_3$ 83 $CH_3CH_2CH=CHCH_2+$ 69

c)
$$R_1-\underset{\underset{\displaystyle}{|}}{\overset{\overset{\displaystyle R_2}{|}}{C}}-R_3$$

Aromatic compounds of this type fragment by losing R_1, R_2, and R_3 to form benzylic carbocations.

162 - 147 = 15 indicating that at least one R is CH_3

162 - 133 = 29 indicating that at least one R is CH_3CH_2

Since there are 12 carbons, the following must be the compound.

$$CH_3CH_2\underset{\underset{\displaystyle Ph}{}}{\overset{\overset{\displaystyle CH_3}{|}}{C}}CH_2CH_3 \longrightarrow \quad {}^{+}\underset{\underset{\displaystyle Ph}{}}{\overset{\overset{\displaystyle CH_3}{|}}{C}}CH_2CH_3 \ (133) \text{ and } CH_3CH_2\overset{+}{\underset{\underset{\displaystyle Ph}{}}{C}}CH_2CH_3 \ (147)$$

d)
$$R_1CHR_2 \atop \ \ \ OH$$

Alcohols fragment by loss of alkyl groups, R_1 and R_2.

116 - 87 = 29 indicating one group is CH_3CH_2.

116 - 59 = 57 indicating one group is $CH_3CH_2CH_2CH_2$.

The alcohol is:

$$CH_3CH_2\underset{\underset{\displaystyle OH}{|}}{C}HCH_2CH_2CH_2CH_3 \longrightarrow {}^{+}\underset{\underset{\displaystyle OH}{|}}{C}HCH_2CH_2CH_2CH_3 \ (87) \text{ and } CH_3CH_2\underset{\underset{\displaystyle OH}{|}}{C}H{}^{+} \ (59)$$

e) $R_1CH_2NHCH_2R_2$ Amines fragment by loss of alkyl groups (R_1, R_2).

87 - 72 = 15 indicating one R is CH_3-

87 - 58 = 29 indicating other R is CH_3CH_2-

The amine is:

$$CH_3CH_2NHCH_2CH_2CH_3 \longrightarrow {}^{+}CH_2NHCH_2CH_2CH_3 \ (72) \text{ and } CH_3CH_2NHCH_2{}^{+} \ (58)$$

f) Esters fragment as follows:

$$R_1COR_2 \longrightarrow R_1{}^+ \quad R_1C^+ \quad {}^+COR_2$$

57 is $CH_3CH_2CH_2CH_2+$, 85 is $CH_3CH_2CH_2CH_2\overset{O}{\overset{\|}{C}}+$

and 115 is $+COCH_2CH_2CH_2CH_2CH_3$

The ester is:

$$CH_3CH_2CH_2CH_2\overset{}{\underset{O}{\overset{\|}{C}}}OCH_2CH_2CH_2CH_2CH_3$$

14.11 Mass Spectrometry: Section 14.6

a) $CH_3(CH_2)_5+$ $CH_3(CH_2)_5\overset{O}{\overset{\|}{C}}+$ $CH_3(CH_2)_3+$ $CH_3(CH_2)_3\overset{O}{\overset{\|}{C}}+$

        85   113   57   85

b) CH_3CH_2+ $CH_3CH_2\overset{O}{\overset{\|}{C}}+$ $+\overset{O}{\overset{\|}{C}}OH$

    29   57   45

c) $CH_3(CH_2)_5+$ $CH_3(CH_2)_5\overset{O}{\overset{\|}{C}}+$ d) CH_3+ $CH_3\overset{O}{\overset{\|}{C}}+$ $+\overset{O}{\overset{\|}{C}}OCH_2CH_3$

    85   113   15   43   73

e) + $-CH_2+$

   77   91

f) + $-\overset{+}{\underset{CH_2CH_2CH_3}{C}CH_2CH_3}$ $-\overset{CH_3}{\underset{CH_2CH_2CH_3}{\overset{|}{C}+}}$ $-\overset{CH_3}{\underset{+}{\overset{|}{C}CH_2CH_3}}$

   77   161   147   133

g) $\overset{\displaystyle CH_3}{+CH}CH{=}CHCH_2CH_2CH_2CH_3$ $CH_3CH_2\overset{+}{C}HCH{=}CHCH_2CH_2CH_2CH_3$

              111   125

$\overset{\displaystyle CH_3}{CH_3CH_2CH}CH{=}CHCH_2+$

            97

h) $+CH_2\overset{\displaystyle CH_3}{C}{=}\overset{\displaystyle CH_3}{C}CH_3$ i) $+CH_2\overset{\displaystyle CH_3}{N}CH_2CH_3$, $CH_3CH_2CH_2\overset{\displaystyle CH_3}{N}CH_2+$

      83   72   86

j) $HOCH_2+$ $+\overset{\displaystyle O}{\overset{\|}{C}}OCH_2CH_2NH_2$ $HOCH_2CH_2\overset{\displaystyle O}{\overset{\|}{C}}+$ $+CH_2NH_2$

    31   88   73   30

15

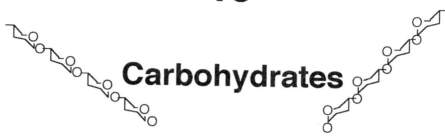

Carbohydrates

CHAPTER SUMMARY

Carbohydrates are a class of organic biopolymers which consist of polyhydroxy aldehydes and ketones, their derivatives and polymers. Other terms for carbohydrates include sugars and saccharides. A single monomer unit is called a **monosaccharide**; several units are referred to as an **oligosaccharide**; larger polymers are called **polysaccharides**.

The nomenclature of carbohydrates usually includes the suffix **-ose**. Monosaccharides may also be identified according to the nature of the carbonyl functional group (**aldose** or **ketose**), the number of carbons in the molecule (tri-, tetr-, pent- ose) or a combination of these two. Monosaccharides also have common names such as **ribose, glucose, galactose**, and **fructose** (four of the most common monosaccharides found in nature).

Monosaccharides have one or more chiral centers and can thereby form **enantiomers** and **diastereomers**. Most common monosaccharides are in the **D-family**. This means that using D-glyceraldehyde as a starting point, other chiral carbons can be inserted between the carbonyl group and the D-carbon, producing families of diastereomers. If two monosaccharides differ in their structures by the configuration at only one chiral center, then they are called **epimers**.

The carbonyl and alcohol groups within the same monosaccharide may react together if the carbon chain is long enough. The result is a **cyclic hemiacetal**. A new chiral center is formed at the carbon which was previously the carbonyl. The two optical isomers that can result are called **anomers**. Five- and six-membered cyclic structures predominate with the alcohol oxygen as the last member of the ring. These are referred to as **furanoses** and

pyranoses, respectively. Cyclic structures exist in equilibrium with the open-chain form.

The method of drawing saccharides in a vertical orientation with the most highly oxidized carbon at the top is called a **Fischer Projection**. It does not represent the real, three-dimensional structure of the molecule. **Haworth Formulas** show the cyclic nature of monosaccharides with the -OH and -CH_2OH groups oriented up and down around a planar ring, that is, above and below the plane of the ring. **Conformational structures** more accurately illustrate the three-dimensional nature of cyclic monosaccharides, especially in the chair conformation of six-membered rings. The -OH and -CH_2OH groups are oriented in axial and equatorial positions around the ring and correspond to the up and down placement in a Haworth Formula. Anomeric isomers are designated as α- if the -OH group is down or axial or as β- if up or equatorial.

In nature monosaccharides can be found in an oxidized state. If the carbonyl of an aldose is oxidized to a carboxyl group the result is called an -**onic acid**. If the primary alcohol group is oxidized to a carboxylic acid the molecule is a **-uronic acid**. For example, glucose would become gluconic acid and glucuronic acids, respectively.

The easy oxidation of the aldehyde group using a mild oxidizing agent such as copper (II) and silver (I) can detect the presence of carbohydrate. The carbohydrates are referred to as **reducing sugars**. This type of test cannot distinguish between aldoses and ketoses, however, because the alkaline conditions of the reaction lead to tautomerization of the ketone and immediate oxidation.

The carbonyl group can be reduced to produce **sugar alcohols** like sorbitol and mannitol. Also the alcohol groups may be esterified with a variety of acids including phosphoric acid. These derivatives are found extensively in metabolism. Reaction between one of the many alcohol groups on a monosaccharide with the hemiacetal group of an adjacent monosaccharide molecule to form an acetal is the method by which carbohydrates polymerize into disaccharides (sucrose and lactose), oligosaccharides, and polysaccharides such as starch, cellulose and glycogen. The α- or β-configuration of the anomeric carbon will be locked into position by this polymerization process. Not only does the diether linkage, called a **glycosidic bond**, resist oxidation by weak oxidizing agents (becoming **nonreducing**

sugars) but it also is metabolically stable. Stereo-specific hydrolysis agents, known as enzymes, are required to cleave the glycoside.

The monosaccharide glucose is the most important source of metabolic energy in living organisms. Panel 15.1 describes the condition known as **diabetes mellitus**, the inability to use glucose. Carbohydrate derivatives are easily found in nature, often functioning as cell-specific markers. Panel 15.2 emphasizes the importance of ascorbic acid, Vitamin C, an essential metabolic product of glucose oxidation while Panel 15.3 outlines our quest for low- and noncaloric sugar substitutes. Panel 15.4 illustrates the semi-synthetic materials rayon and nitrocellulose made from cellulose.

SOLUTIONS TO PROBLEMS

15.1 Optical Isomerism in Carbohydrates: Section 15.3
There are no enantiomers or meso forms in Figure 15.1.
Erythrose and threose are diastereomers. Ribose, arabinose, xylose, and lyxose are diastereomeric with each other. All of the D-aldohexoses are diastereomers also.

15.2 Nomenclature: Sections 15.2, 15.3
Allose is an aldohexose while xylose is an aldopentose.

15.3 Structure: Section 15.3

D-idose   **L-idose**   **D-glucose**   **L-glucose**

15.4 Structures: Section 15.3

a	b	c	d
CH₂OH	CH₂OH	CH₂OH	CH₂OH

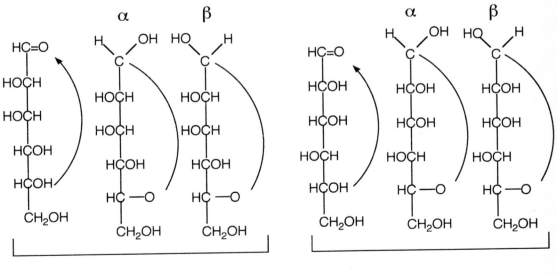

The epimers are a/c, a/d, b/c, and b/d.

15.5 Cyclic Structures: Section 15.3

D-mannose

D-glucose

15.6 Haworth Formulas: Section 15.3

Five-membered rings

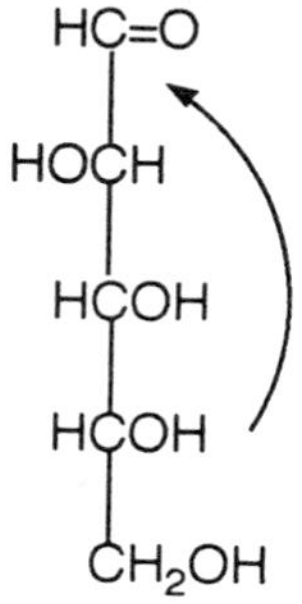

D-arabinose

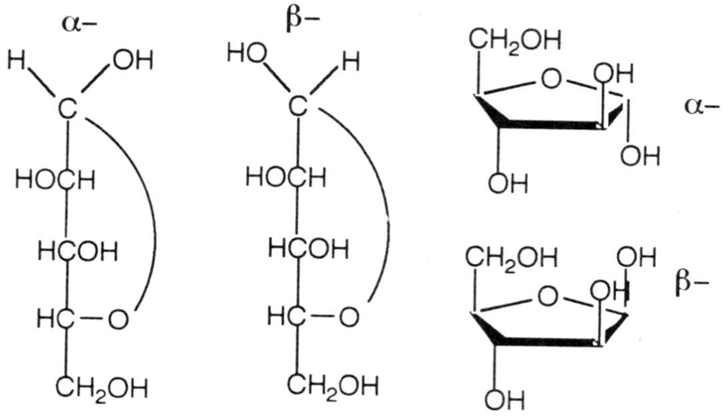

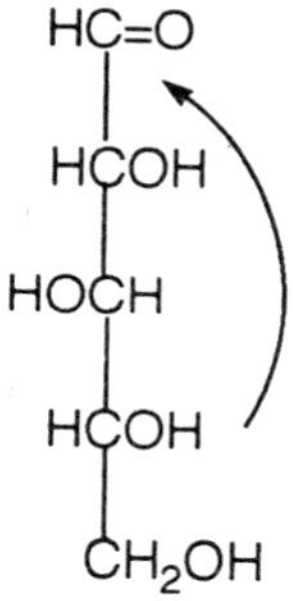

D-xylose

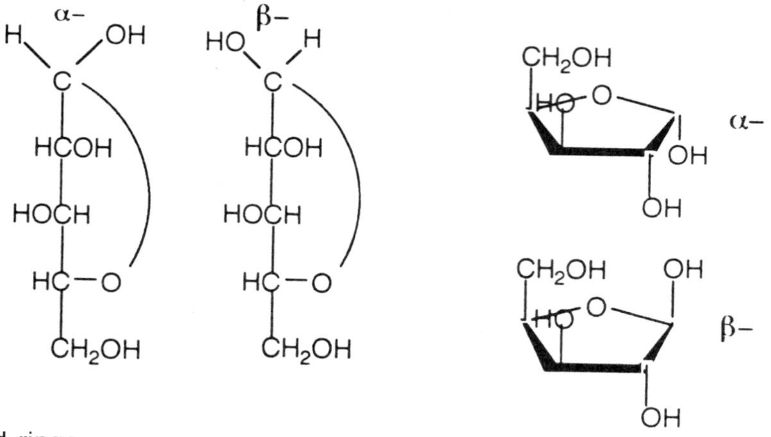

Six-membered rings

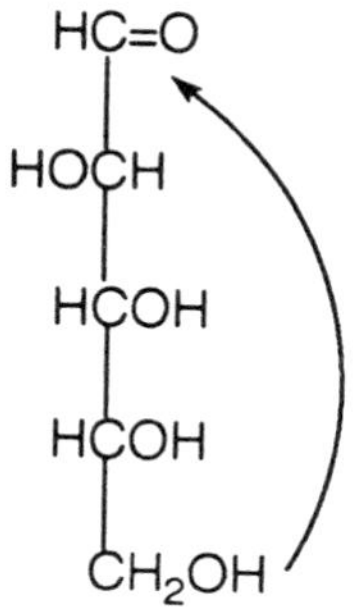

D-arabinose

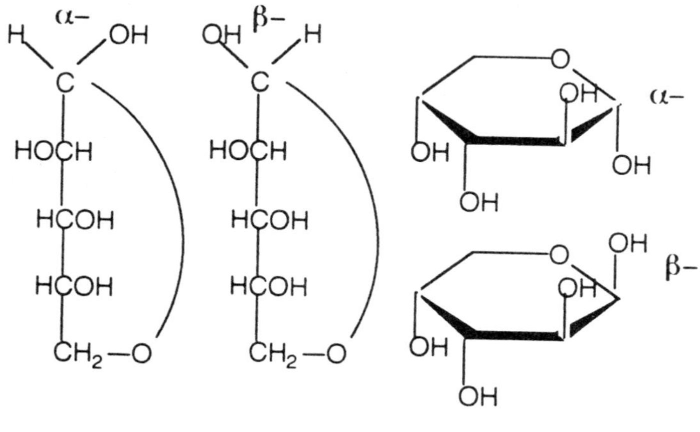

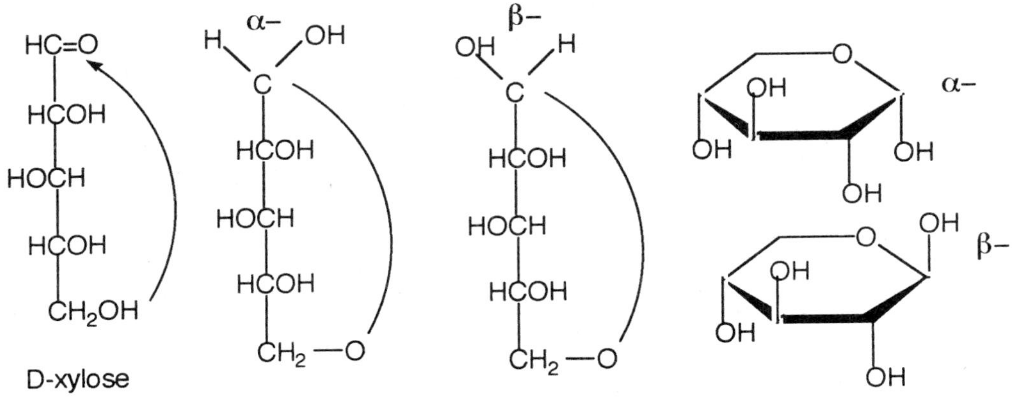

D-xylose

15.7 Cyclic Structures: Section 15.3

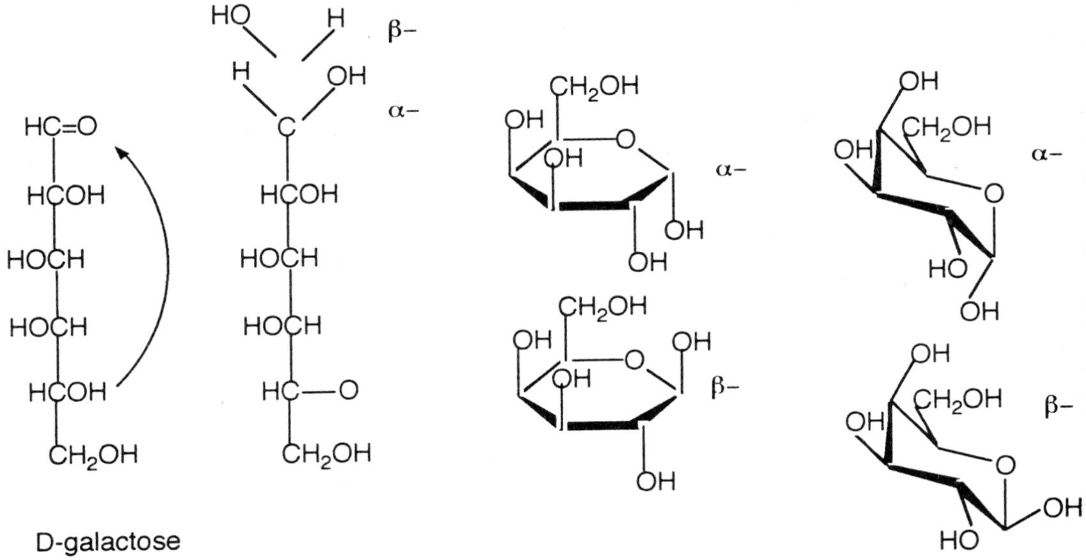

D-galactose

15.8 Cyclic and Open-chain Structures: Section 15.3

a)

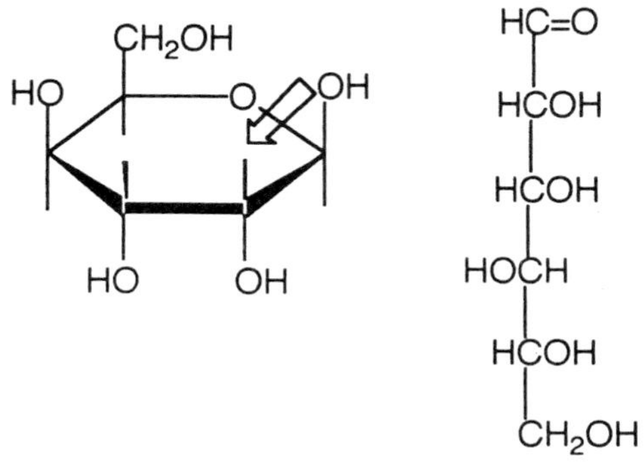

b)

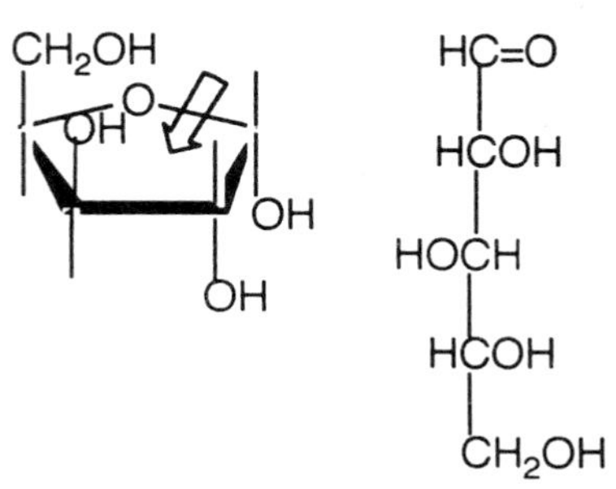

c)

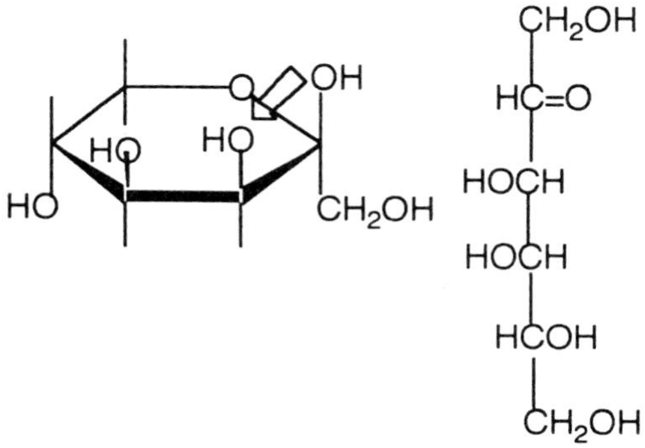

d)

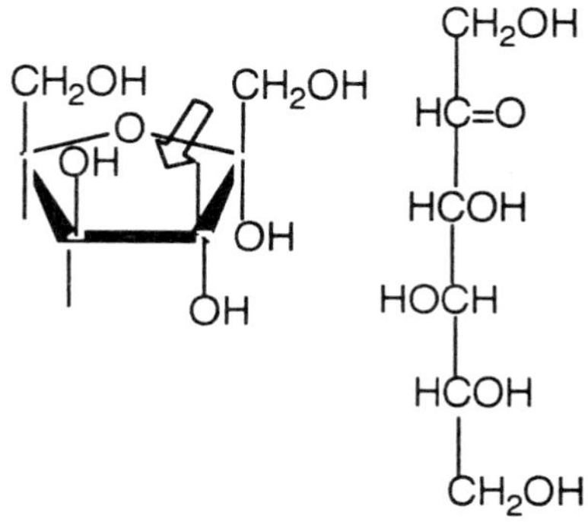

15.9 Oxidation: Section 15.4

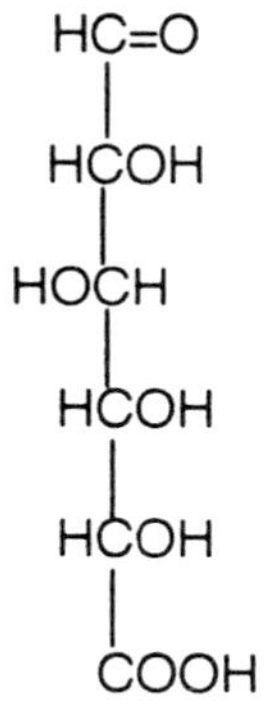

glucuronic
acid

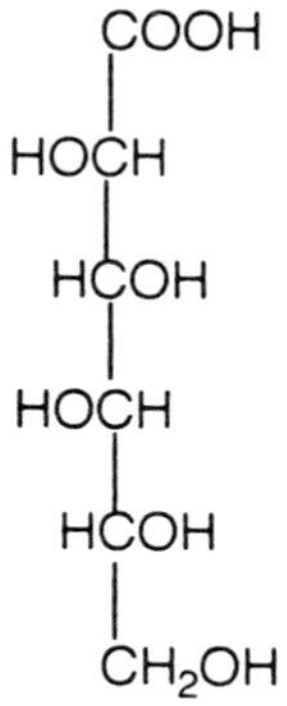

idonic
acid

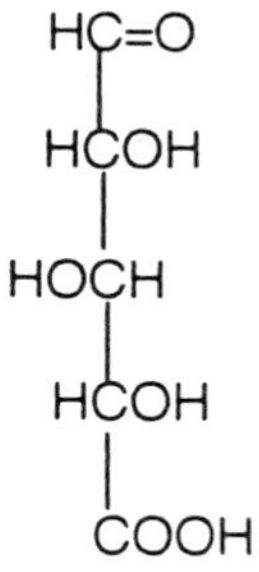

xyluronic
acid

15.10   **Disaccharides:**   Section 15.5

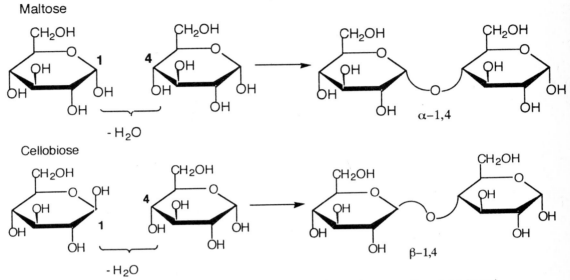

Maltose

Cellobiose

They are both reducing sugars because the anomeric carbon on the right-hand unit can still undergo a ring opening to an aldehyde.

15.11   **Glycosidic Bonds:** Section 15.5
a)   β–1,4   b)   α–1,5   c)   α–1,1

15.12   **Glycosidic Bonds:** Section 15.5

a) α–1,3   b) β,α -1,2

15.13 Conformational Structures/Polysaccharides: Sections 15.3 and 15.5

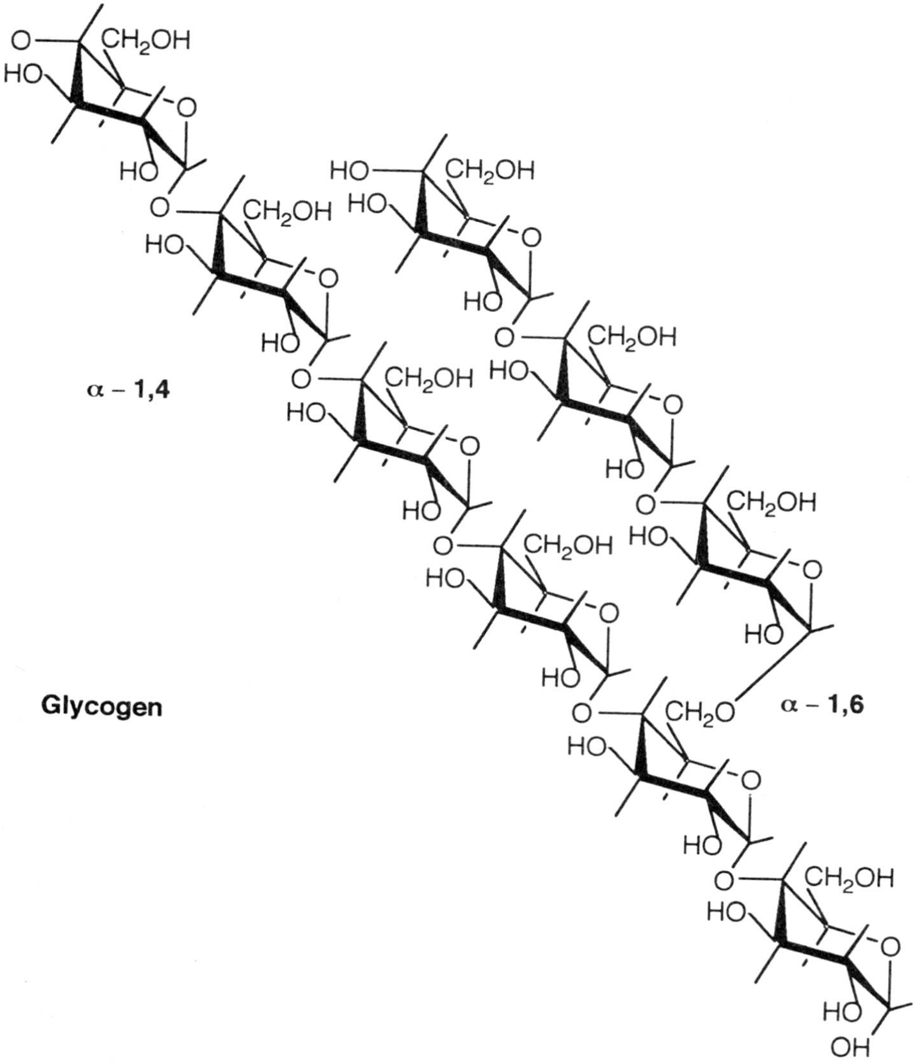

α – 1,4

Glycogen

α – 1,6

15.14 Carbohydrate Derivatives: Section 15.5

Chitin contains monomer units in which the -OH group on C-2 of glucose has been replaced by an acetylated amine group. The glycosidic bond is β -1,4 as is the bond in celluose. The bonds in starch are α-1,4.

15.15   Carbohydrate Derivatives: Section 15.5
Heparin is acidic because of the numerous carboxyl and sulfonic acid groups.

15.16   Terms
a)   A hexose is a six-carbon sugar while a pentose has five carbons.

b)   An aldose has an aldehyde (RCH=O) functional group while a ketose has a ketone (RCOR).

c)   A reducing sugar has an alcohol and ether functional groups on the same carbon. A nonreducing sugar is a diether.

d)   Monosaccharides are small, single carbohydrate units, usually containing five or six carbons. Polysaccharides are polymers of monosaccharides linked by glycosidic bonds.

e) α- and β-D-glucose are anomers, that is, they are both the cyclic forms of glucose with opposite configurations for the -OH group attached to the new chiral center, the former carbonyl group.

f)   Fischer projections are structures drawn vertically with the most oxidized carbon appearing at the top. They are not related spatially to real structures. Haworth formulas are cyclic carbohydrate structures in the form of cyclopentane and cyclohexane-type rings.

g)   Amylose is the "linear" form of starch in which all of the glucose units are linked α-1,4 while amylopectin is branched, its main chain of glucose units linked through an α-1,4 glycosidic bond with α-1,6 bonded-branches about every 25 monomer units.

h)   Glycogen is a polymer of glucose with a main chain having α-1,4 glycosidic bonds and α-1,6 branches every 8-10 units. Cellulose is polyglucose linked β-1,4.

i)   Type I diabetes involves the absence or near absence of insulin to regulate body glucose concentrations. Type II diabetes is a more complicated condition in which insulin is usually present but not functioning properly.

j)   Viscose rayon is a form of cellulose which has been processed by derivatization and reconstitution while acetate rayon is derivatized cellulose.

k)   Fehling's test uses Cu (II) as a weak oxidizing agent while Tollen's test uses Ag (I).

15.17 Structure: Sections 15.3, 15.4 and 15.5

a)   Cellobiose and maltose are both dimers of glucose and both are reducing sugars. Cellobiose has a β-1,4 glycosidic bond while maltose has an α-1,4 bond.

b)   Galactose and glucose, linked β-1,4, are the units in the reducing sugar lactose. Sucrose is nonreducing because of the β,α-2,1 glycosidic bond between fructose and glucose.

c)   α-D-glucose and α-D-galactose are epimers differing in the configuration about C-4. Both are reducing sugars.

d)   α-D-glucose and α-D-fructose are reducing sugars and functional isomers of each other.

e)   α-D-xylose and β-D-ribose are both aldopentoses and reducing sugars. They differ in configuration at C-1 of the cyclic form as well as at C-3.

f)   Maltose has two glucose units linked α-1,4 while lactose is composed of galactose and glucose linked β-1,4; both are reducing.

g)   Cellulose is a linear polyglucose linked β-1,4 while starch consists of amylose (polyglucose α-1,4) and amylopectin (α-1,4 with α-1,6 branches). Both have reducing ends but not much would be seen with Fehling's or Tollen's tests because of the large molecular weight of both polymers.

15.18 Structure: Section 15.4

a) β-D-fructose   b) α-D-idose   c) β-D-talose   d) α-D-lyxose

15.19   **Reactions:** Section 15.4

```
   CH2OH        CH2OH
    |            |
  HCOH         HOCH                    CH2OH
    |            |                      |
  HOCH         HOCH                   HOCH
    |            |                      |
  HCOH         HCOH                   HCOH
    |            |                      |
  HCOH         HCOH                   HCOH
    |            |                      |
   CH2OH        CH2OH                  CH2OH
```

    reduction products of fructose   reduction product of arabinose

15.20   **Terms**

a)   A disaccharide is a dimer of two monosaccharide units linked by a glycosidic bond.

b)   An anomer is an optical isomer which forms at the new chiral center produced when a carbonyl group reacts with an alcohol group.

c)   Lactose intolerance is the inability to digest the lactose in dairy products due to the absence of the enzyme lactase in the small intestine. The result is gas, bloating and diarrhea.

d)   An epimer is an optical isomer which differs from another structure in the configuration at only one chiral carbon.

e)   Invert sugar is the mixture of glucose and fructose resulting from the hydrolysis of sucrose.

f)   A hemiacetal is a compound with an alcohol and an ether group on the same carbon. It results from the reaction of an aldehyde with an alcohol.

g)   Type I diabetes is a condition resulting from the absence of insulin which is necessary to regulate blood glucose concentrations.

h)   Glycoside refers to any polymer of two or more saccharide units linked by a glycosidic bond.

i)   Hyperglycemia is the condition of having an excessive amount of glucose in the blood due to the absence or malfunction of hormones such as insulin.

j)   A reducing sugar is one with a carbon that is an alcohol-ether. It can
exist in equilibrium with an open-chain carbonyl.

15.21   Structure: Section 15.3

a)
```
    HC=O
     |
    HCOH
     |
   HOCH
     |
    HCH  ◄----
     |
    CH₂OH
```
note the deoxy
condition of C-4

b)
```
   CH₂OH
     |
    C=O
     |
    HCOH
     |
    HCOH
     |
    HCOH
     |
   CH₂OH
```

c)
```
    HC=O
     |
   HOCH
     |
   HOCH
     |
   HOCH
     |
    HCOH
     |
   CH₂OH
```

d)
```
    HC=O
     |
   HOCH
     |
    HCOH
     |
    HCOH
     |
   CH₂OH
```

15.22   Optical Isomers: Section 15.3

There are two chiral carbons; four optical isomers are possible.

a
```
   CH₂OH
     |
    HCOH
     |
    C=O
     |
    HCOH
     |
   CH₂OH
```

b
```
   CH₂OH
     |
   HOCH
     |
    C=O
     |
   HOCH
     |
   CH₂OH
```

c
```
   CH₂OH
     |
    HCOH
     |
    C=O
     |
   HOCH
     |
   CH₂OH
```

d
```
   CH₂OH
     |
   HOCH
     |
    C=O
     |
    HCOH
     |
   CH₂OH
```

a/b are identical because you can rotate **b** in the plane of the page and it
will superimpose on **a**. **a** is also obviously meso and is therefore not optically
active.

c/d are enantiomers. **a** is a diastereomer of **c** and **d**.

15.23   Reactions: Section 15.4

In the presence of acid, which also assumes an aqueous solution, the α–
and β– forms of D-glucose will rapidly come into equilibrium with the open-
chain aldehyde. Both the α– and β– anomers can react with methanol to form
the methyl acetal.

15.24 Structure: Section 15.5

Starch contains α– glycosidic bonds which are digestible by enzymes in mammalian gastrointestinal tracts while the β– bonds in cellulose cannot.

15.25 Structure: Section 15.5

a) β–1,4   aldohexose-ketoheptose   b) α–1,5   aldopentose-aldopentose
c) β–2,3 ketoheptose-aldopentose   d) α,β–2,2 ketopentose-ketoheptose

15.26 Structure: Section 15.5

a)

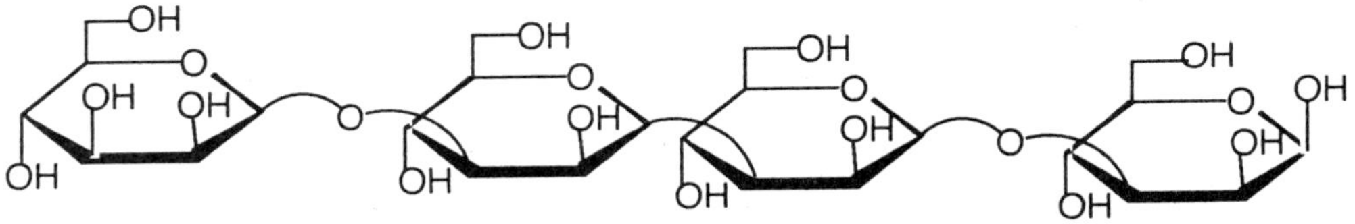

b)

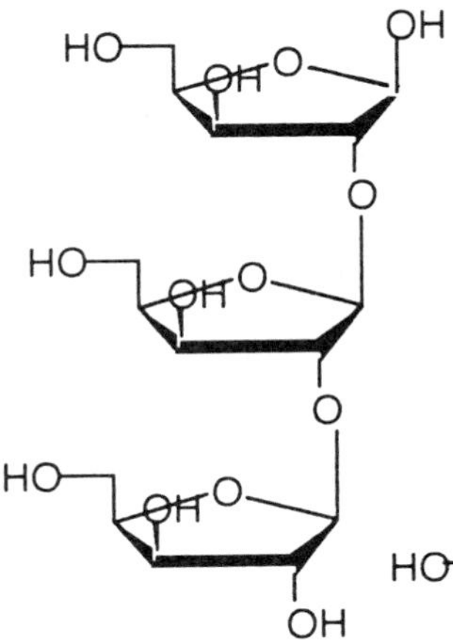

c)

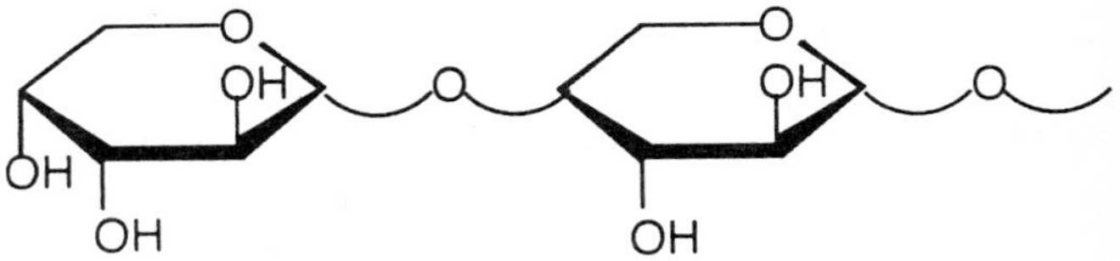

d)

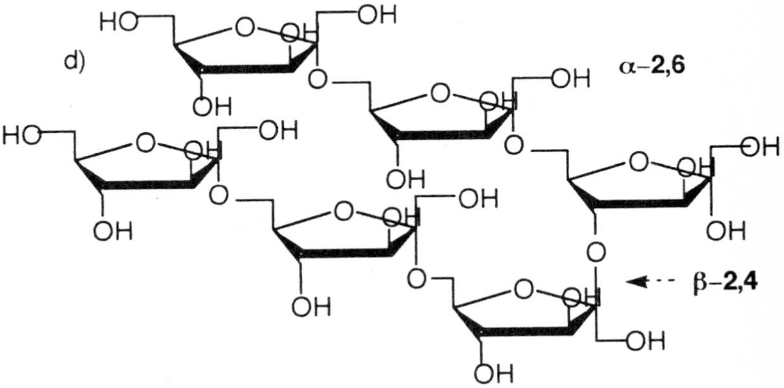

α–2,6

◄-- β–2,4

15.27   Reactions: Sections 15.4 and 15.5

a)

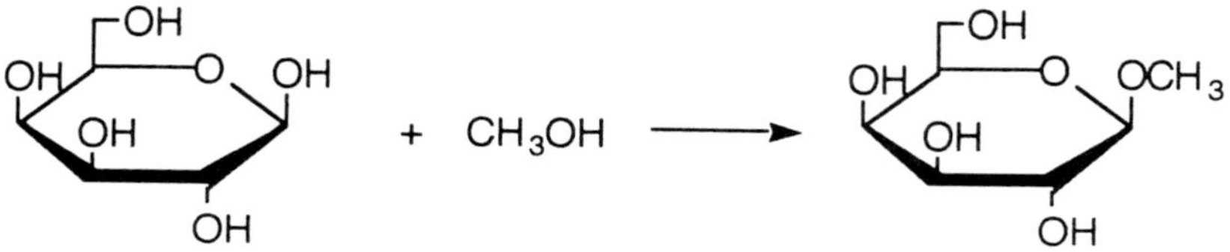

b)

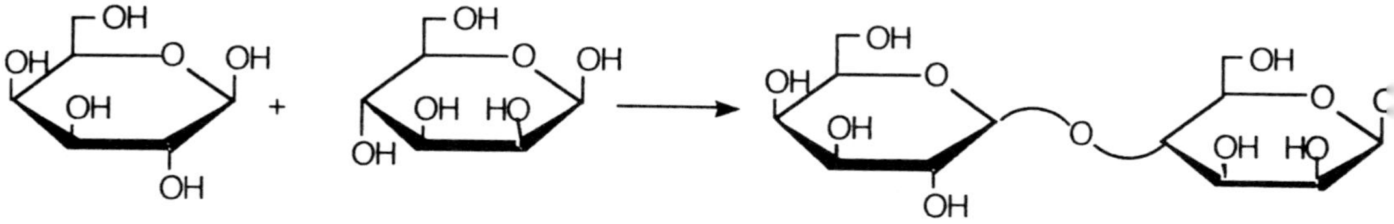

c)

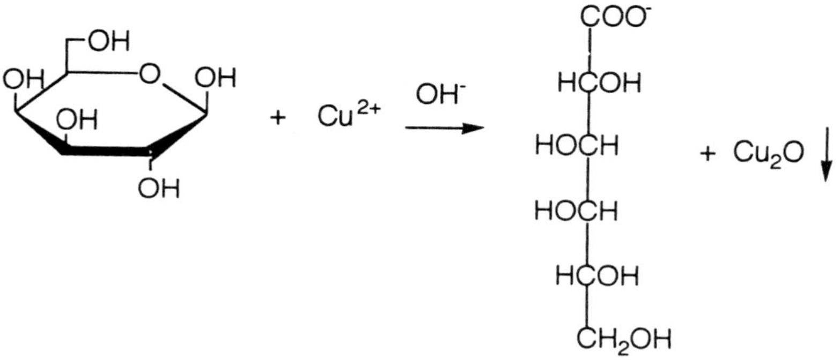

d)

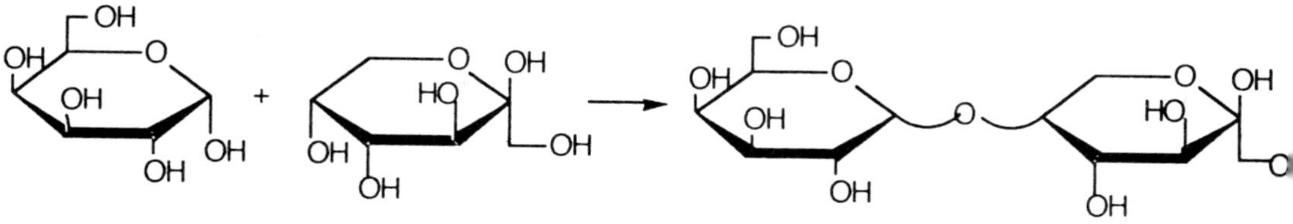

15.28   Reactions: Section 15.4

Both Fehling's and Tollen's tests use a basic solution. Ketoses undergo tautomerism in base to aldehydes and therefore give a positive reaction.

15.29   Reactions: Section 15.4

In base D-glucose will undergo tautomerization. As the reaction reaches equilibrium C-2 can be found as a carbonyl or -CHOH with this chiral center assuming either configuration.

| glucose | enediol | fructose | enediol | mannose |

15.30   Reactions: Section 15.4

Since Fehling's test gives a positive result for any aldose or 2-ketose, it is evidence that there could be a sugar in urine but would not be a definitive test for glucose per se.

16

Lipids

CHAPTER SUMMARY

Lipids can best be defined as biomolecules which are **soluble to a great extent in nonpolar solvents**. In contrast to carbohydrates, proteins and nucleic acids, lipids do not have polymeric forms. By virtue of their **hydrophobic nature** they aggregate into large, nonpolar complexes.

The structures of lipids are quite varied: triacylglycerols (fats and oils), waxes, phospholipids, sphingolipids, steroids, eicosanoids, fat soluble vitamins, and pigments. Some lipids are **simple** in structure while others are more **complex**. Among these molecules are those which are esters in nature and therefore **saponifiable** in aqueous base. Others are **nonsaponifiable**. Many are **completely nonpolar** while others are **amphipathic**, that is, they have a polar/nonpolar nature.

Fats and oils are triesters of glycerol and long-chain fatty acids. The fatty acids are usually **10-24 carbons** in length; they can be **saturated** or **cis-unsaturated**. Saturated triacylglycerols have high melting points and are commonly called fats. Cis-unsaturation leads to a dramatic lowering of melting point and the presence of a liquid, or oil, at room temperature. **Short-hand notations** can be written for the fatty acids which indicate the number of carbons/ number of double bonds/ positions of double bonds from the carboxyl end of the molecule; for example, linoleic acid would be $C_{18:2}{}^{\Delta 9,\ 12}$. Another way to describe unsaturated fatty acids denotes the position of the first double bond from the alkyl end of the molecule; for example, linoleic acid would be $\omega 6$. The double bonds are subject to addition reactions such as **iodination** and **hydrogenation** as well as **oxidative cleavage** and **polymerization**.

The ester bonds in fats and oils can be hydrolyzed in the presence of base to produce **soaps** which are the sodium salts of fatty acids. A soap molecule has a nonpolar, alkyl end and a polar, salt end. Because of this dual polarity, it is called amphipathic. This hydrophobic/hydrophilic nature is essential to the function of such molecules. The **cleaning action of soap** involves lowering the surface tension of water by disrupting hydrogen bonds at the surface and the formation of micelles within the volume of water present. **Micelles** are aggregrations of soap molecules arranged so that the hydrophobic "tails" are oriented towards each other away from the water solvent and the hydrophilic "heads" are pointed into the water.

Detergents are amphipathic molecules which have enhanced solubility and biodegradability properties compared to soaps.

Other amphipathic lipids include the **phospholipids** and **sphingolipids**. The main function of these two subclasses is to produce the semipermeable **lipid bilayer membrane** structure of the cell. The current model of a cell membrane is referred to as **fluid mosaic.** Proteins and cholesterol are also incorporated with the bilayer for purposes of stability, permeability, and cell recognition.

A fused multiple-ring system is the structural framework for **steroids**. **Cholesterol** is the nonpolar, nonsaponifiable progenitor of the metabolic and gonadal hormones such as cortisol, testosterone and estrogen as well as the bile acids used for the intestinal absorption of fats and oils. Many toxins fit into this lipid subclass.

Eicosanoids in the form of prostaglandins, prostacyclins, thromboxanes, and leukotrienes are short-lived metabolites of fatty acids which affect a variety of tissues in the body.

Vitamins A, D, E, and K are called the **fat-soluble vitamins** and must be part of the diet for health and vigor.

Many **pigments** found in algae, bacteria and plants, such as chlorophyll, are lipid in nature. These molecules help to convert light energy to metabolic energy by systems of conjugated bonds.

SOLUTIONS TO PROBLEMS

16.1 Structure: Section 16.3

a) $CH_3(CH_2)_{17}CH=CH(CH_2)_7COOH$

b) $CH_3(CH_2)_{15}CH=CHCH_2CH_2CH=CH(CH_2)_3COOH$

c) $CH_3(CH_2)_4CH=CHCH_2CH=CHCH_2CH=CHCH_2CH=CH(CH_2)_7COOH$

16.2 Structure: Section 16.3

$$
\begin{array}{l}
CH_2OH \quad CH_3(CH_2)_{14}COOH \\
| \\
HCOH \quad + \; CH_3(CH_2)_7CH=CH(CH_2)_7COOH \\
| \\
CH_2OH \quad CH_3(CH_2)_4CH=CHCH_2CH=CH(CH_2)_7COOH
\end{array}
\xrightarrow{-3H_2O}
\begin{array}{l}
CH_2OCO(CH_2)_{14}CH_3 \\
| \\
HCOCO(CH_2)_7CH=CH(CH_2)_7CH_3 \\
| \\
CH_2OCO(CH_2)_7CH=CHCH_2CH=CH(CH_2)_4CH_3
\end{array}
$$

16.3 Structure: Section 16.3

$CH_3(CH_2)_7CH=CH(CH_2)_{11}COOH$ is erucic acid.

16.4 Reactions: Section 16.4

a)   Trimyristin is saturated and therefore has an I_2 number of zero.
Triolein would have one double bond per fatty acid and 3 moles of I_2 would react with it.
Glyceryl oleopalmitostearate has only one double bond (oleo).
The order would be trimyristin<glyceryl oleopalmitostearate<triolein.

b)   Stearic<oleic<linoleic<linolenic

16.5 Biolipids - Structure: Section 16.6

$$
CH_3(CH_2)_4CH=CHCH_2CH=CH(CH_2)_7\overset{\displaystyle O}{\underset{\displaystyle ||}{C}}OCH
\begin{array}{l}
\quad \overset{O}{\underset{||}{}} \\
CH_2OPOCH_2CH_2\overset{+}{N}(CH_3)_3 \\
\qquad\quad | \\
\qquad\quad O^- \\
CH_2OC(CH_2)_{14}CH_3 \\
\qquad || \\
\qquad O
\end{array}
$$

16.6 Biolipids - Structure: Section 16.6

$$R_1COCH \begin{array}{c} O \\ \parallel \end{array} CH_2OPOCH_2CHCOO^- + 4 H_2O \longrightarrow$$

1 mole glycerol

2 moles of fatty acids

serine

phosphate

16.7 Steroids - Structure: Section 16.6

Alcohol groups: estradiol(2), testosterone(1), aldosterone(2), cortisol(3)

Aldehydes: aldosterone(1)

Amines: none

Ketones: testosterone(1), progesterone(2), aldosterone(2), cortisol(2)

Phenols: estradiol(1)

Carboxylic acids: none

16.8 Terms

a)   Amphipathic refers to molecules that have portions which are hydrophilic and hydrophobic.

b)   Emulsification is a process whereby an amphipathic molecule solubilizes a nonpolar solute in a polar solvent or vice versa by forming micelles.

c)   Hydrophobic literally means "water-fearing"; molecules which are insoluble in water.

d)   Iodine number is the number of grams of I_2 that will add to 100 grams of a fat or oil. It is a measure of the degree of unsaturation.

e)   A micelle is a vesicle formed when amphipathic molecules are mixed in a polar solvent. The amphipathic molecules are oriented with their hydrophobic portions inward away from the polar solvent and their hydrophilic portions oriented outward into the polar solvent.

f)   Saponification is the process of hydrolyzing a fat or oil in the presence of excess base to produce glycerol and the sodium or potassium salts of the component fatty acids (soaps).

g)   Saponification number is defined as the number of milligrams of KOH required to saponify one gram of a fat or an oil.

h)   The steroid nucleus is a molecule consisting of four fused rings: three 6-membered and one 5-membered.

16.9 Structures of Fats and Oils: Section 16.3

a)

$$\begin{array}{c} O \\ \| \\ CH_2OC(CH_2)_{10}CH_3 \\ | \\ CH_3(CH_2)_{10}COCH \quad O \\ \| \\ CH_2OC(CH_2)_{10}CH_3 \end{array}$$

b)

$$\begin{array}{c} O \\ \| \\ CH_2OC(CH_2)_{12}CH_3 \quad \text{myristic} \\ | \\ CH_3(CH_2)_{14}COCH \quad O \\ \text{palmitic} \quad \| \\ CH_2OC(CH_2)_{16}CH_3 \\ \text{stearic} \end{array}$$

c)

$$\begin{array}{c} O \\ \| \\ CH_2OC(CH_2)_{12}CH_3 \\ \text{myristic} \\ | \\ CH_3(CH_2)_{12}COCH \quad O \\ \text{myristic} \quad \| \\ CH_2OC(CH_2)_7CH=CH(CH_2)_7CH_3 \\ \text{oleic} \end{array}$$

or

$$\begin{array}{c} O \\ \| \\ CH_2OC(CH_2 \\ \text{my} \\ | \\ CH_3(CH_2)_7CH=CH(CH_2)_7COCH \quad O \\ \text{oleic} \quad \| \\ CH_2OC(CH_2 \\ \text{myris} \end{array}$$

d)

$$\begin{array}{c} O \\ \| \\ CH_2OC(CH_2)_{14}CH_3 \quad \text{palmitic} \\ | \\ CH_3(CH_2)_7CH=CH(CH_2)_7COCH \quad O \\ \text{oleic} \quad \| \\ CH_2OC(CH_2)_7CH=CHCH_2CH=CH(CH_2)_4CH_3 \\ \text{linoleic} \end{array}$$

The fatty acids may appear in other combinations as we For example, an isolated triglyceride molecule might three oleic acids while anot could have three linoleics.

e)   Same as for part d.

16.10 Reactions of Fats and Oils: Section 16.4

The following products will be formed with the glyceride in question:

a)

glyceride
+ 3 H_2O
+ excess NaOH

$\longrightarrow$

CH_2OH
$HOCH$
CH_2OH

$CH_3(CH_2)_7CH=CH(CH_2)_7COO^-Na^+$

$CH_3(CH_2CH=CH)_3(CH_2)_7COO^-Na^+$

$CH_3(CH_2)_{14}COO^-Na^+$

b)

glyceride + 4 H_2 $\longrightarrow$

$$CH_2O\overset{\overset{\displaystyle O}{\|}}{C}(CH_2)_{16}CH_3$$
$$CHO\overset{\overset{\displaystyle O}{\|}}{C}(CH_2)_{16}CH_3$$
$$CH_2O\overset{\overset{\displaystyle O}{\|}}{C}(CH_2)_{14}CH_3$$

c)

glyceride + 4 I_2 $\longrightarrow$

$$CH_2O\overset{\overset{\displaystyle O}{\|}}{C}(CH_2)_7\overset{\overset{\displaystyle I}{|}}{C}H\text{-}\overset{\overset{\displaystyle I}{|}}{C}H(CH_2)_7CH_3$$
$$CHO\overset{\overset{\displaystyle O}{\|}}{C}(CH_2)_7(\overset{\overset{\displaystyle I}{|}}{C}H\text{-}\overset{\overset{\displaystyle I}{|}}{C}HCH_2)_3CH_3$$
$$CH_2O\overset{\overset{\displaystyle O}{\|}}{C}(CH_2)_{14}CH_3$$

16.11 Reactions of Soaps: Section 16.4

a) $2\ CH_3(CH_2)_{16}COO^-Na^+ + Mg^{2+} \longrightarrow (CH_3(CH_2)_{16}COO^-)_2Mg^{2+} + 2\ Na^{1+}$

b) $3\ CH_3(CH_2)_{16}COO^-Na^+ + Fe^{3+} \longrightarrow (CH_3(CH_2)_{16}COO^-)_3Fe^{3+} + 3\ Na^{1+}$

c) $\quad CH_3(CH_2)_{16}COO^-Na^+ + H^+ \longrightarrow CH_3(CH_2)_{16}COOH + Na^{1+}$

16.12 Structure of Fatty Acids: Section 16.3

a) $\quad CH_3(CH_2)_5CH=CH(CH_2)_9COOH$ This is neither $\omega 3$ nor $\omega 6$. It is $\omega 7$.

b) $\quad CH_3(CH_2CH=CH)_6(CH_2)_2COOH$ This is an $\omega 3$.

16.13 Structure of Fatty Acids: Section 16.3

a) $CH_3(CH_2)_3(CH_2CH=CH)_4(CH_2)_7COOH$
The first double bond would appear at position 9 from the carboxyl end.

b) $CH_3(CH_2)_3(CH_2CH=CH)_5(CH_2)_{10}COOH$
The first double bond would appear at position 12 from the carboxyl end.

c) $CH_3(CH_2)_3(CH_2CH=CH)_3(CH_2)_{12}COOH$
The first double bond would appear at position 14 from the carboxyl end.

16.14 Structures of Soaps and Detergents: Section 16.4

a) $CH_3(CH_2)_{14}CO_2^-Na^+$ would be an effective soap because it is the sodium salt of a long chain fatty acid.

b) $(CH_3(CH_2)_{16}CO_2^-)_2Ca^{2+}$ would be insoluble in water and so would not be effective as a soap.

c) $CH_3CH_2CO_2^-Na^+$ does not have a long nonpolar carbon chain and therefore could not make good micelles. It would not be an effective soap.

d) $CH_3(CH_2)_{14}CH_2N(CH_3)_3^+Cl^-$ would be a good detergent because it is a soluble ammonium salt with a long carbon chain.

e) $CH_3(CH_2)_{16}CH_3$ has no polar region, is not amphipathic, and therefore cannot have soap action.

f) $CH_3(CH_2)_{14}CO_2H$ is a neutral molecule. The protonated carboxyl end is not polar enough to counteract the long hydrocarbon chain.

g) $CH_3(CH_2)_{14}CH_2OSO_3^-Na^+$ should be a good detergent.

16.15 Properties of Soaps and Detergents: Section 16.4

a)   d)   g)

polar   *polar*   *polar*

$CH_3(CH_2)_{14}COO^-Na^+$ $CH_3(CH_2)_{14}CH_2N(CH_3)_3^+Cl$ $CH_3(CH_2)_{14}CH_2OSO_3^-Na^+$

nonpolar   *nonpolar*   *nonpolar*

16.16 Consumer Chemistry

This should be carried out in the grocery and/or drug stores.

16.17   Properties of Fats and Oils: Section 16.3
The melting point of a triglyceride decreases with an increase in double bonds
or unsaturation. The iodine number is a measure of unsaturation. Therefore
the higher the iodine number, the lower the melting point.

16.18   Structure: Section 16.4
Detergents, phospholipids and sphingolipids are alike in that they are
amphipathic molecules. In a polar solvent like water they will aggregate so that
their nonpolar portions are away from the solvent.

They differ because phospholipids and sphingolipids have two nonpolar "tails"
while detergents usually have only one. While detergents form micelles with
one layer of molecules shaped into a sphere, the other two types form lipid
bilayers such that a polar solvent can appear within and outside of the structure.

16.19   Structure of Biolipids: Section 16.6
a)

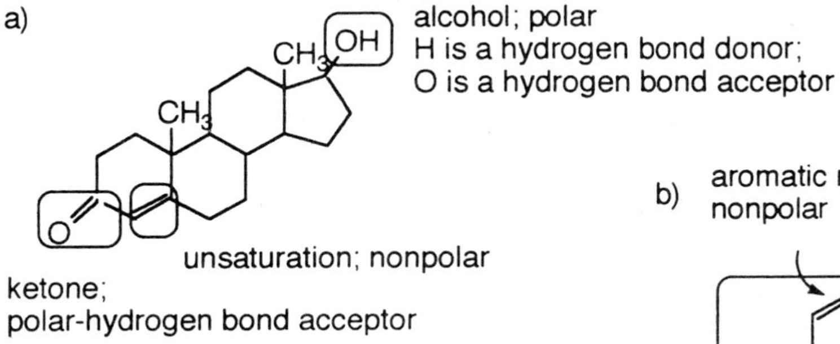

CH₃ OH   alcohol; polar
          H is a hydrogen bond donor;
          O is a hydrogen bond acceptor

CH₃

O=

unsaturation; nonpolar

ketone;
polar-hydrogen bond acceptor

b)   aromatic ring;
     nonpolar   CH₃ OH

HO

phenol; polar
H is a hydrogen bond donor;
O is a hydrogen bond acceptor

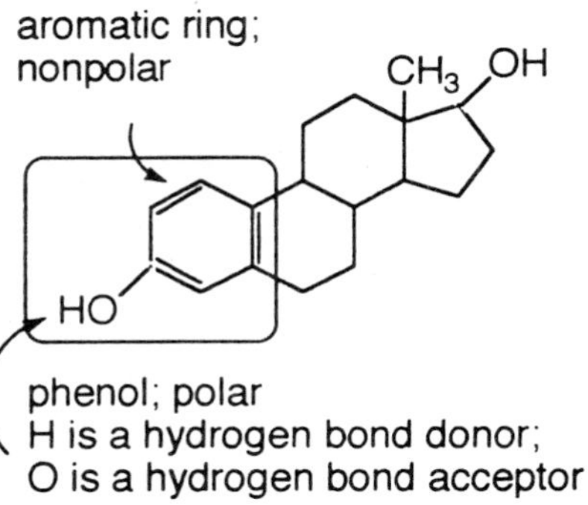

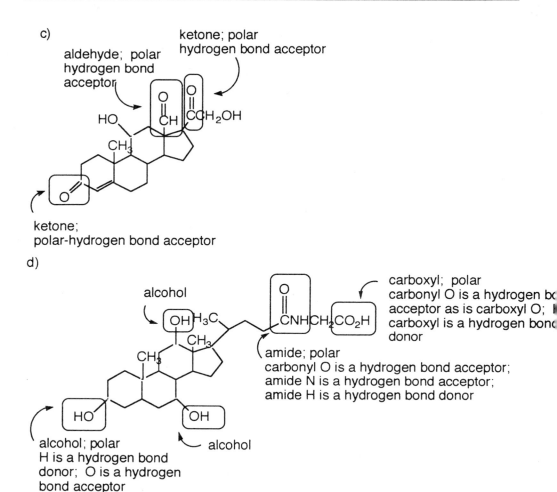

16.20 Functions of Biolipids: Section 16.6

The acidic carboxyl and sulfonic acid portions of the bile acids are polar as are the alcohol groups rimming the steroid nucleus. The fused ring system is nonpolar. Therefore the bile acids are amphipathic and could form micelles which could engulf fats and oils in the intestines.

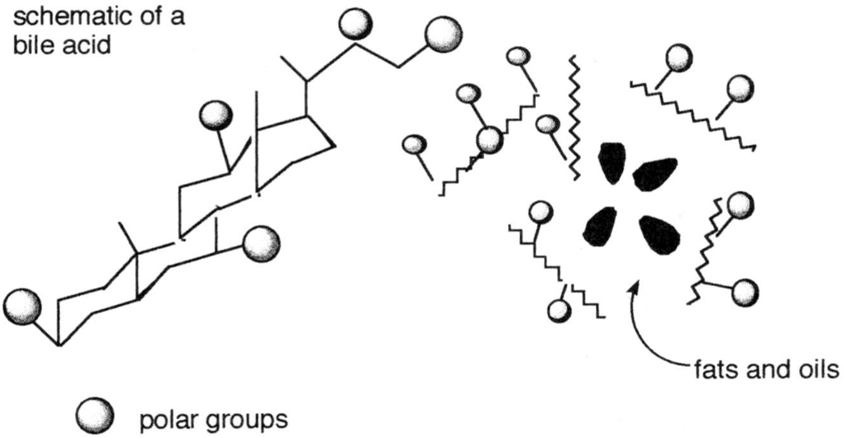

schematic of a
bile acid

polar groups

fats and oils

16.21   **Structure of Biolipids:**   Section 16.6

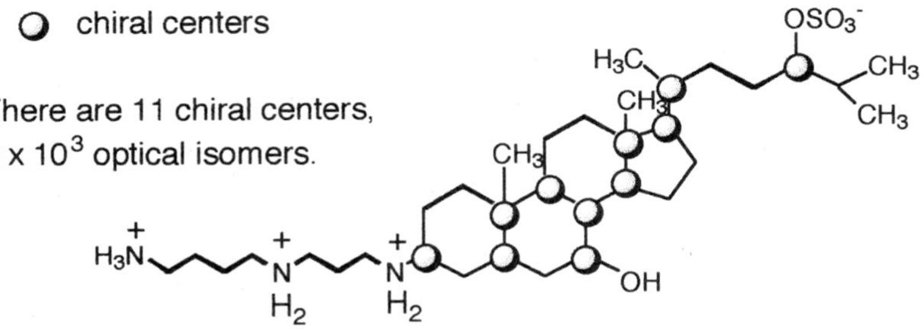

chiral centers

There are 11 chiral centers,
2×10^3 optical isomers.

OSO_3^-

H_3C

CH_3 CH_3

CH_3

CH_3

H_3N^+ $\overset{+}{\underset{H_2}{N}}$ $\overset{+}{\underset{H_2}{N}}$

OH

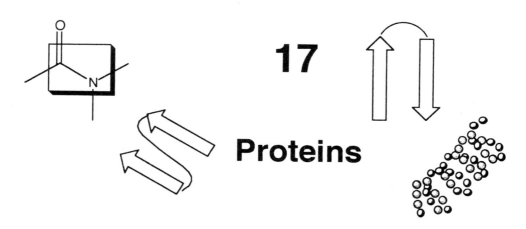

17

Proteins

CHAPTER SUMMARY

Proteins are polymers of amino acids. With **20 different fundamental amino acids** as building blocks, an extraordinarily large variety of proteins can be biosynthesized under the direction of the genetic code. As the term amino acid describes, each monomer has an **amine group** and a **carboxylic acid group** attached to a prochiral carbon. In addition **side chains** can also be present. These range from a simple H to long carbon chains with functional groups.

The amine and carboxyl groups exhibit typical **acid-base behavior** which is pH-dependent. At low pH both groups are protonated: the amine group has a plus (+) charge and the carboxyl is neutral (0). As the pH rises the carboxyl loses its proton becoming negatively charged (-). At higher pH values the amine (+) deprotonates to produce a neutral amine (0). The result of this sequential deprotonation is a series of charged forms ranging from + to 0 to -. If the side chains are capable of acid-base reactions, **the number of possible charged forms depends upon the number and types of amino acids present, the pH, and the pK$_a$ of each ionizable group.** This is true of proteins as well as amino acids. The pH at which the molecule has a net charge of zero, the **zwitterion form**, is called the **pI** or isoelectric (isoelectronic) state. The pI can be calculated by taking the average of the two pK$_a$ values on either side of the zwitterion form. At a pH lower than the pI the molecule will be in a net + charged form while at a pH greater than the pI it will be in a net - charged form.   Charged forms can be separated in an electric field, a process known as **electrophoresis**.

Polypeptides and proteins are the products of amide, or **peptide, bond** formation between the amine group of one amino acid and the carboxyl of another. The sequence of amino acids in the polymer, from the free amino- or N-terminus to the free carboxyl- or C-terminus, is called the **primary (1^0)** **structure** of a protein. A peptide bond has partial double bond character that makes it planar; the geometry is usually trans. As the polypeptide chain grows, the peptide bond can participate in hydrogen bonding - amide hydrogen to carbonyl oxygen. Because of the geometry of the peptide bond, this hydrogen bonding goes on between amino acids which are distant from each other. Organized, folded **secondary (2^0) structures** are formed. The **alpha helix** and **beta pleated sheet** are the two most common secondary structures. In the alpha helix hydrogen bonding usually occurs between the peptide bonds of four amino acids distant from each other.   Beta structure involves the polypeptide chain in its fully extended form coming back on itself to hydrogen bond side-to-side.   The two polypeptide strands in beta structures may be **parallel** or **antiparallel** to each other.

Secondary structures are, in turn, organized into **domains**, or supersecondary structures. Side chains of the amino acids participate in **tertiary (3^0) structure**, that is, they stabilize the overall conformation of the protein molecule.   The forces which hold tertiary structure together include covalent (**disulfide bridges**) and noncovalent (**hydrogen bonding, salt bridge, hydrophobic) interactions**. Shapes of tertiary structure subunits can be **globular** or **fibrous**. Many proteins have more than one folded subunit, linked by the same types of noncovalent forces which hold 3^0 structure together. All of the subunits are needed for the protein to function properly. This is known as **quaternary (4^0) structure**.

Collagen, which is the most abundant protein of the body, has unique primary and secondary structures. A high glycine and proline content leads to fairly rigid, kinked strands which can intertwine in a triple helical structure held together by hydrogen bonding between strands.   The collagen helices aggregate to form skin, bone and connective tissue.

The forces which hold a protein molecule together can be disrupted by changes in temperature and pH as well as by organic solvents and mechanical manipulation. This is known as **denaturation**.

All of the interactions mentioned above are integral parts of the **simple** structure of a protein. In addition proteins may have cofactors such as metal ions, carbohydrates or lipids, and/or organic molecules associated with them. This makes the proteins **complex**. **Myoglobin** and **hemoglobin** are examples of related complex proteins. Myoglobin has a single globular protein subunit complexed with an organic heterocyclic system known as **heme**. The heme in turn holds an iron (II) ion which can bind molecular oxygen, O_2. All of these components contribute to the function of myoglobin: the storage of oxygen in muscle tissue. Hemoglobin is related to myoglobin both structurally and functionally. It contains four myoglobin-type subunits each of which has an iron(II)-heme complex that can bind O_2. However, the four subunits interact cooperatively in order to transport oxygen in the blood from the lungs to the cells. **Sickle cell disease** involves the mutation of hemoglobin such that its function is fatally impeded.

With the great structural versatility available, proteins exhibit a phenomenal breadth of function. Catalysis, protection and regulation were but three discussed in this chapter.

Enzymes are proteins which act as **catalysts** to the complex reactions that occur in the metabolism of living organisms. These reactions include oxidation-reduction, the formation and breaking of carbon-carbon, carbon-nitrogen, and other bonds, hydrolysis, synthesis, group transfer, and isomerization. An enzyme functions by presenting an interactive, three dimensional environment to the reactants (**substrates**). This allows the reaction to be **stereospecific, rapid, and selective**, that is, producing few, if any, spurious by-products. The **active site** of an enzyme has a **substrate binding subsite** and a group of amino acids which effect catalysis, the **catalytic site**. Nonprotein components are common partners in a cooperative catalytic process. The actions of enzymes can be controlled and/or modified by species known as **inhibitors** or an enzyme may be activated/inactivated by **covalent modification**. Most enzymes have precursor forms which are inactive. These are known as **zymogens**.

The complex protective network of higher organisms is called the **immune system**. One part consists of **glycoproteins** (carbohydrate-protein) called **antibodies**. Antibodies bind to foreign substances, **antigens**, and help to mark and destroy the invader. This assault is a key component to the process of **immunization** in which the immune system is trained to respond aggressively to unwanted toxins, bacteria, and viruses. The specificity of antibodies has proven invaluable in diagnostics and has high potential for targeted medications.

The regulation of metabolism is in part due to polypeptide and protein **hormones**, the products of the endocrine system. With the development of recombinant DNA techniques, specific protein hormones can now be made using bacteria and yeast. There has been ongoing discussion and controversy concerning the genetic manipulation of proteins for medical and commercial purposes.

There exists a general concensus that **the primary structure of a protein eventually determines its tertiary structure**. Therefore it is extremely important to be able to study a protein's primary structure. Amino acid content is found by complete hydrolysis of the peptide bonds, separation of the constituent amino acids by column chromatography, and quantitation using reagents such as **ninhydrin** or **dansyl chloride**. However, this gives us no information about the N- to C- sequence. Such sequential analysis can be accomplished by using the Edman technique. Treatment of an intact polypeptide with **phenylisothiocyanate** derivatizes the N- amino acid leaving the rest of the peptide intact for further **Edman degradation**. Large chains must be **fragmented** into shorter peptides, more easy to work with chemically. Cleavage of peptide bonds at specific amino acid residues is accomplished using enzymes such as **trypsin** (Lys, Arg), **chymotrypsin** (aromatics), and **carboxypeptidase** (C-terminus amino acids).

Polypeptides can be produced synthetically by reactions common to organic chemistry. Since both the amine and carboxyl groups are functionally active, a general procedure of functional group blocking, activation of other groups, and coupling of amino acids is carried out. Such a series of reactions is conveniently carried out using the **solid state**, that is, columns to which the growing polypeptide chain is attached while various reagents are washed through.

An understanding of proteins is essential for appreciating the link between organic chemistry and biochemistry.

SOLUTIONS TO PROBLEMS

17.1 Ionized Forms of Amino Acids: Section 17.1

Lysine

pK$_a$ values are boxed.

$$\boxed{8.9} \quad \underset{\substack{+\\H_3NCHCOH\\|\\(CH_2)_4\\|\\NH_3^+\\\boxed{10.3}}}{\overset{O}{\parallel}} \boxed{2.2} \rightleftharpoons \underset{\substack{+\\H_3NCHCO^-\\|\\(CH_2)_4\\|\\NH_3^+}}{\overset{O}{\parallel}} \rightleftharpoons \underset{\substack{H_2NCHCO^-\\|\\(CH_2)_4\\|\\NH_3^+}}{\overset{O}{\parallel}} \rightleftharpoons \underset{\substack{H_2NCHCO^-\\|\\(CH_2)_4\\|\\NH_2}}{\overset{O}{\parallel}}$$

| Net Charge | +2 | +1 | 0 | -1 |

Glutamic Acid

$$\boxed{9.7} \quad \underset{\substack{+\\H_3NCHCOH\\|\\(CH_2)_2\\|\\COOH\\\boxed{4.3}}}{\overset{O}{\parallel}} \boxed{2.2} \rightleftharpoons \underset{\substack{+\\H_3NCHCO^-\\|\\(CH_2)_2\\|\\COOH}}{\overset{O}{\parallel}} \rightleftharpoons \underset{\substack{+\\H_3NCHCO^-\\|\\(CH_2)_2\\|\\COO^-}}{\overset{O}{\parallel}} \rightleftharpoons \underset{\substack{H_2NCHCO^-\\|\\(CH_2)_2\\|\\COO^-}}{\overset{O}{\parallel}}$$

| Net Charge | +1 | 0 | -1 | -2 |

Valine

$$\boxed{9.7} \quad \underset{\substack{+\\H_3NCHCOH\\|\\CH\\H_3C\quad CH_3}}{\overset{O}{\parallel}} \boxed{2.3} \rightleftharpoons \underset{\substack{+\\H_3NCHCO^-\\|\\CH\\H_3C\quad CH_3}}{\overset{O}{\parallel}} \rightleftharpoons \underset{\substack{H_2NCHCO^-\\|\\CH\\H_3C\quad CH_3}}{\overset{O}{\parallel}}$$

| Net Charge | +1 | 0 | -1 |

Tyrosine

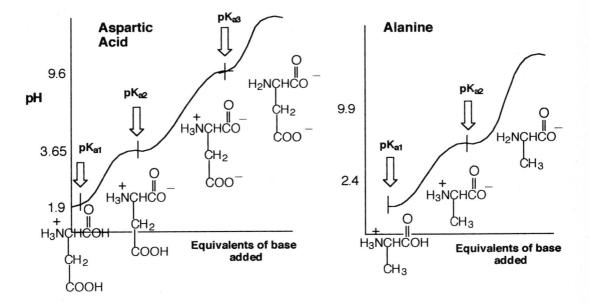

17.2 Acid-Base Behavior of Amino Acids: Section 17.1

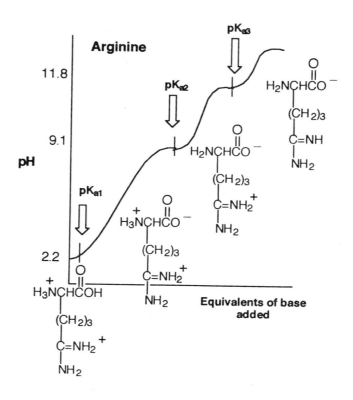

17.3 Ionization of Amino Acids: Section 17.1

glutamic acid	pK_a and charge change of groups			charge at 8.7	net charge at pH 8.7	
α-COOH	2.2	0	→ -1	-1		
α-NH$_2$	9.7	+1	→ 0	+1	-1	Will travel to + pole
R	4.3	0	→ -1	-1		
arginine						
α-COOH	2.2	0	→ -1	-1		
α-NH$_2$	9.1	+1	→ 0	+1	+1	Will travel to - pole
R	11.8	+1	→ 0	+1		

alanine

α-COOH	2.4	$0 \longrightarrow -1$	-1		Will not travel from origin
α-NH$_2$	9.9	$+1 \longrightarrow 0$	+1	0	

tyrosine

α-COOH	2.2	$0 \longrightarrow -1$	-1		Will not travel from origin
α-NH$_2$	9.1	$+1 \longrightarrow 0$	+1	0	
R	10.1	$0 \longrightarrow -1$	0		

cysteine

α-COOH	1.7	$0 \longrightarrow -1$	-1		Will travel to + pole
α-NH$_2$	10.8	$+1 \longrightarrow 0$	+1	-1	
R	8.3	$0 \longrightarrow -1$	-1		

17.4 Ionization of Amino Acids: Section 17.1
histidine

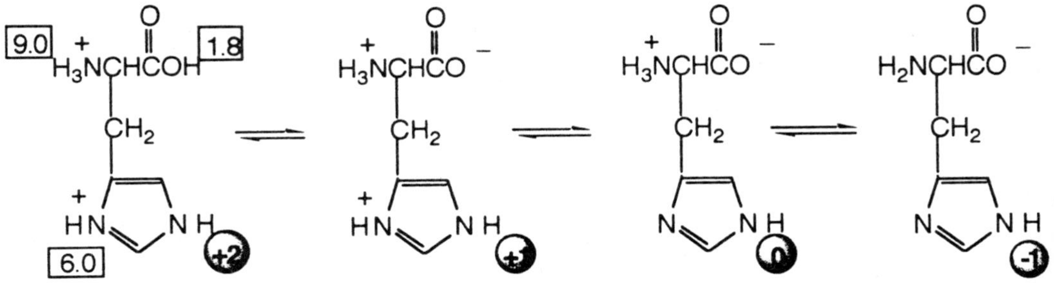

$$pI = \frac{6.0 + 9.0}{2} = 7.5$$

isoleucine

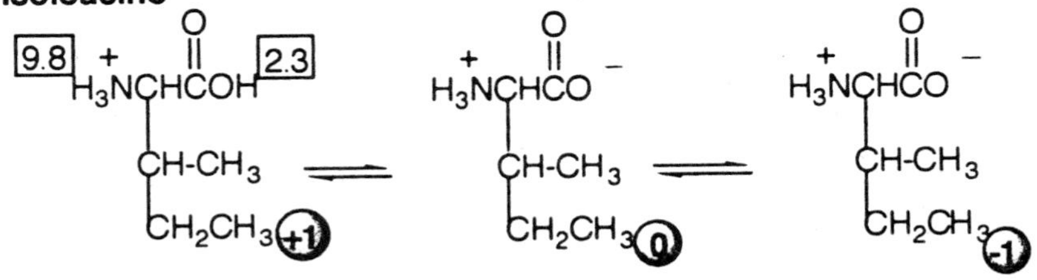

$$pI = \frac{2.3 + 9.8}{2} = 6.05$$

cysteine

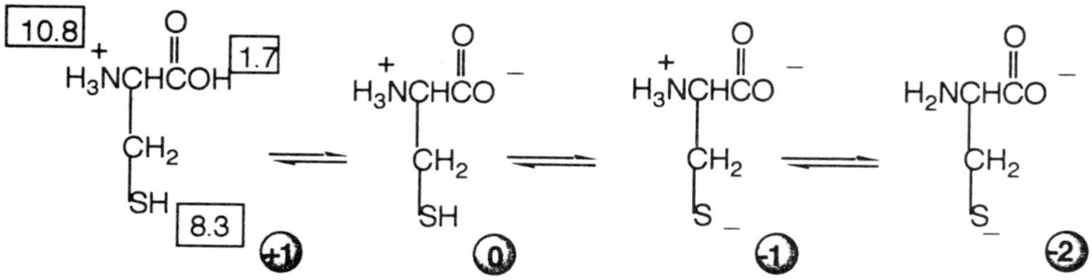

$$pI = \frac{1.7 + 8.3}{2} = 5.0$$

17.5 Ionization of Amino Acids: Section 17.1

glutamine

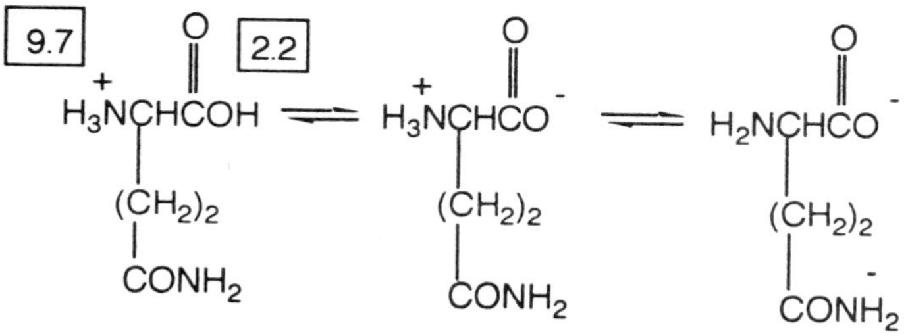

$$pI = \frac{2.2 + 9.7}{2}$$

$$= 5.95$$

Net Charge   +1   0   -1

glutamic acid

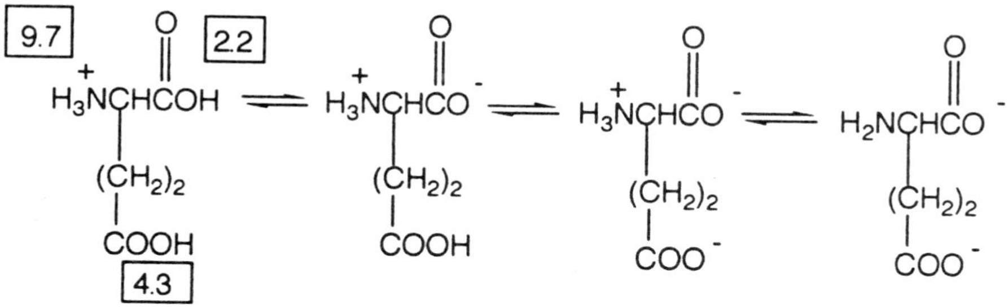

$$pI = \frac{2.2 + 4.3}{2}$$

$$= 3.25$$

Net Charge   +1   0   -1   -2

The pI for Gln is higher than that for Glu due to the loss of ionizability of the side chain carboxyl group.

17.6 Ionization of Amino Acids: Section 17.1

See problems 17.4 and 17.3 for ionization information.
Histidine would most likely be in the 0 or zwitterion form at pH 6.8.
Tyrosine should be in its -1 form at pH 8.5.

17.7 Chirality of Amino Acids: Section 17.1

Threonine and isoleucine each have two chiral centers.

threonine

$$
\begin{array}{cccc}
\text{COOH} & \text{COOH} & \text{COOH} & \text{COOH} \\
| & | & | & | \\
H_3\overset{+}{N}-C-H & H-C-\overset{+}{N}H_3 & H_3\overset{+}{N}-C-H & H-C-\overset{+}{N}H_3 \\
| & | & | & | \\
HO-C-H & HO-C-H & H-C-OH & H-C-OH \\
| & | & | & | \\
CH_3 & CH_3 & CH_3 & CH_3
\end{array}
$$

isoleucine

$$
\begin{array}{cccc}
\text{COOH} & \text{COOH} & \text{COOH} & \text{COOH} \\
| & | & | & | \\
H_3\overset{+}{N}-C-H & H-C-\overset{+}{N}H_3 & H_3\overset{+}{N}-C-H & H-C-\overset{+}{N}H_3 \\
| & | & | & | \\
H_3CH_2C-C-H & H_3CH_2C-C-H & H-C-CH_2CH_3 & H-C-CH_2CH_3 \\
| & | & | & | \\
CH_3 & CH_3 & CH_3 & CH_3
\end{array}
$$

17.8 Chirality of Amino Acids: Section 17.1

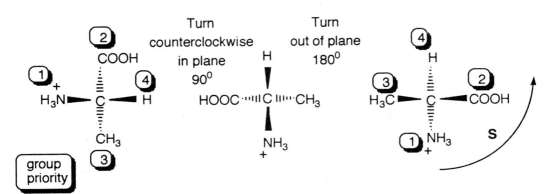

17.9 Ionization of Polypeptides: Sections 17.1 and 17.2

$$\overset{+}{H_3N}\text{---- Ala} \sim \text{Lys} \sim \text{Asp} \sim \text{Tyr} \sim \text{Asp} \sim \text{His} \sim \text{CySH} \sim \text{Leu} \sim \text{Phe} \sim \text{Gln ---COO}$$

Charge at low pH	+	+	0	0	0	+	0	0
pK_a	9.9	10.3	3.65	10.1	3.65	6.0	8.3	2.2
Charge at pH 7.4	+	+	−	0	−	0	0	−

Net charge of polypeptide would be -1.

17.10 Hierarchy of Protein Structure: Section 17.3

At pH 7.4 polyaspartic acid would have a large net negative charge on its side chains while polylysine would have a large net positive charge. This would cause repulsion of the R groups and lead to helix destabilization.

17.11 Hierarchy of Protein Structure: Section 17.3

17.12 Hierarchy of Protein Structure: Section 17.3

a)   Thr and H_2O - hydrogen bonding
b)   Asn and Trp - hydrogen bonding
c)   Asp and Glu - repulsive forces
d)   His and Val - hydrophobic interactions if above pH 6.0

17.13 Hierarchy of Protein Structure: Section 17.3

Since the interior of a water soluble protein has a large degree of hydrophobicity or nonpolarity, nonpolar O_2 and N_2 could stabilize the denaturation of a protein by exposing the nonpolar interior to the air.

17.14 Hierarchy of Protein Structure: Section 17.3

Salt bridges and ion-dipole interactions would be upset by lowering the pH of a protein solution.

17.15 Enzyme Function: Section 17.4

The large-scale industrial use of enzymes must be careful of the possibility of denaturation due to heat, mechanical agitation, the presence of inhibitors, changes in pH, and the presence of organic solvents.

17.16 Determination of Protein Structure: Section 17.5

17.17 Determination of Protein Structure: Section 17.5

Two more cycles of degradation on the polypeptide remaining in Example 17.3 would produce PTH-Tyr, PTH-Gly and free Met.

PTH-Tyr   PTH-Gly   Met

17.18   Determination of Protein Structure: Section 17.5

The theoretical yield for a five-step N-terminal sequential degradation would be

Step 1:	85%
Step 2:	(0.85) * 85% = 72.25%
Step 3:	(0.85) * 72.25% = 61.4%
Step 4:	(0.85) * 61.4% = 52.2 %
Step 5:	(0.85) * 52.2% = 44.4%

17.19   Determination of Protein Structure: Section 17.5

Chymotrypsin digestion of the polypeptide in Example 17.4 would have produced the fragments: Gly ~ His ~ Lys ~ Gly ~ Phe and free Ile.

Trypsin digestion followed by chymotrypsin would produce the following three fragments: Gly ~ His ~ Lys, Gly ~ Phe and free Ile.

17.20 The Organic Synthesis of Polypeptides: Section 17.6

For the hypothetical amino acids - A, B, C, and D - 4! or 24 possible combinations exist.

ABCD	BCDA	CDAB	DABC
ABDC	BCAD	CDBA	DACB
ACDB	BDAC	CABD	DBCA
ACBD	BDCA	CADB	DBAC
ADBC	BACD	CBDA	DCAB
ADCB	BADC	CBAD	DCBA

17.21 Terms

a) An α-amino acid is a molecule with an amine groups on the carbon adjacent to a carboxyl group.

b) An L-amino acid is an amino acid with the amine group on its primary chiral center oriented in the same way as the -OH group in L-glyceraldehyde.

c) A zwitterion is the ionized form of an amino acid or peptide which has a net zero charge.

d) Primary protein structure refers to the linear sequence of amino acids from N- to C-terminus.

e) A basic amino acid is one with a pI greater than 7, more + forms than others, and usually contains one or more nitrogen atoms.

f) A nonpolar amino acid has a side chain which is alkyl and/or aromatic in nature; a hydrophobic amino acid.

g) Tertiary structure refers to the overall structure of a protein subunit, globular or fibrous, which is stabilized by covalent and noncovalent forces.

h) pI is the isoelectric or isoionic pH of an amino acid, polypeptide or protein; for simpler molecules it can be approximated by averaging the pK_a values on either side of the zwitterion form.

i) An antibody is a glycoprotein produced by the B-cells of the immune system.

j) An antigen is a material to which the immune system responds.

17. 22 Structure: Section 17.1

a)   glycine

b)   tyrosine

c)   cysteine

d)   all except Gly, Thr, Ile

e)   proline

f)   serine, threonine,
     asparagine, glutamine,
     histidine, tryptophan, tyrosine

g)   threonine, isoleucine

17.23 Structure: Section 17.2

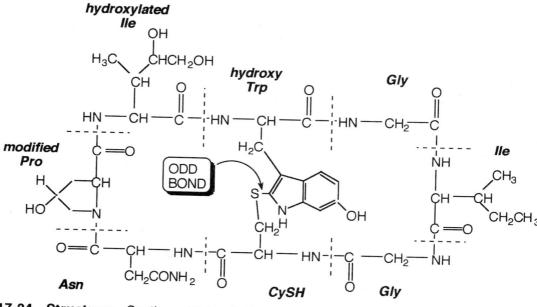

17.24 Structure: Sections 17.1 and 17.5

The amino acids, from N- to C-termini are: Glu, Ile, Thr, Lys.

17.25 Structure: Section 17.1

a)

$$H_3\overset{+}{N}—Tyr \sim Gly \sim Gly \sim Phe \sim Met—COOH$$

$\boxed{9.1}$ $\overset{|}{OH}$ $\boxed{10.1}$ $\boxed{2.3}$

$$pI = \frac{2.3 + 9.1}{2} = 5.7$$

b)

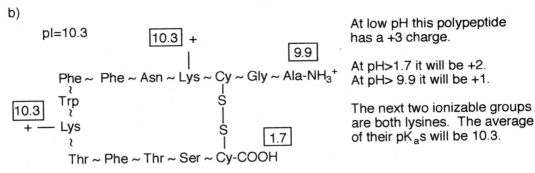

At low pH this polypeptide has a +3 charge.

At pH>1.7 it will be +2.
At pH> 9.9 it will be +1.

The next two ionizable groups are both lysines. The average of their pK_as will be 10.3.

17.26 Structure: Sections 17.1 and 17.2

To associate with the negatively charged nucleic acids, histones would have a net positive charge, that is, they are basic. The basic amino acids are lysine and arginine with some contributions from histidine, depending upon the pH.

17.27 Structure: Sections 17.1, 17.2, and 17.4

Keep in mind that each hemoglobin molecule has two α and two β chains.
Using normal hemoglobin, HbA, as a starting point, find the change in charge which occurs with the change in amino acid.

Hb variant		Changes in Primary Sequence			
	chain	position from N-terminus	amino acid in HbA	amino acid in variant	
S	β	6	Glu	Val	change of +2
C	β	6	Glu	Lys	change of +4
Chesapeake	α	92	Arg	Leu	change of -2
Hasharon	α	47	Asp	His	change of +2
Koln	β	98	Val	Met	no change

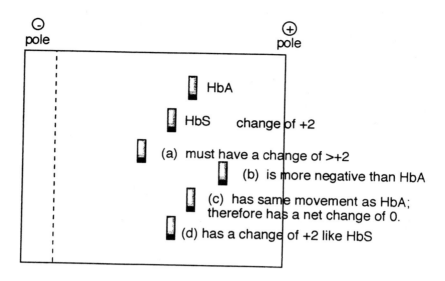

a) is Hb C; b) is Hb Chesapeake; c) is Hb Koln; d) is Hb Hasharon.

17.28 Hierarchy of Protein Structure: Section 17.3

a) 4^0 b) $3^0,4^0$ c) $2^0,3^0,4^0$ d) 1^0 e) $3^0,4^0$ f) 3^0

17.29 Hierarchy of Protein Structure: Section 17.3

a)   hydrogen bonding   b)   hydrophobic interactions
c)   salt bridges   d)   none

17.30 Determination of Protein Structure: Section 17.5

Three cycles of the Edman degradation would produce three PTH - amino acids and a free amino acid.

PTH-His

Ser

17.31   Determination of Protein Structure: Section 17.5

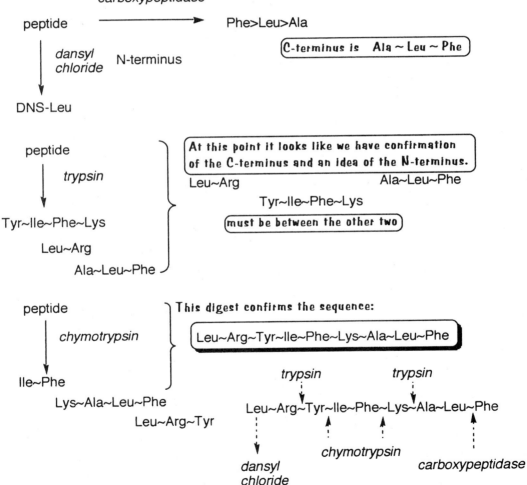

carboxypeptidase

peptide ────────────▶ Phe>Leu>Ala

C-terminus is Ala ~ Leu ~ Phe

dansyl chloride   N-terminus

DNS-Leu

peptide

trypsin

Tyr~Ile~Phe~Lys

Leu~Arg

Ala~Leu~Phe

At this point it looks like we have confirmation of the C-terminus and an idea of the N-terminus.

Leu~Arg   Ala~Leu~Phe

Tyr~Ile~Phe~Lys

must be between the other two

peptide

chymotrypsin

Ile~Phe

Lys~Ala~Leu~Phe

Leu~Arg~Tyr

This digest confirms the sequence:

Leu~Arg~Tyr~Ile~Phe~Lys~Ala~Leu~Phe

trypsin   _trypsin_

Leu~Arg~Tyr~Ile~Phe~Lys~Ala~Leu~Phe

dansyl chloride   _chymotrypsin_   _carboxypeptidase_

17.32 Determination of Protein Structure: Section 17.5

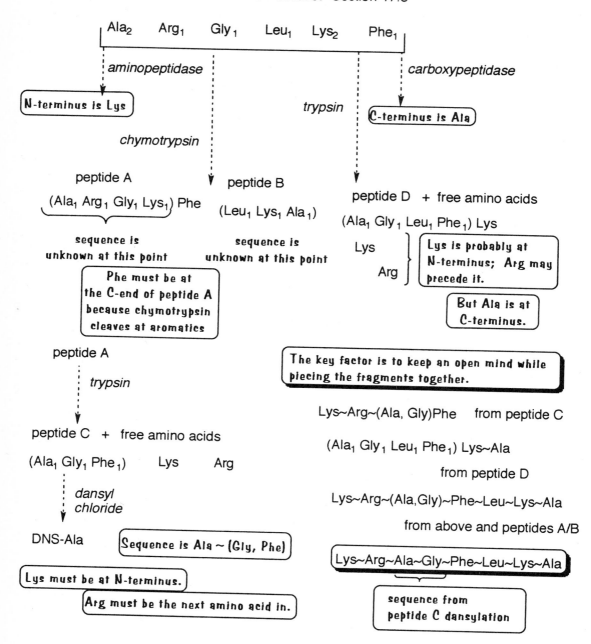

recombinant DNA

genetic code

viruses

AIDS

18

Nucleic Acids

CHAPTER SUMMARY

Nucleic acids are the biopolymers which constitute our genes. The monomer unit is called a **nucleotide.** A nucleotide is composed of a **heterocyclic base**, either a **purine** or **pyrimidine**, a **ribose** or **deoxyribose** sugar unit, and a **phosphate** group.

The two types of nucleic acids are **DNA (deoxyribonucleic acid)** and **RNA (ribonucleic acid)**. These differ in their chemical makeup in the sugar group: deoxyribose in DNA and ribose in RNA; and in the heterocyclic bases: **DNA has adenine(A), guanine(G), thymine(T), and cytosine(C) while RNA has uracil(U) in place of thymine.** The primary structures of DNA and RNA polymers have phosphodiester bridges between (deoxy)ribose units to form a sugar-phosphate backbone. The bases are covalently bound from the hemiacetal group of the sugar to a ring nitrogen. Nucleic acid polymers are usually written from the 5' end (of the sugar unit) to the 3' end, left to right. Often the backbone is represented simply as a horizontal line with the bases protruding. The acidity of the polyprotic phosphate imparts a negative charge and hydrophilicity to the sugar-phosphate backbone at physiological pH. The polymerization of just a few nucleotides produces an **oligonucleotide** while many comprise a **polynucleotide** and a very large number, a nucleic acid.

The secondary structure of nucleic acids involves **hydrogen bonding** between the heterocyclic bases. A and T can form two hydrogen bonds $(A =\!\!= T)$

as can A and U $(A \cdot = = U)$ while G and C form three$(G = = = C)$. In DNA two polynucleotide strands hydrogen bond to each other through their bases in an **antiparallel** fashion. Bond angles in the sugar-phosphate backbone cause the double strand to twist into a helix. This is the classical **double helix structure of DNA** as postulated by Watson, Crick and Wilkinson. DNA is complexed with basic proteins called **histones** forming supercoiled coils. RNA can appear as a double helix but is usually found as a single strand (ss) taking on a variety of secondary structures depending upon its function.

Mononucleotides and dinucleotides are important in metabolism. Adenosine tri-, di- and mono-phosphates, **ATP, ADP and AMP**, are energy intermediates while nicotinamide adenine dinucleotide, **NAD**, and the flavin mono- and di- nucleotides (**FMN, FAD**) are oxidation-reduction cofactors. Cyclic adenosine monophosphate (**cAMP**) acts as a second messenger within cells.

The main function of DNA is to **store genetic information in its nucleotide sequence**. The genetic code consists of base triplets (**codons**) most of which correspond to one of the 20 fundamental amino acids in proteins. **Replication** or duplication of DNA is a **semiconservative** process which depends upon **base pairing**, that is, hydrogen bonding. The **transcription and translation** of the DNA code to protein products proceeds through a complicated series of steps first involving the formation of a **messenger RNA (mRNA)** having a base sequence complementary to that of the parent DNA strand. The mRNA then associates with **ribosomal RNA (rRNA)** - protein complexes. **Transfer RNA (tRNA)** molecules bearing specific amino acids are then base-paired with the mRNA. Many enzyme-catalyzed reactions later, a protein product is formed.

A higher order (eukaryotic) gene contains both coding (**exon**) sequences and intervening (**intron**) or noncoding sequences. Therefore the transcription and translation process is also one of cutting and splicing the exon sequences for the production of a functional protein.

Although the replication and transcription/translation processes occur with high fidelity, occasionally **mutations** can occur. These can lead to death, predisposition to disease, congenital malformations or syndromes, or evolutionary progress.

Viruses are species consisting of nucleic acids, usually ssRNA, encased in a **protein coat** and require a host organism for their replication. Once a virus invades a host cell, it uses its own **reverse transcriptase** enzyme to encode its genome into the host DNA thereby ensuring its survival. **AIDS, acquired immune deficiency syndrome**, is produced by a **retrovirus** that attacks the immune system.

Oncogenes are those genes which are believed responsible for uncontrolled, cancerous cell growth. Cancer can be due to the production of growth factors or the inhibition of growth suppressors.

Manipulation of the genetic code through **recombinant DNA** allows molecular biologists to modify and transfer genes both for the study of disease and the production of new cellular characteristics.

SOLUTIONS TO PROBLEMS

18.1 Structure: Section 18.1
Purines and pyrimidines contain nitrogen which has a nonbonding electron pair. This pair acts as a Lewis base, that is, an electron-pair donor.

18.2 Polynucleotide Structure: Section 18.1

18.3 Structure: Section 18.2

18.4 Structure: Section 18.2

18.5 Structure: Section 18.2

This molecule should be acidic because it still has an ionizable H on the phosphate group.

18.6 Nucleic Acid Structure: Section 18.3

Histones should have amino acids with basic side chains, that is, lysine, arginine and histidine. These amino acids would interact with the negatively charged phosphates in the sugar-phosphate backbone of the nucleic acid.

18.7 The Genetic Code: Section 18.4

coding (sense) *strand*	5'	GGT ACT CCC TGA	3'
antisense strand	3'	CCA TGA GGG ACT	5'

codon	5' - GGU ACU CCC UGA - 3'

anticodon	3'	CCA UGA GGG ACU	5'

peptide	Gly Thr Pro Stop

18.8   The Genetic Code: Section 18.4

a)

⇓

modified codon	5' - GGU A**A**CU CCC UGA - 3'
	GGU AAC UCC CUG A

anticodon	3'	CCA UUG AGG GAC U	5'

peptide	Gly Asn Ser Leu

⇓

modified codon	5' - GGU -CU CCC UGA - 3'
	GGU CUC CCU GA

anticodon	3'	CCA GUG CCA CU	5'

peptide	Gly Leu Pro

b)

modified codon	5' - GGU ACU CCC GUG UGA - 3'

anticodon	3'	CCA UGA GGG CAC ACU	5'

peptide	Gly Thr Pro Val Stop

18.9 Terms

a)   An anticodon is a three-base polynucleotide sequence of tRNA which base-pairs with a specific mRNA codon.

b)   Base-pairing is the hydrogen bonding which occurs between the heterocyclic bases in nucleic acids: A to T, A to U, G to C.

c)   A codon is a three base polynucleotide sequence of mRNA corresponding to an amino acid or protein synthesis directive, including start or stop.

d)   The degeneracy of the genetic code refers to the fact that there is more than one three base codon for most amino acids.

e)   A deoxyribonucleotide is a monomer unit of DNA containing a heterocyclic base, a deoxyribose sugar and a phosphate group.

f)   An exon is a DNA sequence of nucleotides which is expressed as a protein product.

g)   The genetic code is the nucleotide base triplet pattern present in DNA which codes for the amino acids that make up the proteins in an organism.

h)   A genome is the entire genetic makeup of an organism.

i)   Heterocyclic bases are organic ring systems containing nitrogen.

j)   Histones are the basic proteins associated with nucleic acids.

k)   An intron is an intervening sequence, that is, a portion of DNA (and its complementary mRNA) which is not transcribed and translated into protein.

l)   A nucleoside is a combination of a heterocyclic base and a sugar group.

m)   Oncogenes are genes associated with cancer.

n)   The primary transcript is the first DNA-directed mRNA product containing both introns and exons.

o)   Recombinant DNA is a modified gene sequence which has be spliced into a foreign host.

p)   Retroviruses are RNA viruses which use the enzyme reverse transcriptase to transfer their genetic codes into the host genome.

q)   Semiconservative replication is the process followed in DNA replication; the two daughter strands produced by base-pairing with the original DNA strands are combined so that there is a parent DNA strand associated with a daughter strand.

r)   Transcription is the process of producing a mRNA with a base sequence complementary to that of the (-) sense strand of DNA and analogous to the (+) coding or antisense strand of DNA.

s)   Translation is the process wherein mRNA becomes associated with a ribosome and base pairs with tRNAs which carry specific amino acids, followed by protein formation.

t)   The triplet code is the three-base nucleotide sequence in DNA (mRNA) which eventually is processed into an amino acid in a protein product.

u)   A virus is an entity which requires a host cell for its propagation; it consists of a nucleic acid genome (usually RNA), a few enzymes and a protein coat.

18.10   Structure: Section 18.1

	DNA	RNA
Sugar	deoxyribose	ribose
Bases	A,T,G,C	A,U,G,C
Secondary Structure	double helix	single stranded

18.11   Structure: Section 18.3

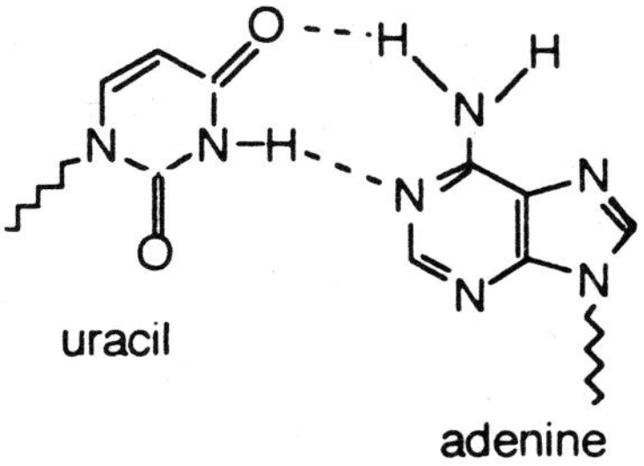

uracil

adenine

18.12 Genetic Code: Section 18.4

sense ***DNA***	5′	G T A A C G T C G C T T 3′
antisense ***DNA***	3′	C A T T G C A G C G A A 5′
mRNA	5′	G U A A C G U C G C U U 3′
mRNA as ***triplet code*** ***(codon)***		GUA ACG UCG CUU
tRNA ***(anticodon)***		CAU UG C AGC GAA
peptide		Val Thr Ser Leu

18.13 Structure: Section 18.1

One mole of the polynucleotide sequence in problem 18.12 would produce the following upon hydrolysis:

3	moles	G
4	moles	T
2	moles	A
3	moles	C
12	moles	deoxyribose
12	moles	phosphate

18.14 Structure: Section 18.1

$$10^6 \text{ nucleotides} \left(\frac{\text{helix turn}}{10 \text{ nucleotides}} \right) \left(\frac{34 \, \overset{\text{o}}{\text{A}}}{\text{helix turn}} \right) \left(\frac{10^{-10} \text{ meters}}{\overset{\text{o}}{\text{A}}} \right) = 3.4 \times 10^{-4} \text{ meters}$$

18.15   Energy-Related Nucleotides: Section 18.2

Species	Bases		Number of moles	
			Ribose	Phosphate
ATP	adenine	1	1	3
FAD	flavin	1	2	2
	adenine	1		
NADH	nicotinamide	1	2	2
	adenine	1		
FMN	flavin	1	1	1

18.16   Genetic Code: Section 18.4

Glucagon would require a minimum of (37 x 3) + 6 (start/stop) nucleotides, that is, 117 nucleotides.

18.17   Genetic Code: Section 18.4

For hemoglobin E the amino acid substitutions are lysine for glutamic acid. The codons for Lys are AAA and AAG while those for Glu are GAA and GAG. The difference is A - G in the first nucleotide of the triplet. Both of these bases are purines and would fit about the same in the helix of DNA. The hydrogen bonding patterns are different, with A involved in 2 while G is involved in 3, but hydrogen bonding to a lesser extent could still occur.

For hemoglobin M_{Boston} the tyrosine would come from UAU and UAC codons while the normal hemoglobin's histidine is derived from CAU and CAC. Again we see the substitution of U for C (T for C in the parent DNA). Both are pyrimidines. U (T) usually forms 2 hydrogen bonding pairs while C forms 3.

Mutations could change the bases in the DNA complementary strand such that only 2 hydrogen bonds were available and in the correct orientation. Such a change would cause the aberrant base to be paired during replication.